STUDIES IN MODERN THERMODYNAMICS 8

BIOCHEMICAL THERMODYNAMICS
(second edition)

STUDIES IN MODERN THERMODYNAMICS

STUDIES IN MODERN THERMODYNAMICS 8

BIOCHEMICAL THERMODYNAMICS
(second edition)

Edited by

M.N. JONES

Department of Biochemistry
University of Manchester
Manchester M13 9PL, U.K.

ELSEVIER
Amsterdam – Oxford – New York – Tokyo   1988

ELSEVIER SCIENCE PUBLISHERS B.V.
Sara Burgerhartstraat 25
P.O. Box 211, 1000 AE Amsterdam, The Netherlands

Distributors for the United States and Canada:

ELSEVIER SCIENCE PUBLISHING COMPANY INC.
52, Vanderbilt Avenue
New York, NY 10017, U.S.A.

LIBRARY OF CONGRESS
Library of Congress Cataloging-in-Publication Data

Biochemical thermodynamics / edited by M.N. Jones. -- 2nd ed.
      p.   cm. -- (Studies in modern thermodynamics ; 8)
   Bibliography: p.
   Includes index.
   ISBN 0-444-42943-3 : fl 280.00
   1. Thermodynamics. 2. Bioenergetics. 3. Biochemistry.
I. Jones, Malcolm N. II. Series.
QP517.T48B56 1988
574.19'283--dc19   88-416
                                                                 CIP

ISBN 0-444-42943-3 Vol. 8)
ISBN 0-444-41762-1 (Series)

v

EDITOR'S PREFACE

Since the first edition of Biochemical Thermodynamics in 1979 the application of thermodynamics to the study of biochemical and biological systems has continued to advance and has given rise to an expanding body of literature. These advances have arisen largely for two reasons, firstly the development of more sensitive and more versatile calorimetric equipment which has extended the range of direct enthalpy measurements to smaller heat effects in a wide variety of systems and secondly because of the rapid and significant advances which have taken place in the field of molecular biology. Such developments have prompted the design of new and sophisticated experiments and also led to a better understanding of thermodynamic data in molecular terms.

In such a developing science it is constantly necessary to order, review and analyse the expanding literature. This volume has been prepared with these objectives in mind and I am greatly indebted to all the contributors, who are experts in various areas of biochemical thermodynamics, and who have willingly given their time and effort to make the production of this volume possible.

As with the First Edition the volume starts with studies on peptides and model systems and passes on to macromolecules and systems of increasing complexity. Apart from sections dealing with basic thermodynamics most of the chapters are completely new and relate to literature which has appeared since the First Edition. Despite every effort to produce perfect camera-ready copy I beg the readers indulgence with regard to any remaining errors in the text for which I take responsibility. While all the contributors are well aware of the desirability to use SI units, because of the use of literature material it has not always been possible. However, such problems are largely confined to the presentation of some data in calories instead of joules.

Finally I would like to express my appreciation and thanks to Miss Margaret Barber who has not only typed Chapters 5 and 7 but has once again given me invaluable secretarial assistance and displayed great patience with the numerous typographical problems during the preparation of this volume.

Malcolm Norcliffe Jones,
November, 1987.

CONTRIBUTORS

B. Crabtree — Rowett Research Institute, Aberdeen, Scotland, AB2 9SB, U.K.

S.J. Gill — Department of Chemistry & Biochemistry, University of Colarado, Boulder, Colorado, 80309, U.S.A.

M.N. Jones — Department of Biochemistry & Molecular Biology, University of Manchester, Manchester, M13 9PT, U.K.

H.H. Klump — Institut fur Physikalische Chemie der Universitat Freiburg, Freiburg, Albert str. 23a, Fed. Rep. Germany.

T.H. Lilley — Chemistry Department, The University, Sheffield, U.K.

A.G. Lowe — Department of Biochemistry & Molecular Biology, University of Manchester, Manchester, M13 9PT, U.K.

B.A. Nicholson — Rowett Research Institute, Aberdeen, Scotland, AB2 9SB, U.K.

W. Pfeil — Akademie der Wissenschaften der DDR, Zentralinstitut fur Molekularbiologie.

C.H. Roberts — Department of Chemistry & Biochemistry, University of Colardo, Boulder, Colorado, 80309, U.S.A.

I. Wadso — Division of Thermochemistry, Chemical Centre, University of Lund, P.O. Box 124, S-221 00 Lund, Sweden.

A.R. Walmsley — Department of Biochemistry, University of Leicester, Leicester, U.K.

J. Wyman — Department of Chemistry and Biochemistry, University of Colarado, Boulder, Colarado, 80309, U.S.A.

CONTENTS

Chapter 7 — Conformational Changes in Proteins in Relation to Energy Transduction —
A. G. Lowe and A.R. Walmsley

Chapter 8 — Thermodynamics of Metabolism — R. Crabtree and B.A. Nicholson

CHAPTER 1

THERMODYNAMICS OF PEPTIDES AND MODEL SYSTEMS

T.H. LILLEY

1.1 INTRODUCTION

There is currently a considerable thrust in both the chemical and biochemical areas towards achieving greater understanding of the underlying chemical and physical phenomena involved in biological processes. In this chapter we will be concerned particularly with describing studies which are directed towards some problems in protein chemistry but many of the conclusions which are drawn have relevance to the behaviour of systems containing other macromolecules and certainly much of the methodology which is outlined is applicable to systems other than those considered here. It should be stated at the outset that we will not, in the main, be concerned with the problems associated with the formation of covalent bonds and the chemistry of metabolism will not be directly addressed. Rather, attention will be directed towards the weak, so-called 'non-bonding' interactions which are important in macromolecular assembly and molecular recognition processes such as enzyme active site – substrate interactions and macromolecule – macromolecule associations such as those involved in antibody – virus problems. As might be expected both from the title of this book and this chapter attention will be directed towards thermodynamic investigations but it should be stressed that such studies cannot be viewed in isolation and complementary work using, for example, theoretical methods (both quantum and statistical mechanics) and structural and spectroscopic techniques is also necessary if a complete understanding and description of the molecular processes is to be achieved. Much of the work which will be described will be taken from either published or as yet unpublished studies which have been performed in recent years in the Sheffield laboratories although we shall when it is considered appropriate or pertinent draw on the very extensive studies of others to illustrate particular points.

In essence the programme of work with which we have been involved has been directed towards two major problems in protein chemistry. These are:

(i) the propensity of some oligopeptides to fold spontaneously, when present in a suitable environment, and give rise to biologically useful and/or enzymatically active macromolecular structures (refs.1,2,3)

and (ii) the tendency for some polypeptide substrates to interact in a pharmacologically useful manner with the active sites of some enzymes (ref.4).

Both of these areas have attracted considerable attention from scientists in various disciplines over the years but the first is now of increased importance because of the rise of the relatively new and important area of protein engineering. It is apparent that there is little point in modifying nucleic acids so that proteins with changed sequences result if the final protein product does not adopt a biologically active 'shape' or has unsuitable physical and mechanical properties. The second of the two problems referred to above is part of a much more general area namely the delineation of the factors pertinent to the design of drugs.

Most of the thermodynamic studies which we and others have performed using model compounds to address problems in protein chemistry have used water as the solvent and this is of course understandable in view of this solvent being the biological milieu. However as will become apparent from some work which will be presented later in this chapter, the medium in which the model compounds are immersed has a significant effect on their thermodynamic properties (and presumably on other properties also). In view of this and the fact that there are some macromolecular phenomena which occur in either partly aqueous or non-aqueous environments, we would suggest that when one is considering model systems, the model must include, perhaps in some 'model' way, an appropriate environment for the central molecule or molecules.

Thermodynamic studies of the properties of model compounds in condensed media fall into two distinct although related categories. Firstly, there are those investigations which are directed towards what can be termed the 'standard state' properties of the solute and the information obtained reflects, at least to a good approximation, how the solute interacts with the solvent although, and this sometimes seems to be either forgotten or neglected, there must also be a concomitant perturbation of the solute because of its interaction with the solvent. Frequently studied properties which fall into this category are partial molar volumes and partial molar heat capacities at 'infinite dilution' and a considerable amount of information is available on these for amino acids and peptides (see *e.g.* refs.5,6). The second category of investigation are those studies which are concerned with evaluating the interactions which occur between (solvated) molecules in a particular solvent and as examples of these types of investigation one can mention the information which can be obtained from the determination of osmotic and activity coefficients and enthalpies of dilution (see *e.g.* refs.7-9). From both a biochemical and a chemical viewpoint studies in this latter category are certainly more important than those in the first category but it is nevertheless undoubtedly true that since the solute-solute interactions of concern occur between solvated species, from an interpretational view, a knowledge of the solvation properties is extremely useful. Later in this chapter it will be shown that the relationship between experimental measures of solute-solute interactions and solvation is closer than has been generally appreciated. Before embarking on considerations of the properties of particular systems it is appropriate to outline the thermodynamic formalism and methodology which is in current use.

1.2 BACKGROUND THERMODYNAMICS

We begin by considering a solution containing n_s moles of a single non-electrolytic solute s and n_w moles of a single component solvent which is denoted by w. At constant temperature and pressure the Gibbs energy (G) of the system is

$$G = n_w \mu_w + n_s \mu_s \qquad (1)$$

where μ_w and μ_s are the chemical potentials of the solvent and solute respectively. The former is usually written in terms of the molar Gibbs energy of the pure solvent ($\mu_w^{\ominus}$) and the activity (a_w) of the solvent in the solution *i.e.*

$$\mu_w = \mu_w^{\ominus} + RT \ln a_w \qquad (2)$$

Most commonly a mole fraction defined ideal state is adopted for the solvent and consequently we can also write equation (2) as

$$\mu_w = \mu_w^{\ominus} + RT \ln x_w + RT \ln f_w \qquad (3)$$

where x_w is the mole fraction of the solvent and f_w is the rational activity coefficient. The solvent activity can also be expressed in terms of its molal osmotic coefficient (ϕ)

$$a_w = \exp -(n_s/n_w)\phi = \exp -(m_s M_w \phi) \qquad (4)$$

where m_s is the molality of the solute and M_w is the (kg) R.M.M. (molecular weight) of the solvent. Combination of equations (3) and (4) gives the following alternative expression for the chemical potential of the solvent

$$\mu_w = \mu_w^{\ominus} - RT M_w m_s \phi \qquad (5)$$

It is often convenient when using the molal scale to consider the partial specific Gibbs energy (G_w) of the solvent (*i.e.* the property for 1 kg of solvent) rather than the chemical potential and we define

$$G_w = M_w^{-1} \mu_w = G_w^{\ominus} - RT m_s \phi \qquad (6)$$

where $G_w^{\ominus}$ is the Gibbs energy of 1 kg of pure solvent. The chemical potential of the solute could be expressed using an equation analogous to equation (3) but it is more usual to write it in terms of a standard state chemical potential ($\mu_s^{\ominus}$) pertaining to a molality $m_s^{\ominus}$. This standard chemical potential refers to a hypothetical 1 molal solution which reflects in large part interactions occurring between the solute and the solvent only. The chemical potential is given by

$$\mu_s = \mu_s^{\ominus} + RT \ln (a_s/m_s^{\ominus}) \qquad (7)$$

which in turn can be expressed as

$$\mu_s = \mu_s^{\ominus} + RT \ln (m_s/m_s^{\ominus}) + RT \ln \gamma_s \qquad (8)$$

where γ_s (the molal activity coefficient) is related to the molality and the solute activity (a_s) by

$$\gamma_s = a_s/m_s \qquad (9)$$

This activity coefficient is defined so that $\gamma_s \rightarrow 1$ as $m_s \rightarrow 0$. A solution is considered to be ideal when both ϕ and γ_s are unity and accordingly we can write using these conditions and equations

4

(5),(6) and (8)

$$\mu_w^{id} \;=\; \mu_w^{\ominus} \;-\; RT \; M_w \; m_s \tag{10}$$

$$G_w^{id} \;=\; G_w^{\ominus} \;-\; RT \; m_s \tag{11}$$

$$\mu_s^{id} \;=\; \mu_s^{\ominus} \;+\; RT \; \ln \, (m_s/m_s^{\ominus}) \tag{12}$$

Excess functions are defined as the differences between the values which pertain in a real solution and in an ideal solution. The excess Gibbs energy per kg of solvent is therefore given by

$$G^{ex} \;=\; G \;-\; G^{id} \tag{13}$$

which can be written as

$$G^{ex} \;=\; G_w^{ex} \;+\; m_s \, \mu_s^{ex} \tag{14}$$

where

$$G_w^{ex} \;=\; [\, \partial (G^{ex}/m_s)/\partial \, m_s^{-1}\,] \;=\; RT \; m_s \; (1-\phi) \tag{15}$$

and

$$\mu_s^{ex} \;=\; [\, \partial G^{ex}/\partial m_s\,] \;=\; RT \, \ln \, \gamma_s \tag{16}$$

Other thermodynamic functions can be defined similarly and for example using the links between the Gibbs function and temperature or pressure

$$[\, \partial (G/T)/\partial \, T^{-1}\,]_{P,composition} \;=\; H \tag{17}$$

$$[\, \partial G/\partial P\,]_{T,composition} \;=\; V \tag{18}$$

where H and V are respectively the enthalpy and volume of the system considered, We thus get

$$
\begin{aligned}
H^{ex} \;=\; H \;-\; H^{id} &\;=\; H_w^{ex} \;+\; m_s \, H_s^{ex} \\
&\;=\; R \; m_s \; [\, \partial \{(1-\phi) \;+\; \ln \, \gamma_s\}/\partial \, T^{-1}\,]
\end{aligned} \tag{19}
$$

and

$$
\begin{aligned}
V^{ex} \;=\; V \;-\; V^{id} &\;=\; V_w^{ex} \;+\; m_s \, V_s^{ex} \\
&\;=\; RT \; m_s \; [\, \partial \{(1-\phi) \;+\; \ln \, \gamma_s\}/\partial \, P\,]
\end{aligned} \tag{20}
$$

These excess functions can be related to other thermodynamic functions which are frequently used to express solution behaviour such as for example the relative apparent molar enthalpy and volume of the solute. The apparent molar property ($X_{s,app}$) of the solute can be defined by

$$X = X_w^{id} + m_s X_{s,app} \qquad (21)$$

i.e.

$$X_{s,app} = (X - X_w^{id})/m_s \qquad (22)$$

which should be compared with the expression for the corresponding partial molar property X_s

$$X_s = (\partial X/\partial m_s) = X_{s,app} + (\partial X_w^{id}/\partial m_s) + m_s (\partial X_s^{app}/\partial m_s) \qquad (23)$$

This can be related to the excess property by

$$X_{s,app} = X_s^{id} + X^{ex}/m_s \qquad (24)$$

and

$$X_s = X_s^{id} + m_s (\partial X_s^{id}/\partial m_s) + (\partial X_w^{id}/\partial m_s) + (\partial X^{ex}/\partial m_s) \qquad (25)$$

Consequently using equations (24) and (25) we have

$$\mu_{s,app} = \mu_s^{id} + G^{ex}/m_s = \mu_s^{\ominus} + RT \ln (m_s/m_s^{\ominus}) + (G^{ex}/m_s) \qquad (26)$$

$$H_{s,app} = H_s^{\ominus} + (H^{ex}/m_s) \qquad (27)$$

$$V_{s,app} = V_s^{\ominus} + (V^{ex}/m_s) \qquad (28)$$

$$\mu_s = \mu_s^{\ominus} + RT \ln (m_s/m_s^{\ominus}) + (\partial G^{ex}/\partial m_s) \qquad (29)$$

$$H_s = H_s^{\ominus} + (\partial H^{ex}/\partial m_s) \qquad (30)$$

$$V_s = V_s^{\ominus} + (\partial V^{ex}/\partial m_s) \qquad (31)$$

The apparent molar free energy of the solute is a quantity which is seldom considered as such but the corresponding enthalpy and volumes are frequently used because they are directly related to experiment. The apparent molar volume (usually given the symbol ϕ_V) can be evaluated in an absolute sense but for the enthalpy only relative values can be obtained and the difference

$$H_{s,app} - H_s^{\ominus} = H^{ex}/m_s \qquad (32)$$

is called the relative apparent molar enthalpy and is often written as ϕ_L. It is clear from equation (27) that it can alternatively be called the molar excess enthalpy ($H_{m,s}^{ex}$).

If one now considers the partial specific properties of the solvent then from equations (13) and (14) we have for the Gibbs function

$$G_w = G_w^{id} + [\partial(G^{ex}/m_s)/\partial m_s^{-1}] \tag{33}$$

and analogous expressions for say the partial specific enthalpy and volume. These latter can be written as

$$H_w = H_w^{\ominus} + [\partial(H^{ex}/m_s)/\partial m_s^{-1}] \tag{34}$$

$$V_w = V_w^{\ominus} + [\partial(V^{ex}/m_s)/\partial m_s^{-1}] \tag{35}$$

As is usual for enthalpies only relative values are obtainable from experiment. Equations (34) and (35) can all be expressed as partial molar properties of the solvent by division by the number of moles of the solvent in 1 kg.

1.3 SOLVATION PROPERTIES

It is worth stressing at the outset that by their very nature the types of molecules with which we shall be concerned are polyfunctional and this necessarily means that their solvation regions will be heterogeneous. Even at a relatively crude level, if we consider the simplest of the amino acids, glycine, dissolved in a solvent, then the solvation region will consist of three sub-regions, one associated with the positively charged amino-group, one associated with the negatively charged carboxylate group and the third arising from the presence of the apolar methylene group. It should also be remembered that one is dealing with a dynamic situation and there is therefore not only continuous interchange on a relatively short time scale between the solvent in its possible microenvironments (including as one of these bulk solvent) but also both the solute and solvent molecules will be undergoing vibrational, rotational and librational motions all of which can contribute to experimentally determined properties. [Some aspects of these problems have been commented on earlier (refs.6,10).] The nature of the solvent consequently can have a marked effect on thermodynamic properties and to illustrate this two examples will be presented now although some others will be referred to later.

In Table 1.1 the standard state partial molar volumes of urea in several solvents are given and the relatively large variations exhibited clearly indicate that the different solvents interact with this solute in markedly different ways and to different extents.

The second example also considers volumetric information and Fig.1.1 shows some recent (ref.12) apparent molar volume results for the amide N-methylacetamide in water and the 'mixed-solvent', aqueous 6 molar guanidinium chloride. Now given that the apparent molar volume can be expressed as a function of solute molality by (see section 1.4.2)

$$V_{s,app} = V_s^o + v_{ss} m_s + v_{sss} m_s^2 + \cdots \cdot \tag{36}$$

TABLE 1.1

Standard state partial molar volume ($V^{\ominus}$) at 25°C of urea in six solvents (ref.11)

Solvent	$V^{\ominus}$ (cm^3 mol^{-1})
Water	44.24
Methanol	36.97
Ethanol	40.75
Formamide	45.34
N,N–Dimethylformamide	39.97
Dimethylsulphoxide	41.86

then the point which emerges is that solvation, reflected in the standard state volume, and solvated solute–solvated solute interactions, expressed by the molality dependence of the apparent molar volumes, both depend on the nature of the solvent medium.

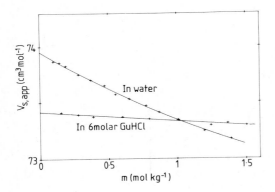

Fig.1.1 The apparent molar volume of N–methylacetamide in water and in 6 molar guanidinium chloride (GuHCl) solutions at 25°C as a function of the amide molality

One of the difficulties in attempting to get molecular rationalisations of the properties of amino acids and peptides arises from their structural diversity (see Table 1.3) and other problems come from the fact that the naturally occurring compounds are zwitterionic in nature. There are many ways in which this charge distribution manifests itself in protein chemistry and by way of illustration we will consider some examples which utilise recent literature information.

The standard state volume and heat capacity changes associated with the condensation reactions for the formation of peptides

$$n \{ \overset{+}{H_3N}-CH(R)-COO^- \} \rightarrow \overset{+}{H_3N}-[-CH(R)-CONH-]-_{n-1}CH(R)COO^- + (n-1)H_2O \qquad (37)$$

are shown as a function of the number of peptide groups formed in Figs.1.2 and 1.3. (The changes were calculated using the information given in refs.5,13,14,15).

It is apparent that there are reasonable although not perfect linear relationships and the

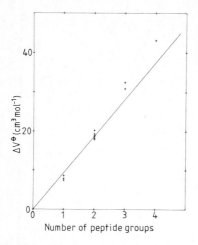

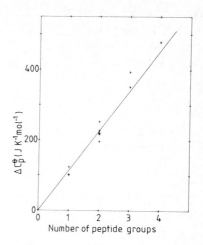

Fig.1.2 The standard state volume changes for the formation of peptides in water at 25°C against the number of peptide groups formed.

Fig.1.3 The standard state heat capacity changes for the formation of peptides in water at 25°C against the number of peptide groups formed.

volume and heat capacity changes for the formation of one peptide group are about 9.4 cm^3mol^{-1} and 113 J $K^{-1}mol^{-1}$ respectively. It is possible to rationalise a large proportion of the observed results without recourse to discussions of 'structural' effects either of the solutes or from the solvent simply by assuming that their origin is primarily electrostatic and arises because during the condensation reaction charges are 'destroyed' when the peptides are formed. Using the classical Born dielectric continuum approach one can predict that the volume change associated with this nett charge reduction will be proportional to the function $\{(\partial \ln \epsilon/\partial P)/\epsilon\}$, where ϵ is the relative permittivity of the solvent, and to a function of the sizes and shapes of the ionic groups. If the charged groups are assumed to be spherical and are of radius r_+ and r_- for the amino and carboxylate groups respectively and are sufficiently far apart on the molecules to act independently, then the standard state volume changes for the formation of peptides will be given by

$$\Delta V^\ominus = K(V) \{n-1\} \{r_+^{-1} + r_-^{-1}\} \qquad (38)$$

For water at 25°C, the constant K(V) which depends only on solvent dielectric properties, has the value 4.17×10^{-16} m^4mol^{-1}. Using this and the slope in Fig.1.2 leads to a value for the term reflecting ionic size in equation (38) of 2.24×10^{10} m which corresponds to a harmonic mean radius for the charged groups of 89 pm. The standard state heat capacity changes can be treated in the same sort of way and the expression corresponding to equation (39) is

$$\Delta C_p^\ominus = K(C_p) \{n-1\} \{r_+^{-1} + r_-^{-1}\} \qquad (39)$$

where $K(C_p)$ depends on the temperature, the relative permittivity and the first and second derivatives of the permittivity with respect to temperature and, for water at 25°C has the value 5.41×0^{-9} J $K^{-1}mol^{-1}m$. Using this and the mean radius deduced from the volume changes leads to a predicted value for the slope of the heat capacity plot (Fig.1.3) of 121 J $K^{-1}mol^{-1}$ which is close to that observed. It is apparent therefore that most if not all of the effects seen in the volume and heat capacity changes come from electrostatic sources and for these examples at least specific effects are relatively minor. Specific effects are clearly evident for even relatively simple zwitterionic peptides and as an illustration, Table 1.2 gives some results for three isomeric peptides which differ in sequence.

TABLE 1.2

Standard state molar volumes and heat capacities of some isomeric peptides in water at 25°C (ref.15)

Peptide	$v^{\ominus}(cm^3\ mol^{-1})$	$C_{p,s}^{\ominus}(J\ K^{-1}\ mol^{-1})$
Glycylglycylalanine	128.79	296.0
Glycylalanylglycine	129.67	289.8
Alanylglycylglycine	130.16	286.2

Before moving away from this area it seems worth mentioning that some fairly comprehensive volumetric and heat capacity data have been obtained where concentrated urea solutions are used as the solvent (refs.5,16,17) and these have been used to evaluate the standard volume and heat capacity changes for the formation of peptides in 7 molar urea solutions. The results obtained for the standard volume changes are shown in Fig.1.4 and it is apparent that an approximately linear relationship exists, so by inference from the results obtained in water, one can conclude that much of what is observed in this solvent also comes from electrostatic sources and one need not invoke 'solvent structural' effects to account for at least most of the observations. It is not possible to construct a theoretical line for the urea-water solvent system because appropriate permittivity data are not available but assuming the effects seen are entirely electrostatic one can conclude that the function $\{(\partial \epsilon / \partial P)/\epsilon\}$ for 7 molar urea solutions is about 3×10^{-16} m^4 mol^{-1} which does not seem unreasonable given that urea-water mixtures are accepted (ref.18) as being more structurally disordered that water itself.

It was because of our awareness of the perturbing influence of the ionic groups on the naturally occurring compounds that we decided to use electrostatically neutral compounds to probe molecular interactions and in most of our recent work (refs.19-31) we have used terminally substituted amino acids and peptides. In view of the fact that we shall refer to them quite frequently, the compounds and their abbreviated names which will be mentioned are given in Table 1.3.

Confirmation that specific effects in solvation properties are present is indicated by the results (ref.31) which are given in Table 1.4 for the formation of peptides from uncharged amino acid derivatives. Here the reactions considered are of the type

TABLE 1.3

Structures and abbreviations of amino acid and peptide N-acetyl amides.

$$CH_3CON(R_1)CH(R_2)CONH(R_3)$$

R_1	R_2	R_3	Abbreviation
H	H	H	GLY
H	CH_3	H	ALA
H	$CH(CH_3)_2$	H	VAL
H	$CH_2CH(CH_3)_2$	H	LEU
H	$CH_2C_6H_5$	H	PHE
H	H	CH_3	GLYMe
H	CH_3	CH_3	ALAMe
H	$CH_2CH(CH_3)_2$	CH_3	LEUMe
CH_3	H	H	SAR
CH_3	CH_3	H	MAL
CH_3	H	CH_3	SARMe

$$CH_3CO-N\underline{\quad}CONH(R_3)$$

		R_3	Abbreviation
		H	PRO
		CH_3	PROMe

$$CH_3CONHCH(R_1)CONHCH(R_2)CONH_2$$

R_1	R_2		Abbreviation
H	H		GG
H	CH_3		GA
CH_3	H		AG
CH_3	CH_3		AA
H	$CH_2C_6H_5$		GF
$CH_2C_6H_5$	H		FG
CH_3	$CH_2C_6H_5$		AF
$CH_2C_6H_5$	CH_3		FA

$$CH_3CO-N\underline{\quad}CONHCH(R_2)CONH_2$$

	R_2		Abbreviation
	H		PG
	CH_3		PA
	$CH(CH_3)_2$		PV
	$CH_2CH(CH_3)_2$		PL

$$CH_3CONHCH(R_1)CO-N\underline{\quad}CONH_2$$

R_1			Abbreviation
H			GP
CH_3			AP
$CH_2CH(CH_3)_2$			LP

In addition:

	Abbreviation
$CH_3CO-N\underline{\quad}CO-N\underline{\quad}CONH_2$	PP
$CH_3CO-N\underline{\quad}CO-CH[CH_2CH(CH_3)_2]CONHCH_2CONH_2$	PLG
$CH_3CO[NHCH_2CO]_3NH_2$	GGG
$CH_3CO[NHCH(CH_3)CO]_3NH_2$	AAA

$$CH_3CONHCH(R_1)CONH_2 + CH_3CONHCH(R_2)CONH_2$$
$$= CH_3CONHCH(R_1)CONHCH(R_2)CONH_2 + CH_3CONH_2 \qquad (40)$$

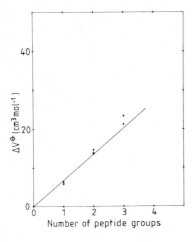

Fig.1.4 The standard state volume changes for the formation of peptides in 7 molar urea solutions at 25°C against the number of peptide groups formed.

TABLE 1.4

Standard state volume changes for the formation of peptide amides in water at 25°C [see equation (41)].

Compound formed	$V^{\ominus}/cm^3 \ mol^{-1}$
GLYGLY	1.86
ALAALA	1.28
PROPRO	3.52
GLYPRO	0.31
PROGLY	2.77
PROALA	1.67

At the moment only volumetric information is available although a study of the heat capacities is currently the subject of a collaborative investigation between G.R. Hedwig and our laboratory. Two features are apparent from the results given in Table 1.4. Firstly, the volume changes observed for the formation of both secondary and tertiary amide bond are considerably smaller than those found for the formation of the zwitterionically charged molecules, which indirectly confirms the above electrostatic arguments and secondly, that specific effects are clearly evident. It is of course these which require interpretation in terms of the structural characteristics of the molecules concerned. In Table 1.5 are given the standard state partial molar volumes of some amino acid and peptide derivatives (we have included in this table the volumetric virial coefficients also). Now there have been considerable efforts in the past to find predictive schemes for representing the molar volumes of polyfunctional compounds and these usually (ref.32) although not always (refs.33,34) have taken the form

$$V^{\ominus} = \Sigma_i n_i V_i^{\ominus} \qquad\qquad (41)$$

where n_i is the number of defined groups of type i and $V_i^{\ominus}$ is the group molar volume. For example Millero and his coworkers (ref.32) used equation (41) for the amino acids and established a table of group volumes of various functional groups in water at 25°C and there have been other examples where the same sort of approach has proved useful (refs.13,16)

TABLE 1.5

Volumetric results for some N–acetyl amino acid and peptide amides in water at 25°C

Solute (A)	$V_A^{\ominus}(cm^3 mol^{-1})$	$v_{AA}(cm^3 kg mol^{-2})$
GLY	90.56	0.23
ALA	108.06	−0.12
VAL	139.00	−1.15
PRO	126.51	−0.43
SAR	107.14	−0.06
GLYMe	108.93	−0.30
ALAMe	122.87	−0.59
LEUMe	174.75	−1.60
GG	127.37	−0.54
AA	161.79	0.32
PP	200.93	−1.37
GP	161.77	−0.37
PG	164.23	−0.31
PA	180.67	−0.80
P–D–A	180.63	−0.78

The abbreviations used are given in Table 1.3 In the last entry 'D' is added to stress that the D–isomer is used rather than the L–isomer.

The simplest application of equation (41) would be to assume that each molar volume is atom additive and a direct test of this possibility is provided by the three isomeric compounds GLYMe, ALA, and SAR which differ from each other only in the dispositions of hydrogen atoms and methyl groups on the backbone (see Fig.1.5). It is clear from the information given in Table 1.5 that there are however significant differences in the standard state volumes of these three compounds and this observation immediately negates the possibility of atom additivity. In view of this the next level of approximation would be to assume that the defined groups are the amino acid residues but even using this a volume additivity scheme does not apply very well and this is strikingly evidenced from the fact that for the two dipeptides GP and PG, the change in sequence results in a volume change of about 2.5 cm^3 mol^{-1}.

Notwithstanding this it is possible to construct a crude additivity scheme and the results obtained from the present systems are given in Table 1.6. The data set used is rather limited but our general feeling about the usefulness of such approaches is that although they can be used to estimate volumes they yield little molecular information and are quite inadequate for addressing the subtle problems of peptide–solvent interactions. Some examples were given earlier (Table 1.1 and Fig.1.1) which illustrated that the partial molar volume depends on the solvent considered and some further examples (ref.35) are given in Table 1.7 where a comparison is made of information on some amidic solutes in water and in the solvent N–methylacetamide. The superficial picture which emerges from these results is that whereas the simple amides generally have molar volumes which are greater in NMA than in water, if anything for the amino acid amides, the converse is so. It is also worth stressing that partial molar volumes can depend quite markedly on temperature (see Fig.1.6 for an example of this) and this dependence indicates that the solute–solvent interactions are also sensitive to temperature variations.

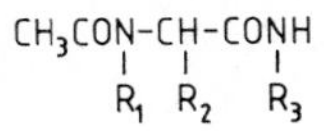

		R_1 R_2 R_3	
	GLYMe	H H CH_3	
$CH_3CON-CH-CONH$ 　　　$|$　$|$　$|$ 　　R_1 R_2 R_3	ALA	H CH_3 H	
	SAR	CH_3 H H	

Fig.1.5   Structures   of   some isomeric amino amides.

An important point which is often overlooked is that a relatively direct measure of solute–solvent interactions can be obtained from standard state partial molar volumes. The link between molecular events and this property was made some years ago (refs.40–42) but seems to have been largely ignored but as has been shown the following relationship

$$V_s^{\ominus} - RTK = L_A \int_0^{\infty} <1 - exp - w_{sw}(r)/kT> 4\pi r^2 \, dr \qquad (42)$$

is strictly correct. In this equation L_A is Avogadro's constant, K is the isothermal compressibility of the solvent and $w_{sw}(r)$ is the potential of mean force between one solvent molecule and one solute molecule in the pure solvent at a separation r and the brackets indicate that rotational averaging must be performed. At the moment evaluation of the functional dependence of $w_{sw}(r)$ is prohibitively difficult for molecules of any complexity but examination of the temperature dependence of the nett solute– solvent interaction for different compounds indicates some quite marked variations. In view of some comments which will be made later it is convenient to consider the function

$$\frac{\partial}{\partial T^{-1}} (V_s^o - RTK) = L \int_0^{\infty} <\{\frac{\partial}{\partial T^{-1}} w_{sw}(r)/kT\} exp - w_{sw}(r)/kT> 4\pi r^2 \, dr \qquad (43)$$

and Figs.1.7 and 1.8 illustrate this for the solutes urea (ref.11) and sucrose (ref.43) respectively . It is apparent from both of these diagrams that as the temperature is raised the slopes of the curves become less negative which is indicative of a gradual weakening in the solute solvation. In contrast to these, Fig.1.9 shows the corresponding information for some

TABLE 1.6

Approximate group contributions to the standard state partial molar volumes in water at 25°C.

Group	V (cm^3 mol^{-1})	
	Ref.31	Ref.32
$-NH-CH_2-CO-$	36.8	35.7
$-NH-CH(CH_3)-CO-$	53.7	51.2
$-NH-CH[CH(CH_3)_2]-CO-$	85.1	82.6
$-NH-CH[CH_2CH(CH_3)_2]-CO-$	104.1	98.4
$-N(CH_3)-CH_2-CO-$	53.3	51.5
$-N\text{---}CH-CO$	73.1	70.3

TABLE 1.7

Volumetric properties of some amidic solutes in N–methylacetamide at 32.6°C and in water at 25°C.

Solute	$V^\ominus$(cm^3 mol^{-1})	
	NMA(ref.35)	Water
F	38.88	38.51(ref.36)
NMF	59.32	56.87(ref.37)
DMF	77.45	74.50(ref.38)
A	55.67	55.60(ref.39)
DMA	93.24	89.65(ref.36)
GLY	90.65	90.56(ref.31)
ALA	107.57	108.06(ref.31)
VAL	138.40	139.00(ref.31)
PRO	124.93	126.51(ref.31)
PHE	170.11	–

alcohols in water (refs.44,45) and these exhibit quite different behaviour in that minima are clearly evident which correspond to the situation where the function given in equation (43) is zero. It is apparent that at this temperature the solvation as expressed by the exponential function in equation (42) is at a maximum.

The marked dissimilarity in the shapes of the curves shown in Figs.1.7 and 1.8 and Fig.1.9 indicates that the nature of the solvation is quite different and one can suggest that for urea and sucrose the solvation is rather specific and directional (this will be discussed later for saccharides) and that increasing the temperature simply overcomes these predominantly hydrogen–bonding interactions between these solutes and the solvent. The results for the molecules containing apolar residues do seem to be compatible with the idea that solvation of solutes containing apolar residues is qualitatively different and that an indirect contribution to the potential of mean force comes from the solvent structure. Of course, for polyfunctional molecules there will be contributions to the total effect from different groups and this is illustrated in Fig.1.10 for N–methylformamide in water which shows intermediate behaviour with a slight tendency for 'direct' interactions to predominate. It is unfortunate that there are

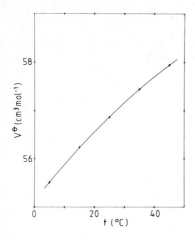

Fig.1.6 Temperature dependence of the partial molar volume of N—methylformamide in water.

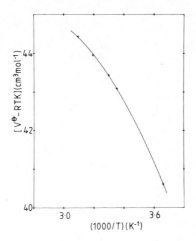

Fig.1.7 Plot of the function given in equation (43) against reciprocal temperature for aqueous urea solutions.

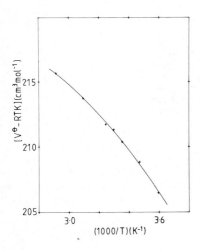

Fig.1.8 The same plot as in Fig.1.7 for aqueous sucrose solutions.

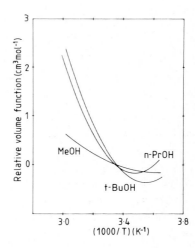

Fig.1.9 Corresponding plot to those in Fig.1.7 and 1.8 for some alcohols in water. The information is presented relative to the results at 25°C.

relatively few data available in the literature on the temperature variation of partial molar volumes since from the few examples we have chosen, these can serve to highlight solvation differences (see Table 1.8). The idea that the expansibility of solutes can give useful solvation information was suggested some years ago (refs.46,47) and used in a qualitative way but to the best of our knowledge has never been expressed from a rigorous viewpoint.

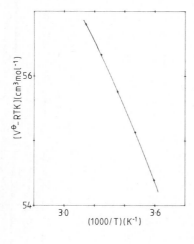

Fig.1.10 The same plot as Fig.1.7 for N-methylformamide solutions.

TABLE 1.8

Values for the function given in equation (44) for some solutes in water at 25°C.

Solute Function ($cm^3 \ mol^{-1}K$)	Sucrose	Urea	NMF	MeOH	EtOH	PrOH
	131	116	42	−46	−93	−219

1.4 SOLUTE–SOLUTE INTERACTIONS

We now turn to consider the behaviour of solutions where the solute concentrations are finite. The approach which is adopted in treating such systems from a thermodynamic viewpoint is based on the excess function idea which was initially proposed in an empirical way but has since been justified from statistical mechanical arguments.

1.4.1 Thermodynamic Formulation of Molal Excess Functions

As was mentioned above the concept of excess functions is used fairly extensively in various areas of thermodynamics and the approach which is most used in solution chemistry stems from some early work of Scatchard (refs.48,49) and has been popularised by Friedman (ref.50). The excess property (X^{ex}) of a solution is simply the difference between the value the property takes (X) for the real solution and the value it would have for the ideal solution (X^{id}). A pictorial representation of an ideal solution is shown in Fig.1.11a where we have a given number of solute molecules in a solvent and the solute molecules interact with peripheral solvent molecules to form the solvation regions but these solvation regions do not interact. The 'real' solution of the same solute is illustrated in Figure 1.11b and now for various reasons the

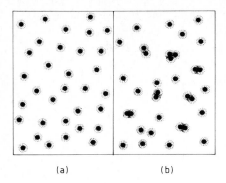

(a)   (b)

Fig.1.11 Schematic representation
of ideal and non-ideal solutions.

solvated solute species are interacting. The excess functions for this system can be imagined
to be the differences between the real and the ideal situations and they will clearly reflect not
only direct solute–solute but also solvation interactions. As might be expected for non–
cooperative situations in relatively dilute solutions, pairwise events would be more prevalent
than higher order contacts but these latter would become more important as the concentration
of the solute increases. It therefore seems reasonable to write the excess functions as
polynomials in the solute molality and if we consider the excess functions per kg. of solvent
then generally we have

$$X^{ex} = X - X^{id} = x_{AA} \, m_A^2 + x_{AAA} \, m_A^3 + \cdots \cdots = \Sigma \, x_{nA} \, m_A^n \qquad (44)$$

For the excess Gibbs function this becomes

$$G^{ex} = g_{AA} \, m_A^2 + g_{AAA} \, m_A^3 + \cdots \cdots = \Sigma \, g_{nA} \, m_A^n \qquad (45)$$

where g_{AA}, g_{AAA} *etc.* are molal free energetic virial coefficients. In experimental practice free
energetic virial coefficients are obtained from the molality dependence of either activity
coefficients or, more usually for non electrolytes, osmotic coefficients. The expressions for the
activity and osmotic coefficients for a single solute containing solution are obtained from
equation (45) and (15) and (16) and the results are

$$\mu_A^{ex} = RT \ln \gamma_A = 2g_{AA} \, m_A^2 + 3g_{AAA} \, m_A^3 + \cdots \cdots = \Sigma \, ng_{nA} \, m_A^{n-1} \qquad (46)$$

$$RT \, (\phi - 1) = g_{AA} \, m_A + 2g_{AA} \, m_A^2 + \cdots \cdots = \Sigma \, (n-1)g_{nA} \, m_A^{n-1} \qquad (47)$$

The most frequently used procedures for non–electrolytes is to use the colligative properties
of either vapour pressure or freezing temperature depression. Most of the available information
has been obtained using the isopiestic vapour pressure technique (ref.51) which has the

disadvantage that relatively concentrated solutions are required. There have been several studies in recent years (refs.7,52–58) which have used freezing temperature depressions and a suggested iterative procedure (ref.59) for determining solute concentrations and Fig.1.12 shows some illustrative results (ref.54). The difficulty with the colligative methods is that they always contain a very large contribution from the 'ideal' state and consequently the precision required in measurements is high if the excess properties are to be obtained with any accuracy.

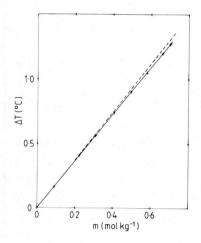

Fig.1.12 Freezing temperatures of aqueous acetamide solutions.

Enthalpic virial coefficients can be obtained from enthalpy of dilution measurements and using equation (45) it can be shown that the molar enthalpy of dilution, H^{ex} for the process of diluting a solution from an initial molality m_A to a final molality m_A' is

$$H_m^{ex} = (m_A' - m_A) \{h_{AA} + (m_A' + m_A) h_{AAA} + \cdots \cdots\} \qquad (48)$$

and Fig.1.13 shows some recent results (ref.12) for amidic solutes in water.

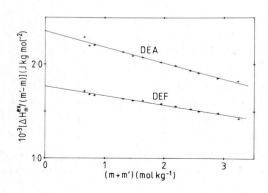

Fig.1.13 Illustration of the applications of equation (48) to two aqueous systems at 25°C.

Volumetric virial coefficients are obtained *via* precise density measurements and in recent years there have been a considerable number of systems investigated largely because of the comparative ease with which such measurements can now be made. Incidentally unlike the above energy functions the standard state function for the solute ($V_A^{\ominus}$) – the partial molar volume at infinite dilution (often written symbolically as $\phi_V^{\ominus}$) can also be obtained. An illustration of some results was given earlier in Fig.1.1.

Other thermodynamic virial coefficients (such as those associated with partial molar heat capacities, compressibilities *etc.*) are also accessible *via* appropriate measurements but we will not discuss these here although reference will be made to some of these a little later.

Before leaving this section one point should be made regarding the ideal state which is chosen from which to represent excess properties. As was mentioned above we adopted a molal scale representation of this and this is of course arbitrary and we could equally will have used say a molarity scale and obtained the analogous expression to equation (46)

$$G^{ex} = g'_{AA} c_A^2 + g'_{AAA} c_A^3 + \cdots \cdots \qquad (49)$$

where now c_A is the molarity of the solute and the g' terms are molar virial coefficients. Alternatively one can adopt a mole fraction definition of ideality for both the solute and the solvent and if this is done a difference results such that if we had a solution which was ideal on this scale but chose to represent ideality on the molal scale the excess Gibbs function takes the form

$$G^{ex} = -RT \, M_w \, m_A^2 \, \{1/2 - M_w \, m_A/6 + \cdots \cdots \} \qquad (50)$$

In other words there are terms present which are not removed by extrapolation to infinite dilution. The point which must be stressed however is that even though the molality definition of ideality is arbitrary so too is the mole fraction definition and not only is there no good reason to 'correct' free energetic coefficients, if this is done one must be careful when transposing the results to more fundamental states otherwise erroneous conclusions can be drawn.

1.4.2 Virial expansions of thermodynamic properties (ref.60)

In view of the fact that virial expansions of various forms are used fairly extensively at the present time for representing the thermodynamic behaviour of solutions containing both (or either) non–electrolytes, it is worthwhile justifying them from a statistical mechanics viewpoint and pointing out the relationships the virial coefficients have to molecular events.

The simplest point to start from is to consider the behaviour of a single component real (*i.e.* one in which the molecules are interacting) gas. The justification for beginning here is that the statistical mechanics is easier for gaseous systems than for solutions but the forms of the equations obtained are identical even though there are some important differences which will be alluded to later.

The Grand Canonical Partition Function (GCPF) for a system can be expressed, in terms of the Canonical Partition Function (CPF) and the absolute activity (λ) of the species in the system, as

$$GCPF = 1 + \Sigma \; (CPF)_N \; \lambda^N \tag{51}$$

Where N is the number of molecules in the system and λ is given by

$$\lambda = \exp \; (\mu/kT) \tag{52}$$

μ being the chemical potential (per molecule). Equation (51), which is a power series in the absolute activity, may be transformed into a corresponding power series in the activity (z) to give

$$GCPF = \Sigma \; (CI)_N z^N/N! \tag{53}$$

where $(CI)_N$ is the configurational integral for the N molecules and the activity used is defined by

$$z = (CI)_1 \lambda \; /V \tag{54}$$

V being the volume of the system. The Grand Canonical Partition Function can also be written in terms of the pressure of the system *i.e.*

$$GCPF = \exp \; (pV/kT) \tag{55}$$

and so combination of equations (53) and (55) eventually, after some manipulations, leads to an expression for the pressure (p) of the gas in terms of the activity. The resulting expression is

$$p/k1 = z + \Sigma \; b_n \; z^n \tag{56}$$

and the various virial coefficients (b_n, n>1) in this are well-defined cluster integrals which contain information on the intermolecular forces present in the system. The pressure and the activity are also linked to each other and to the number density (ρ) of the molecules by the thermodynamic relationship

$$\rho = [\; \partial(p/kT)/ \partial \ln z]_T \tag{57}$$

and so we can write the number density as

$$\rho = z + \Sigma \; n \; b_n \; z^n \tag{58}$$

i.e. we now have a virial expansion of the number density in terms of the activity. However since activities are not usually known from experiment but densities and pressures are, it is possible by using a straightforward series reversion procedure to obtain a different virial expansion *viz.*

$$p/kT \quad = \quad \rho \; + \; \Sigma \; B_n \; \rho^{\,n} \tag{59}$$

where now we have a direct link between the experimental quantities pressure and number density and some virial coefficients.

It is convenient at this point to express equations (58) and (59) in terms of molar rather than molecular properties and so by writing

$$K_n \quad = \quad b_n \; L_A^{\,n-1} \tag{60}$$

L_A being Avogadro's constant, and by defining a new activity

$$a \quad = \quad z/L_A \tag{61}$$

we obtain

$$p/RT \quad = \quad a \; + \; \Sigma \; K_n \; a^n \tag{62}$$

or in terms of molar concentrations (*i.e.* the molarity, c, of the gas)

$$p/RT \quad = \quad c \; + \; \Sigma \; B_n \; c^n \tag{63}$$

It should be noted that the sums in equations (62) and (63) are taken over all 'clusters' of two or more molecules and that when the gas is ideal then all K_A's and B_n's are zero and so the ideal gas law results.

The preceding is general and applies to all gaseous systems but for purposes of illustration let us suppose that we have a system in which interaction occur between pairs of molecules but interactions between more than two molecules are absent. Under this condition then equation (62) becomes

$$p/RT \quad = \quad a \; + \; K_2 \; a^2 \tag{64}$$

and only the second (activity) virial coefficient is needed to represent the system. However given that the activity may be related to the molarity through an activity coefficient (f) by

$$a \quad = \quad c \; f \tag{65}$$

The expression for the molarity becomes

$$c \quad = \quad a \; + \; 2K_2 \; a^2 \tag{66}$$

and that for the activity coefficient becomes

$$f = 1 - (2K_2 \, a^2/c) = 1 - 2K_2 \, c \, f^2 \tag{67}$$

Consequently the ratio of the experimental pressure to the pressure which would pertain if the gas was ideal is

$$p/p^{id} = \phi = f + K_2 \, c \, f^2 \tag{68}$$

This should be contrasted with the expression which is obtained from the concentration expansion, equation (63),

$$p/p^{id} = \phi = 1 + B_2 \, c + B_3 \, c^2 + \cdots \cdots \tag{69}$$

Using equations (66),(67) and (68) it can be shown (ref.60) that the activity expansion virial coefficient is related to the concentration expansion coefficients by $B_2 = -K_2$, $B_3 = 4K_2^2$, $B_4 = -20K_2^3$ *etc.* In other words for the particular system we have chosen to consider where at the molecular level only pairwise interactions occur, if one uses a concentration expansion then virial coefficients other than B_2 will be required. This is quite an important point since in the literature in many places statements are made on the 'chemical' significance of for example B_3 or coefficients which are related to this and most if not all of these are completely ill-conceived. It should of course be mentioned that for systems where triplet interactions do occur at the molecular level *i.e.* when K_3 is finite then the links with the concentration virial coefficients are more complex. Another point which is worth making is that using the concentration expansion leads to an expression in which when K_2 is positive (which as will be mentioned later corresponds to nett attractions between the species then the coefficients alternate in sign. This is commonly observed in solution situations.

So far we have obtained equations for representing the properties of an interacting system but have not given any quantitative links with molecular events. However it has been shown (ref.60) that the excess number ($\bar{N}_2$) of pairs of molecules which are in proximity is

$$\bar{N}_2 = V \, L_A \, K_2 \, a^2 \tag{70}$$

(where by excess we mean the difference between the real and ideal situations) and the excess molar concentration ($\bar{c}_2$) is

$$\bar{c}_2 = K_2 \, a^2 \tag{71}$$

Consequently we can write, using equation (69)

$$p/RT = \bar{c}_1 + \bar{c}_2 \tag{72}$$

and this equation states quite simply that the ideal gas law will *always* be obeyed provided *excess concentrations* are used. Now equation (72) is precisely what one would expect if on had relatively strong and localised associative interactions occurring between molecules and under these particular circumstances $\bar{c}_1$ can be identified with the concentration of monomeric molecules and $\bar{c}_2$ with the concentration of dimers. In this situation the virial coefficient, K_2, has the meaning of an association constant. However the major advantage of the approach, given above is that unlike equilibrium constants the various K_n's can adopt negative values in these situations where there are nett repulsions between the interacting species. Hence this virial expansion idea has more generality than the more commonly used equilibrium constant formalism in that it more closely reflects molecular events. Some years ago Guggenheim (ref.61) suggested that the term *sociation* should be used to represent the interaction between species so that when nett attractions occur one has negative sociation. The expression for the pairwise 'sociation constant', which is identical to the second virial coefficient in equation (69), is

$$K_2 = -B_2 = 2\pi L_A \int_0^\infty <(\exp -w(r)/kT) - 1> r^2 \, dr \qquad (73)$$

where the brackets (< >) denote that orientational averaging over the coordinates of the interacting molecules is to be performed and $w(r)$ is the intermolecular potential at a distance of separation r. It is worth noting that the corresponding expression for the pairwise *association* constant (K_a) of like molecules is

$$K_a = 2\pi L_A \int_0^? <\exp -w(r)/kT> r^2 \, dr \qquad (74)$$

and that there is necessarily some uncertainty about the upper limit of the integral since one can never state unambiguously the distance at which the interaction between species is zero. When sociation is strong and there is a relatively large interaction over a relatively short distance range then the upper limit of the integral in equation (74) is unimportant and to all intents K_2 and K_a are identical as implied above. However for weak positive sociation this is not so and the association constant can be arbitrarily set by the choice of the integral limit (or in experiment by the method of measurement) but this is not so for the sociation constant which is well-defined and retains its rigorous interpretation as a measure of excess pair concentrations which we stress again can be positive or negative.

The ideas presented above can be directly carried over to deal with solutions of a nonelectrolyte in a solvent (electrolytes add a further complication which owes from the long range nature of the interionic forces). In this situation the pressure in equations (62) and (63) must be replaced by the osmotic pressure (π) and the intermolecular potential in equation (73) now becomes a solvent averaged pair potential. The only other difference is that the equations for solutions

$$\pi/RT = a + \Sigma K_n a^n \qquad \text{[analogue of equation (62)]} \qquad (75)$$

and

$$\pi/RT = c + \Sigma B_n c^n \qquad \text{[analogue of equation (63)]} \qquad (76)$$

apply under conditions where the solution being investigated is held at its osmotic pressure. Now of course this is not the usual experimental situation where most frequently the system would be at a constant external pressure. However the transposition from what has been termed (refs.62–64) the McMillan–Mayer (MM) conditions to the usual Lewis–Randall (LR) conditions can be performed provided that appropriate ancillary volumetric data are available. We will return to this point later but it should be stressed that there can be very significant differences between the MM and the LR quantities and by way of illustration in Table 1.9 we give the osmotic coefficients on the two scales at some round values of molality for urea solutions in water at 25°C.

TABLE 1.9

Lewis–Randall (LR) and McMillon–Mayer (MM) osmotic coefficients at 25°C for aqueous urea solutions.

m(mol kg^{-1})	ϕ(LR)	c^{*}(mol dm^{-3})	ϕ(MM)
1	0.9624	0.9558	1.0047
2	0.9331	1.8349	1.0157
4	0.8904	3.3951	1.0488
6	0.8607	4.7357	1.0957
8	0.8338	5.8979	1.1441
10	0.8299	6.9148	1.1921

* The molarity is at a pressure equal to the osmotic pressure of the solution.

Superficial examination of these results using the LR information would lead one to conclude that because the osmotic coefficients are less than one fairly extensive association must be occurring between urea molecules in solution whereas if one considers the MM information the indications are that nett repulsions occur between urea molecules!

Theoretical approaches to solution thermodynamic properties such as those of McMillan and Mayer (ref.65) and Kirkwood and Buff (ref.40) use the molar scale (or the closely related number density scale) to express solution composition whereas as has been mentioned in experimental situations it is more customary to use the molality scale. The link between the two scales is straightforward to make since the molarity is related to the molality by

$$c_A = m_A / (V_w^{\ominus} + m_A V_{A,app}) \tag{77}$$

and as a consequence of this the molal activity coefficient of the solute is related to its molar activity coefficient by

$$\ln \gamma_A = \ln f_A - \ln \{(V_w^{\ominus} + m_A V_{A,app})/V_w^{\ominus}\} \tag{78}$$

and if we use a virial expansion for the apparent molar volume then this becomes

$$\ln \gamma_A = \ln f_A - (V_A^{\ominus}/V_w^{\ominus}) m_A + \cdots \tag{79}$$

The molar activity coefficient can also be expressed using the Kirkwood-Buff (refs.40,42) formalism as

$$\ln f_A = (2L_A/V_w^{\ominus}) (B_{AA} - B_{Aw})m_A + \quad \cdots \cdots \tag{80}$$

where B_{AA} and B_{Aw} are virial coefficients representing solute-solute and solute-solvent interactions. In the limit of dilute solutions equations (79) and (80) can be combined to give

$$\ln \gamma_A = \{(2L_A/V_w^{\ominus}) (B_{AA}^* - B_{Aw}^*) - (V_A^{\ominus}/V_w^{\ominus})\}m_A \tag{81}$$

where the asterisks on the B terms indicate the dilute solution situation. Since, using the usual molality expansion the activity coefficient is given by [see equation (47)]

$$\ln \gamma_A = (2g_{AA}/RT) m_A + \quad \cdots \cdots \tag{82}$$

we therefore have the following expression for the molal free energetic virial coefficient

$$g_{AA} = (RT/V_w^{\ominus}) \{L_A (B_{AA}^* - B_{Aw}^*) - (V_A^{\ominus}/2)\} \tag{83}$$

Now the coefficient $L_A B_{Aw}^*$ is simply one half of the right hand side of equation (42) i.e.

$$L_A B_{Aw}^* = (V_A^{\ominus} - RTK)/2 \tag{84}$$

and so incorporation of this into equation (83) gives

$$g_{AA} = (RT/V_w^{\ominus}) \{L_A B_{AA}^* - V_A^{\ominus} + RTK/2\} \tag{85}$$

The first term in the second parentheses is simply the McMillan-Mayer virial coefficient [see equation (76)] and equation (85) has been used quite frequently (see e.g. refs.60,66-68) for LR scale MM scale conversions i.e. to relate the experimentally deduced quantity g_{AA} to molecular events as expressed through $L_A B_{AA}^*$. The formal expression for this coefficient is

$$L_A B_{AA}^* = L_A \int_0^{\infty} <1 - \exp - w_{AA}(r)/kT> 2\pi r^2 \, dr \tag{86}$$

where $w_{AA}(r)$ is the potential of mean force between two solute molecules in the pure solvent.
     The molality virial coefficient can be expressed somewhat differently by combining equations (83) and (84) as

$$g_{AA} = (RT/V_w^{\ominus}) \{L_A(B_{AA}^* - 2B_{Aw}^*) - RTK/2\} \tag{87}$$

and if we recognise that the compressibility term is usually relatively small then

$$g_{AA} \approx (RT/V_w^{\ominus}) \{L_A(B_{AA}^* - 2B_{Aw}^*)\} \tag{88}$$

It is apparent therefore that the experimentally based LR free energetic virial coefficient is to all intents determined by the difference between a term which reflects solvated solute–solvated solute interactions and a term reflecting how the solute interacts with the solvent. In other words the LR coefficient g_{AA} directly monitors the *compromise* which is reached between solute–solute interactions and solute solvation and this does seem to be eminently sensible.

Manipulation of equation (88) gives the following expressions for the enthalpic and virial coefficients

$$h_{AA} = g_{AA} T\alpha + (R L_A/V_w^{\ominus}) \{ \partial/\partial T^{-1} (B_{AA}^* - 2B_{Aw}^*) \} \tag{89}$$

and

$$v_{AA} = - g_{AA} K + (RT L_A/V_w^{\ominus}) \{ \partial/\partial p (B_{AA}^* - 2B_{Aw}^*) \} \tag{90}$$

where α and K are the expansibility and compressibility coefficients for the solvent. It is apparent from these last two equations that these virial coefficients also represent balances between interaction and solvation but of course since these are now derivative quantities the weightings can be different not only from each other but also from the free energetic term. There are some interesting analogies, which will not be explored here, between the present formulation of the properties of solutions and the concept of the Boyle temperature for gases and the behaviour of polymers in different solvents.

1.4.3 Experimental results on N–acetyl amino acid amides in water

The experimental information which is currently available on free energetic, enthalpic and volumetric second virial coefficients for terminally substituted amino acids is given in Table 1.10. For many of the systems studied third and sometimes fourth virial coefficients are required to represent the experimental data but given the problems alluded to earlier on the meaning of higher coefficients they have not been included in the table. Perusal of the data presented shows that the following qualitative features are evident.

(a) The free energetic coefficients are all negative and generally become more so as the amino acid side chain becomes more apolar.

(b) The enthalpic coefficients exhibit both positive and negative values, the latter occurring for the more polar residues.

(c) The limited information on volumetric coefficients shows that both positive and negative values occur and in contrast to the enthalpic terms the more apolar the solute the more negative is the coefficient.

(d) All three types of coefficients show a wide range of values. This last observation is important since it allows one quite categorical point to be made which is that the interactions which are occurring between molecules containing amide (peptide) groups are not dominated by hydrogen–bonding interactions of the sort shown in Fig.1.14. The first three points implicate the solvation characteristics of the apolar side–chains. A further

TABLE 1.10

Virial coefficients for amino acid amides in water at 25°C. The units of g and h are J kg mol^{-2} and v cm^3 kg mol^{-2} respectively.

Solute species		$-g_{AB}$	h_{AB}	v_{AB}
A	B			
GLY	GLY	83	−220	0.23
ALA	ALA	114	269	−0.12
VAL	VAL		1259	−1.15
LEU	LEU	732	1714	
PHE	PHE		1049	
GLY	ALA	186	86	
GLY	VAL		385	
GLY	LEU	217	547	
ALA	VAL		591	
ALA	LEU	280	899	
VAL	LEU		1486	
PHE	GLY		−175	
PHE	ALA		356	
PHE	VAL		1392	
PHE	LEU		1960	
GLYMe	GLYMe	77	585	−0.30
ALAMe	ALAMe	119	1181	−0.59
LEUMe	LEUMe	681	3420	−1.60
GLYMe	ALAMe	28	959	
GLYMe	LEUMe	229	1392	
ALAMe	LEUMe	381	1988	
PRO	PRO	111	660	−0.43
MAL	MAL	171	587	
SAR	SAR	90	145	−0.06
PROMe	PROMe	174	1763	
SARMe	SARMe		1417	
MAL	PRO	108	608	
MAL	SAR	68	514	
PRO	SAR	173	367	
PRO	GLY	100	226	
SAR	GLY	122	18	
PRO	ALA	100	412	
SAR	ALA	164	264	
PRO	VAL		882	
SAR	VAL		684	
PRO	LEU	386	1273	
SAR	LEU	204	851	
PRO	GLYMe	114		
PRO	ALAMe	180		
PRO	LEUMe	449		

The abbreviations used for the compounds are given in Table 1.3.
The data sources for this table are refs.19–22;24–29;31;69,70.

inference which can be drawn from closer inspection of the results is that the interactions which are occurring, as reflected in the virial coefficients, cannot be represented by simple additive functions.

Most treatments of molecular interactions adopt what can be termed 'chemical' approaches and either implicitly or explicitly assume that some particular effect is dominant. To illustrate the basis of such treatments and how they would manifest as virial coefficients we

28

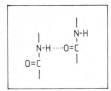

Fig.1.14 'Direct' hydrogen bonding of amide groups.

consider the situation shown in Fig.1.15 where two molecules both of type A and which contain only the functionalities i and j are interacting. If the i–i interactions are predominant then the molecules will tend to associate in the way shown and the nett energetics of the pairwise interaction would be composed of two i–i interactions and three j–j interactions so for example the enthalpic virial coefficient would be of the form

$$h_{AA} = 2H_{i-i} + 3H_{j-j} \tag{91}$$

where the H terms represent the interactions of the subscripted functions and the virial coefficient depends linearly on the number of functional groups present. In contrast to this 'chemical' formulation Savage and Wood (ref.9) in an important and seminal paper, suggested that an alternative view could be taken. Their approach, which we have given the acronym 'SWAG' (Savage and Wood Additivity of Groups) has been used fairly extensively (see *e.g.* ref.10 for references to earlier work) to represent and correlate molecular interactions in solution.

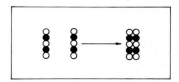

Fig.1.15 Schematic illustration of 'direct' interactions.

1.4.4 Group additivity

The SWAG approach in essence is similar to an earlier idea put forward by Schrier and Schrier (ref.71) when they considered the behaviour of aqueous solutions containing amides and salts and attempted to use this to model some aspects of salt–induced protein denaturation. The idea was that if one considers a series of amides containing apolar (say $-CH$, $-CH_2$ and $-CH_3$) and amide (CONH) functionalities interacting with an ion then if each interaction is weak *i.e.* energetically small compared to thermal energies, then the total interaction of an amidic molecule with the ion will depend simply on the numbers of each of the groups on the amide and how each group interacts with the ion. Fig.1.16 illustrates the basis of the approach and some recent results (ref.72) which seem to confirm the ideas are shown in Fig.1.17.

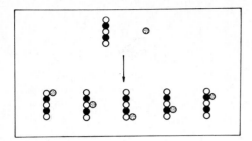

Fig.1.16 Schematic representation of 'weak' interaction between a solute and an ion.

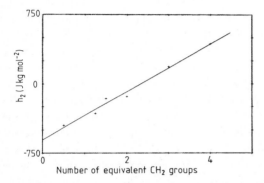

Fig.1.17 The variation of the heterotactic virial coefficients for the interaction of some amides with KCl in water at 25°C against the number of equivalent methylene groups on the amide.

In the SWAG formulation the conceptual basis of this treatment was extended to consider the interactions occurring between, for example, two amide molecules in solution and then if no one group interaction is dominant, every functionality on one molecule 'samples' every functionality on the other. Consequently for the situation shown in Fig.1.18 the nett enthalpic virial coefficient will take the form

$$h_{AA} = 4H_{i-i} + 9H_{j-j} + 12H_{i-j} \qquad (92)$$

In other words, in contrast to the expression given in equation (91), which applies for 'strong' interactions, the virial coefficient depends *quadratically* on the numbers and natures of the groups present. This approach is easily generalised for the interaction of any two species to give for the enthalpic coefficient

$$h_{AB} = \Sigma \Sigma n_i{}^A n_j{}^B H_{i-j} \qquad (93)$$

if all of the functionalities are all mutually completely accessible. In this equation $n_i{}^A$ denotes the number of i groups on molecule A and similarly for j on B and the H_{i-j}'s are parameters representing the interaction between a single i group and a single j group. There are several

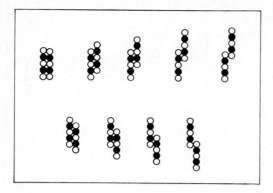

Fig.1.18 Schematic representation of 'weak' interactions occurring between two solutes.

ways of testing and illustrating the SWAG idea and we will take four examples drawn from work from our laboratory. Before beginning this however, one problem of the approach needs to be mentioned which is that it tends to be rather demanding of data and when relatively limited data sets are considered some simplifying approximations are necessary if one is not to have an unacceptably large number of H_{i-j} coefficients. Considerable simplification results if for systems containing aliphatic groups, one adopts the procedure suggested by Friedman and adopted by Savage and Wood (ref.9) that methyne and methyl groups be approximated by one half and one and one half methylene groups respectively. In what follows we have adopted this approximation.

We consider first homotactic and heterotactic enthalpic virial coefficients for some aliphatic amino acid amides and some simple amides. Assuming that primary and secondary amide groups are equivalent to each other then equation (93) becomes

$$h_{AB} = n_{CH_2}^A n_{CH_2}^B H_{CH_2-CH_2} + (n_{CH_2}^A n_{pep}^B + n_{CH_2}^B n_{pep}^A) H_{CH_2-pep}$$
$$+ n_{pep}^A n_{pep}^B H_{pep-pep} \qquad (94)$$

where now $n_{CH_2}^A$ for example denotes the number of equivalent methylene groups on molecule A and the abbreviation 'pep' denotes both primary and secondary amide functions. Fig.1.19 shows the result (ref.26) of fitting the results of a considerable number of systems and the values obtained for the group coefficients were $H_{CH_2-CH_2}=25$; $H_{CH_2-pep}=81$ and $H_{pep-pep}=-292$ (all with units of J kg mol^{-2}). It is apparent from this figure that these three coefficients represent the enthalpic information for a large number of systems well if not perfectly. Superficially it might seem that the relatively large value for the $H_{pep-pep}$ term would lead to this dominating the interactions but that this is not so can be seen from Fig. 1.20 where we have shown the contributions the three terms in equation (94) make to the homotactic interactions of the N-acetylamides of glycine and leucine. It is worth noting the relative magnitudes of the contributions arising from the apolar and polar interactions in the two cases. We now address the heterotactic interactions occurring between N-acetyl-L-phenylalaninamide (PHE) and the corresponding glycyl, L-alanyl, L-valyl and L-leucyl compounds (ref.27). If we take the aromatic residue, C_6H_5-, to be a distinct group (phe) and apply equation (93) then the following

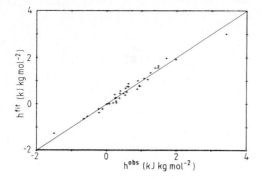

Fig.1.19 Correlation of experimental and fitted enthalpic virial coefficients for amidic solutes in water at 25°C.

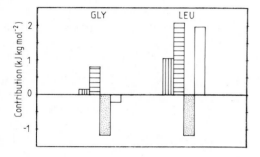

Fig.1.20 Group additivity components for the homotactic enthalpic virial coefficients for N-acetylglycinamide and N-acetyl-L-leucinamide.

expression results:

$$h_{PHE-B} - [3n_{CH_2}^B H_{CH_2-CH_2} + 4H_{pep-pep} + 2(3 + n_{CH_2}^B) H_{CH_2-pep}]$$
$$= n_{CH_2}^B H_{CH_2-phe} + 2H_{phe-pep} \qquad (95)$$

The left hand side of this equation is calculable using experimental results and the previous H_{i-j} coefficients and since the numbers of equivalent methylene groups ($n^B_{CH_2}$) on the amino acid amides are known from their molecular constitution, the unknown coefficients can be obtained. Fig.1.21 illustrates this and from it one gets $H_{CH_2-phe}=-295$ and $H_{phe-pep}=-410$ (both with units of J kg mol^{-2}). The value obtained from the first of these is not unexpected given the value obtained for the CH_2-CH_2 interactions but the second is both unexpected and striking and suggests a relatively strong interaction occurring between the osmotic residue and the peptide group. There is some earlier evidence in the same regard from intramolecular studies on cyclic peptides (refs.73–75) and both this and the present observation suggests that aromatic group–peptide group interactions might well play a significant role in protein architecture.

Our third illustration considers some information obtained on the imino acid amide of proline. This compound contains the tertiary amide functionality (–CON–) and also is severely

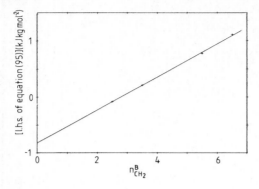

Fig.1.21 Plot of equation (95).

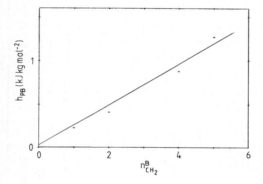

Fig.1.22 Plot of equation (96).

restricted in its motions because of the presence of the closed pyrrolidine ring. If we consider that this tertiary amide group (ipep) is distinct from both the primary and secondary amide functionalities and look at the heterotactic interactions of PRO with the same amino acid amides as for the phenylalaninamide above   then using the SWAG formulation we get the two term expression

$$h_{PB} = \{7.5H_{CH_2-CH_2} + 2H_{pep-pep} + 11.5H_{CH_2-pep} + 2H_{pep-ipep}$$
$$+ 1.5H_{CH_2-ipep}\} + \{5H_{CH_2-CH_2} + H_{CH_2-pep} + H_{CH_2-ipep}\} n_{CH_2}^B \quad (96)$$

Fig.1.22 illustrates the use of this expression and it is apparent that a tolerably good correlation is obtained. this and the earlier values for the coefficients $H_{CH_2-CH_2}$, $H_{pep-pep}$ and H_{CH_2-pep} the unknown terms can be evaluated. The final example which we will mention comes from some recent work from our laboratory (ref.76) but also utilises information obtained by others (ref.9) and addresses the heterotactic interactions occurring between the protein denaturant urea and some simple primary, secondary and tertiary amides containing aliphatic residues. The reason for pursuing this investigation was simply to see if some light could be

TABLE 1.11

Heterotactic enthalpic virial coefficients for amides with urea.

AMIDE	$h_{AB}(J\ kg\ mol^{-2})$
Formamide	−261
Acetamide	−142
NMF	−132
NMA	15
DMF	−155
DMA	−70
DEF	36
DEA	135
NMP	180
NBA	264

thrown on the problem of how urea interacts with apolar residues and peptide groups. A summary of the information which has been obtained is given in Table 1.11. When adopting a SWAG approach it is convenient to consider the urea molecule as being a distinct group although others have used different approximations and given this the heterotactic amide– urea interaction will be of the form

$$h_{Au} = n_{CH_2}^A\ H_{CH-u} + H_{pep'-u} \tag{97}$$

where $n_{CH_2}^A$ is the number of equivalent methylene groups on the amide and the abbreviation pep' refers to the amide group being primary, secondary or tertiary. The results obtained are shown in Fig.1.23 and it seems apparent from this that they fall into two distinct sets, one coming from the primary and secondary amides and the other from the tertiary amides. We analysed the information assuming this was so and the lines drawn in Fig.1.23 show the results obtained. The most striking feature which comes from this analysis and which is clearly apparent from the figure is that from an enthalpic viewpoint the tertiary amide function interacts with urea much more favourably than either the primary or secondary amide functions. This seems to be quite contrary to 'chemical intuition' and serves to illustrate the problem of using such thought processes for the weak interactions which occur in condensed media.

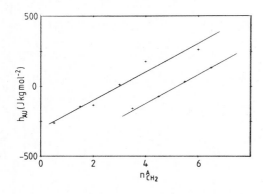

Fig.1.23 The homotactic interactions of some amides with urea in water at 25°C [see equation (97)].

Table 1.12 contains a collection of the free energetic and enthalpic interaction coefficients which we have obtained from group additivity analysis of the data on amide and amino acid (plus some simple peptide) systems. For completeness the results of SWAG analyses on the available free energetic data are included although as will be mentioned later we have misgivings about the validity of such a method of analysis for free energetic virial coefficients. The values obtained for the various interaction coefficients differ somewhat from those obtained by others using a much less coherent data set containing a wider range of functionalities and the indications are that the best fits to data are obtained when the molecules considered have structural and functional similarity. This is confirmed by some recent work by Barone and his coworkers (refs.77,78).

TABLE 1.12

Group interaction coefficients of functional groups in water at 25°C. The units of each are J kg mol^{-2}.

Groups		G_{i-j}	H_{i-j}	TS_{i-j}
i	j			
CH_2	CH_2	−20	25	45
pep	pep	−50	−292	−242
ipep	ipep	−62	−319	−257
CH_2	pep	22	80	58
CH_2	ipep	37	49	12
pep	ipep	−62	−319	−257
CH_2	phe	−	295	−
phe	pep	−	−410	−
phe	phe	−	898	−
CH_2	urea	−	101	−
pep	urea	−	−296	−
ipep	urea	−	−517	−

If we take the information given in Table 1.12 as it stands then several features emerge some of which have been referred to earlier. The CH_2–CH_2 interactions have a negative free energetic coefficient which is composed of positive enthalpic and entropic terms and this finding does seem to be in accord with the general view of the factors involved in the interactions of hydrophobic moieties in water. The pep–pep interaction is likewise negative but this also has large negative enthalpic and entropic components which indicate that the nature of the favourable interaction occurring between these groups is quite different to that occurring between the apolar residues. The tertiary amide–tertiary amide functions self–interaction does seem to be a little different to the pep–pep interaction in that it has more negative enthalpic and free energetic coefficients. The indications are therefore that when the hydrogen–bond denoting propensity of amide groups is removed then they interact together more favourably. This is analogous to the situation referred to above when amide–urea interactions were discussed and using a much over–simplified representation of what must be occurring, it seems that the nett process

$$
\begin{array}{ccc}
\overset{|}{-}\text{C}\overset{|}{-}\text{N}- & & \overset{|}{-}\text{C}\overset{|}{-}\text{N}- \\
\underset{\vdots}{\overset{\parallel}{\text{O}}} & + & \underset{\vdots}{\overset{\parallel}{\text{O}}} \\
\text{H}_2\text{O} & & \text{H}_2\text{O}
\end{array}
\longrightarrow
\begin{array}{c}
\overset{|}{\text{C}}{=}\text{O}\cdots\overset{|}{\text{N}}- \quad + \quad \text{H}_2\text{O} \\
-\overset{|}{\text{N}} \qquad \overset{|}{\text{C}}{=}\text{O}\cdots\text{H}_2\text{O}
\end{array}
$$

is more favoured than

$$
\begin{array}{c}
\overset{|}{\text{C}}{=}\text{O}\cdots\text{H}_2\text{O} \quad + \quad \text{H}_2\text{O}\cdots\text{H}\overset{|}{-}\text{N} \\
\text{H}_2\text{O}\cdots\text{H}\overset{|}{-}\text{N} \qquad\qquad \overset{|}{\text{C}}{=}\text{O}\cdots\text{H}_2\text{O}
\end{array}
\longrightarrow
\begin{array}{c}
\overset{|}{\text{C}}{=}\text{O}\cdots\text{H}\overset{|}{-}\text{N} \\
\text{H}_2\text{O}\cdots\text{H}\overset{|}{-}\text{N} \qquad \overset{|}{\text{C}}{=}\text{O}\cdots\text{H}_2\text{O}
\end{array}
\quad + \quad \text{H}_2\text{O}
$$

in water. The apolar–polar CH_2–pep and CH_2–ipep coefficients do suggest repulsive interactions but it is interesting that the latter are more repulsive which also seems to be the corollary of the polar–polar interactions. We should also stress once again the marked difference between the enthalpic coefficient for aromatic–peptide group interactions and aliphatic–peptide group interactions. This is a convenient point at which mention should be made of some recent work (ref.76) in which, rather than use water as the solvent, mixed solvent containing urea and water was used. The objective of this study was to see how denaturing media modify the solute–solute interactions and since there is contention regarding the major contributions involved in such media to protein stability, a series of amides were investigated which had greater and lesser polar–apolar ratios. A summary of the results obtained in 8 molar urea is given in Table 1.13. (It should be mentioned that information is available for other urea concentrations also). Included in this table is information drawn from the literature for the corresponding virial coefficients in water and it is apparent that the homotactic interactions are generally more positive in the urea–water mixture than in water. Since the increasingly positive values obtained as apolarity increases in water arise in large part from hydrophobic interactions superficially it appears that such interactions are more prevalent in the mixture than in water which does of course contradict a whole mass of evidence. An alternative and more satisfactory conclusion emerges, however, if the previously mentioned information on the heterotactic interactions which occur between amides and urea *in water* is considered. The clear indication from this is that there is a relatively strong interaction occurring between urea and the peptide (primary, secondary and tertiary amidic) groups and so one imagines that in the concentrated urea solution such groups will be effectively saturated by urea molecules [or preferentially solvated (refs.79,80) to use a different terminology]. Now if we apply the group additivity approach to the homotactic interactions occurring in urea solution we get

$$
h_{AA} = n_{CH_2}^A \, H_{CH_2-CH_2} + 2n_{CH_2}^A \, H_{CH_2-pep} + H_{pep-pep} \tag{98}
$$

where the H terms will be different to those obtained in water since the solvent is now different. Assuming that the amidic groups will have no tendency to associate since they are completely 'satisfied' by their peripheral urea one is led to the conclusion that the enthalpic

TABLE 1.13

Homotactic enthalpic virial coefficients of some amides in water and in 8 molar aqueous urea. (Units are J kg mol^{-2}).

	Water	8 molar urea
Formamide	−115	32
Acetamide	1	157
NMF	272	352
NMA	235	538
DMF	737	810
DMA	1081	1060

virial coefficient will be dominated by the first term on the right hand side of equation (98) $i.e.$ the enthalpic virial coefficient will vary with the square of the number of equivalent methylene groups on the amides. That this is indeed the situation which prevails is illustrated in Fig.1.24 which has a slope of about 60 J kg mol^{-2} corresponding to a $H_{CH_2-CH_2}$ value which is more than twice that which is found in water (precise statements about the relative effects are not possible because of the solvent media being different). The question one could ask is "If positive enthalpic coefficients arise in water because of hydrophobically dominated interactions between species does the more positive value obtained in urea solutions mean that the interactions occurring are more hydrophobic in origin?" Our view is that such a superficial view is untenable in view of the large amount of evidence which is available suggesting that urea disrupts the solvent structure and that the value for the CH_2-CH_2 interaction coefficient in urea-water mixtures arises largely from solvation effects.

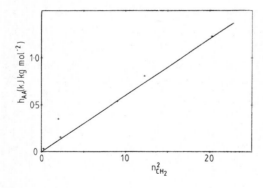

Fig.1.24 The homotactic enthalpic virial coefficients of some amides in 8 molar urea solutions at 25°C.

It was mentioned above that we have some doubts (ref.29) about the applicability of the SWAG idea to free energetic virial coefficients and these can be formulated as follows. Equation (85) gives the link between the Lewis-Randall and McMillan-Mayer homotactic virial coefficients

$$g_{AA} = (RT/V_w^\ominus) \{_{LR}B_{AA}{}^* - V_A^\ominus + KRT/2\} \qquad (85)$$

and the fundamental assumption of the group additivity approach is that the virial coefficients can be written in the form

$$g_{AA} = \Sigma \Sigma n_i^A n_j^A G_{ij} \tag{99}$$

Now it is well known that the molar volumes of non–electrolytes in water are at least approximately linear in the number and types of group on the solute. In other words we have [see equation (42)]

$$V_A^{\ominus} \approx \Sigma n_i^A V_i^{\ominus} \tag{100}$$

It is therefore clear that incorporation of equations (99) and (100) into equation (85) leads to an expression for the experimental virial coefficient

$$g_{AA} \approx (RTL_A B_{AA}^*/V_w^{\ominus}) - (RT/V_w^{\circ}) \Sigma n_i^A V_i^{\ominus} \tag{101}$$

which contains a term linear in the number of groups whereas the SWAG formulation assumes an overall quadratic dependence. It is possible of course that the volumetric contribution to the virial coefficient is sufficiently small so that the B_{AA}^* term is dominant but calculation for some illustrative systems (see Table 1.14) shows that this is not so. It this appears that the underlying physical basis for the treatment of free energetic data by the group additivity approach is weak but this does not exclude its empirical utility for the compact representation of bodies of data.

If we now consider enthalpic coefficients in the same light then the following equation results by appropriate differentiation of equation (85)

$$h_{AA} = g_{AA}T\alpha + \{(RL_A/V_w^{\ominus}) (\partial B_{AA}^*/\partial T^{-1})\} + (RT^2 E_A^{\ominus}/V_w^{\ominus}) - (KR^2T^2/2V_w^{\ominus}) \tag{102}$$

where $E_A^{\ominus} = (\partial V_A^{\ominus}/\partial T)$ and the isothermal compressibility of the solvent is assumed to be independent of temperature. The only amino acid amide for which expansibility data exist is N–acetylglycinamide and calculation of the various terms on the r.h.s. of eq (102) gives them as being −6, −271, 58 and 1 (all in units of J kg mol^{-2}) reading from left to right. The experimental virial coefficient is −220 J kg mol^{-2} and since this is not too dissimilar to the second term of equation (102), the indication is that the SWAG postulate regarding the quadratic nature of group interactions may be applicable to enthalpic coefficients. Some confirmation of this is also obtained if one considers the available information on aqueous urea (ref.11) and aqueous N–methylformamide (refs.19,37) solutions. The enthalpic virial coefficient for urea is −359 J kg mol^{-2} and the second term in equation (102) is −399 J kg mol^{-2}. Of course whether a true group quadratic dependence does result depends on a large number of factors including and involving the form of the solute–solute potential of mean force and the solvation characteristics of the defined groups. It has been suggested (ref.81) that the solvation perturbations by different proximate groups could significantly effect the constancy of parameters reflecting group interactions but although this is eminently reasonable it does not in itself negate SWAG

TABLE 1.14

Components of free energetic virial coefficients for some amidic compounds in water at 25°C.

Solute species		$(V_w^{\ominus} g_{AB}/RT)$	$L_{AB}{}^{*}{}_{AB}$
A	B	$(cm^3 mol^{-1})$	$(cm^3 mol^{-1})$
GLY	GLY	-34^a	57
GLY	ALA	-75^a	23
GLY	LEU	-88^a	36
GLY	PRO	-40^b	68
GLY	SAR	-49^b	49
ALA	ALA	-58^a	49
ALA	LEU	-113^a	19
ALA	PRO	-40^b	76
ALA	SAR	-66^b	41
LEU	LEU	-296^a	-139
LEU	PRO	-156^b	-14
LEU	SAR	-83^b	50
PRO	PRO	-45^c	81
PRO	SAR	-70^c	46
SAR	SAR	-36^c	70
GLYMe	GLYMe	-31^d	77
GLYMe	ALAMe	-11^d	104
GLYMe	LEUMe	-93^d	49
GLYMe	PRO	-46^c	71
ALAMe	ALAMe	-48^d	74
ALAMe	LEUMe	-154^d	-6
ALAMe	PRO	-73^c	51
LEUMe	LEUMe	-276^d	-101
LEUMe	PRO	-182^c	-32
GG	GG	-49^e	78
GG	GLY	-52^e	56
GG	PRO	-66^e	60
AA	AA	-81^e	80
AA	ALA	-7^e	127
PP	PP	-98^e	103
GP	GP	-137^e	25
PG	PG	-93^e	70
PG	GLY	-86^e	41
PG	PRO	-76^e	69
PA	PA	-42^e	138
A	A	-59^f	-2
DMA	DMA	-72^g	17
F	F	-13^h	25
DMF	DMF	-52^h	22

The volumetric information on the aminoacid and peptide amides are taken from ref.31 and that for the simple amides from ref.36–38. The free energetic virial coefficients were obtained from : (a) ref.21; (b) ref.28; (c) ref.29; (d) ref.26; (e) ref.69; (f) ref.7; (g) ref.54; (h) refs.7,54.

approaches since the semiquantitative nature of such treatments was clearly recognised in the seminal paper (ref.9).

1.4.5 Comments on solutions containing saccharides

One group of compounds for which the SWAG idea is of little utility is the saccharides and these have been extensively investigated by the Naples group led by Barone. It is quite clear from the extensive amount of information which has been accrued that the conformational

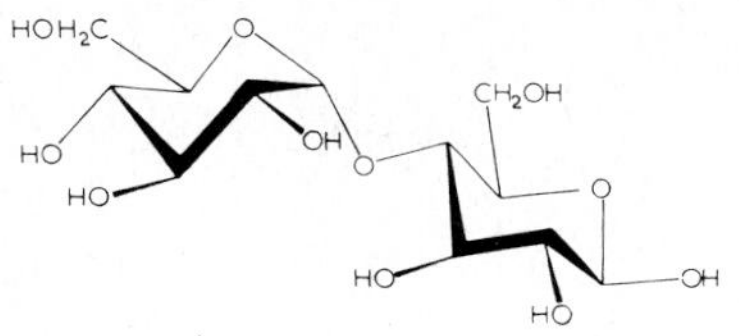

Maltose

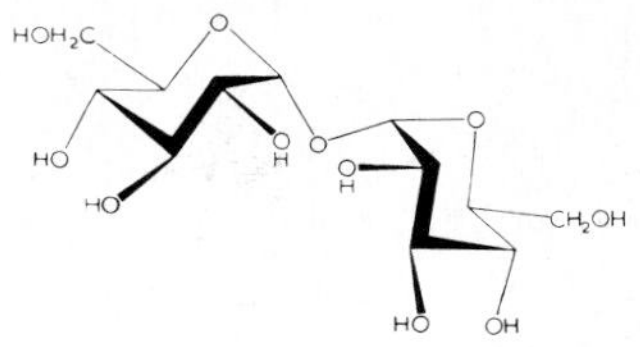

Trehalose

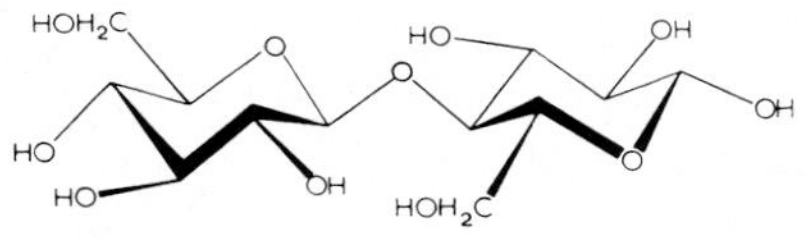

Cellobiose

Fig.1.25 The structures of some isomeric disaccharides.

TABLE 1.15

Homotactic and heterotactic enthalpic second virial coefficients for three disaccharides.

	h_{AB}(J kg mol^{-2})		
	Maltose	Cellobiose	Trehalose
Maltose	571	—	
Cellobiose	734	689	—
Trehalose	574	703	795

properties as well as the group contribution plays a significant role in determining the interactive energetics. We will not discuss this work in detail since it has been reviewed quite recently (ref.82) but rather will illustrate some at least of the problems involved by using recent information (ref.83) on some isomeric disaccharides (see Fig.1.25 and Table 1.15). It is apparent from Table 1.15 that rather significant variations in the virial coefficients are observed which indicate that the SWAG approach is inapplicable for such systems. A general feature which the information presented does indicate is that the enthalpic virial coefficients for saccharides in water are invariably positive and although relatively few free energetic coefficients are currently available (ref.82) these also generally exhibit this feature. Some years ago it was suggested (ref.84) that the hydration of monosaccharides is dependent not only on the number of hydroxyl groups the saccharides have but also on their positions and orientations. It was

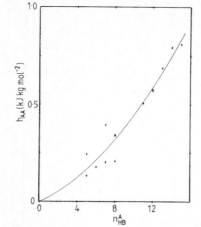

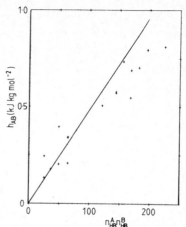

Fig.1.26 Correlation of enthalpic virial coefficients of some saccharides with the hydrogen bonding numbers (see text).

Fig.1.27 Plot of equation (103) for some saccharides.

proposed that equatorial hydroxyl groups should have a greater propensity for hydrogen-bonding to the solvent (which was assumed to have a significant structural remnant from the tridymite ice–lattice) than axial groups. Later studies (see $e.g.$ ref.85) incorporated this idea into a specific hydration model and the suggestion was made that the compatibility of a monosaccharide with a structured water lattice has a major effect on the extent of hydration. This approach has recently been extended (ref.83) and a range of saccharides of known structures have been incorporated into an extended water (ice) lattice using model–building techniques so that hydrogen bonding was optimised whilst maintaining sterically and electronically acceptable conformations for each saccharide. The numbers of hydrogen–bonds each saccharide makes with the lattice was then determined. Fig.1.26 shows that for a range of saccharides there is a reasonable albeit a non–linear correlation between the enthalpic virial coefficients and the number of hydrogen bonds the solutes can form. The correlation can be at least approximately linearised (see Fig.1.27) by assuming that the interaction energetics of two saccharidic species in water depends on the product of the numbers of hydrogen bonds the two species can make with the solvent $i.e.$

$$h_{AA} = n_{HB}{}^{A}\, n_{HB}{}^{B}\, H_{HB-HB} \tag{103}$$

with n^{A}_{HB} and n^{B}_{HB} denoting the 'hydrogen bond' numbers for A and B respectively and H_{HB-HB} is an enthalpic coefficient which represents interactions between the hydrogen bonded species. It is perhaps worth mentioning that this correlation rationalises an observation made some time ago (ref.86) and which has been used and discussed by others (see $e.g.$ ref.87) namely that the heterotactic virial coefficient is approximately given by the square root of the product of the component homotactic coefficients. The information summarised in Fig.1.27 if nothing else serves to highlight the role of solvation in solute–solute energetics.

TABLE 1.16

Components of the homotactic free energetic virial coefficients for some saccharides in water at 25°C.

Solute	$(V_w^\ominus g_{AA}/RT)$ (cm^3 mol^{-1})	$L_{AB}^* {}_{AA}$ (cm^3 mol^{-1})
Xylose	23	117
Glucose	37	148
Sucrose	83	293
Lactose	79	285
Maltose	40	248
Raffinose	143	443

TABLE 1.17

Components of the homotactic enthalpic virial coefficients for some saccharides in water at 25°C.

Solute	h_{AA} (J kg mol^{-2})	I	II	III	IV
			Terms in equation (102) (J kg mol^{-2})		
Glucose	343	7	248	89	−1
Galactose	133	5	18	111	−1
Ribose	202	2	116	85	−1
Sucrose	577	14	440	123	−1

The terms in equation (102) should be read from left to right.

TABLE 1.18

Components of the homotactic volumetric virial coefficients for some saccharides in water at 25°C.

Solute	v_{AA} (cm^3 kg mol^{-2})	I	II	III
			Terms in equation (104) (cm^3 kg mol^{-2})	
Ribose	0.46	0.01	0.14	.−0.31
Xylose	0.68	0.03	0.33	−0.32
Galactose	0.52	0.03	−0.03	−0.52
Glucose	0.68	0.04	0.22	−0.42

The terms in equation (104) should be read from left to right.

Table 1.16 gives the free energetic virial virial coefficients for some saccharides in water and it is apparent from this that, in contrast to the results obtained for amino acid derivatives (see Table 1.10), all are positive. It is also evident that the McMillan–Mayer solute–solute terms are also large and positive and although much of the terms come from excluded volume effects closer examination of the results indicated that when compared with the amino acid compounds the balance of the contributions is more in favour of solvation for the saccharides. One sees this also for enthalpic virial coefficients by applying the same approach as that used earlier [see equation (102)] to four systems for which appropriate data exist (see Table 1.17) the sign of the second term is opposite to that which was deduced for N-acetylglycinamide and so for the saccharides it seems clear that solvation is playing a much more important role. Further confirmation of this comes from a consideration of the volumetric virial coefficients which can be written as

$$v_{AA} = - g_{AA}K + \{(RTL_A/V_w^{\ominus}) \; (\partial B_{AA}^*/\partial P)\} - RTK_A^{\ominus}/V_w^{\ominus} \qquad (104)$$

where $K_A^{\ominus}$ is the partial molar compressibility of the solute. This expression has been applied to some representative data which are given in Table 1.18 and it is apparent from these that a significant proportion of the volumetric virial coefficients comes directly from the pressure variation of the partial molar volumes and this again implicates solvation playing a major if not completely dominant role.

1.4.6 Peptide amides in water

One of the advantages of investigating peptides and their derivatives is that relatively subtle changes in the molecular framework can be introduced such as changes in chirality and changes is amino acid residue sequence and in this section we will present some largely unpublished and recent information (refs.10,24,69,70) which explores sequence dependence. Table 1.19 gives a summary of the available information representing interactions between solute molecules. The first point which should be made for this more complex group of compounds is that if we consider the enthalpic information and apply the group additivity parameters obtained from simpler compounds to them then although there is some broad predictability possible there are also some very marked deviations (see Fig.1.28). This is also clearly evident from even a casual inspection of the results on isomers with different sequences. There are several reasons why this could happen and some of these have been alluded to earlier but an extra important feature becomes increasingly important as the molecules increase in complexity.

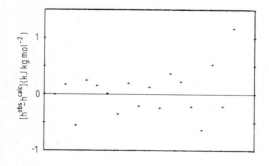

Fig.1.28 Comparision of predicted and experimental virial coefficients for some peptide amides.

It has been known for many years that polypeptides can exist in 'folded' conformations

TABLE 1.19

Virial coefficients for peptide amides in water at 25°C.

Solute species		$-g_{AB}$	h_{AB}	v_{AB}
A	B	(J kg mol^{-2})	(J kg mol^{-2})	(cm^3 kg mol^{-2})
FG	FG		854[b]	
GF	GF		411[b]	
FG	GF		743[b]	
FG	GLY		−215[b]	
FG	PHE		1028[b]	
GF	GLY		−457[b]	
GF	PHE		991[b]	
FA	FA		1223[b]	
AF	AF		2383[b]	
FA	AF		2218[b]	
FA	ALA		314[b]	
FA	PHE		1603[b]	
AF	ALA		906[b]	
AF	PHE		1666[b]	
PG	PG	231[a]	619[a]	−0.31[d]
GP	GP	338[a]	255[a]	−0.37[d]
PA	PA	105[a]	1629[a]	−0.80[d]
AP	AP	118[a]	1226[a]	
PV	PV		3369[a]	
PL	PL	680[a]	3979[a]	
LP	LP		4592[a]	
PP	PP	241[a]	2010[a]	−1.37[d]
GA	GA		190[a]	
AG	AG	59[a]	284[c]	
GG	GG	121[a]	−646[c]	−0.54[d]
AA	AA	200[a]	939[c]	0.32[d]
GGG	GGG		−1499[c]	
AAA	AAA		4881[c]	
PA	ALA	176[a]	769[a]	
AP	ALA	70[a]	750[a]	
PG	GLY	213[a]		
PG	PRO	188[a]	597[a]	
AA	ALA	18[a]		
GG	GLY	129[a]	−211[c]	
GG	PRO	163[a]		
GGG	GLY	−544[c]		
PLG	PLG	642[a]	5866[a]	

(a) ref.69; (b) ref.70; (c) ref.21; (d) ref.31.

and the tendency for such folding to occur will depend on a balance between intramolecular and intermolecular (*i.e.* solvation) interactions. If a molecule is 'well' solvated then there will be little tendency for it to fold and obviously poor solvation will tend to aid what are in effect intramolecular association processes. the consequence of this is that if one has two isomeric molecules, one of which 'folds' more than the other, then notwithstanding their equivalence from a group additivity viewpoint their nett interactions as reflected in say free energetic virial coefficients will be different. In other words one would expect intramolecular events to influence intermolecular events. It is instructive to consider the two prolyl containing peptides N-acetylglycyl–L–prolinamide (GP) and its isomer N–acetyl–L–prolylglycinamide (PG) since for these we have free energetic, enthalpic and volumetric information. If we look at the standard state partial molar volumes then that for GP is smaller than that for PG and using equation (43) one might conclude that GP interacts more strongly with the solvent than PG. However, if this were so then we would expect the free energetic virial coefficient for GP would be more positive than that for PG and this is contrary to observation. In fact if the GP adopted a more compact configuration than its isomer then not only would its molar volume be smaller but also the excluded volume contribution in equation (88) would also be smaller and consequently its g_{AA} would have a more negative value. This is indeed the case and the other virial coefficients support the idea that folding to at least some extent occurs for GP since its homotactic enthalpic coefficient is less positive and its volumetric coefficient is more negative than those for PG and this is what one would expect if intramolecular folding reduced accessibility of hydrophobic portions of the molecule. Supporting non–thermodynamic evidence (ref.69) comes from an n.m.r. study of amide proton exchange rates where it was found that GP exchanged its protons some 2–3 times more slowly than did PG. The experimental evidence for other isomeric dipeptides is less complete but if we use the available rather fragmentary information on prolyldipeptide derivatives it would seem that PA is probably more 'folded' than AP whereas PL is probably less compact than its isomer. These conclusions are however tentative and further experimental work is required.

If we consider the phenylalanyl containing peptides then notwithstanding the fact that we have only enthalpic information (generally these peptides are relatively insoluble and the determination of free energetic and enthalpic information will require some very careful and precise experimental work) one can nevertheless see patterns of behaviour. There is no doubt that the average conformation of isomeric peptides in this series are different since their n.m.r. coupling constant patterns clearly indicate this (ref.70). The homotactic interaction coefficient for FG is more positive than that for GF which suggests that the latter is better folded and the former has more hydrophobic surface exposed. This suggestion is in accordance with the facts that the heterotactic interactions of FG with GLY and PHE are more positive than those with GF. In a similar way the homotactic coefficient for AF suggests that this molecule has more exposed hydrophobic regions than its isomer and in turn AF has more positive heterotactic coefficients than does FA.

Obviously a great deal of further work of both a thermodynamic and a spectroscopic nature needs to be done but this should be rewarding since it seems clear that even for relatively small molecules rather specific and characteristic effects are evident. A striking example of this is the trialanyl peptide derivative AAA which has an enthalpic virial coefficient which is some three times more positive than would be predicted from group additivity and although one does not know the reason for this one would conclude that intramolecular folding must be occurring to produce perhaps a hydrophobic 'face' which in turn gives rise to very significant hydrophobically induced interactions.

1.4.7 Chiral Effects

As was mentioned above because of the nature of the amino acids all except glycine contain chiral centres at the α–carbon and this gives one the opportunity to explore how the presence of such centres affect interactions. A short time ago Watson and his co–workers (ref.88) from measurements of enthalpies of dilution of L–alanylglycine and D–alanylglycine

TABLE 1.20

Enthalpic virial coefficients for chiral solutes.

Solute A	Solute B	h_{AB}(J kg mol^{-2})
L−ALA	L−ALA	269
D−ALA	D−ALA	277
D−ALA	L−ALA	337
L−ALA−L−ALA	L−ALA	488
L−ALA−L−ALA	D−ALA	578
L−PRO−L−ALA	L−PRO−L−ALA	1629
L−PRO−D−ALA	L−PRO−D−ALA	1399
L−PRO−L−ALA	L−PRO−D−ALA	1515
L−PRO−L−ALA	L−ALA	769
L−PRO−L−ALA	D−ALA	748
L−PRO−D−ALA	L−ALA	803

concluded that chiral effects in aqueous solutions of small peptides are negligible but it is possible that the Coulombic effects in such systems mask the rather subtle effects one is looking for. There is some evidence (refs.89−92) which indicates that stereochemical recognition might will operate at the level of peptide dimers or even monomers so a study was performed (ref.24) on the excess enthalpic properties of some systems which incorporated the D−alanyl residue. The results obtained for the enthalpic virial coefficients are given in Table 1.20 and it is clear from these that energetic stereoselectivity does occur. The homotactic virial coefficients of L−ALA and D−ALA are, within experimental error, the same which is what one would expect in an achiral solvent but the heterotactic interaction between L−ALA and D−ALA is different to both of these and is in fact somewhat more positive. The dimeric peptides P−D−A and P−L−A are diastereoisomeric and hence one might expect that both the homotactic and heterotactic interactions of these would be different and this is indeed found. The volumetric information (ref.31) which is available on P−D−A and P−L−A suggests that their solvation properties are essentially the same and so the different enthalpic virial coefficients seem to indicate that it is the solute−solute terms in equation (89) which are different. The implications of the results given in Table 1.20 have been discussed elsewhere (ref.24) and we will not pursue these here except to point out that the effects observed indicate that the weak intermolecular interactions which are being assessed from the experimental measurements do seem to be remarkably sensitive to the stereochemistry of the solutes.

1.4.8 Interactions in non−aqueous solvents

It should by now be clear that the problems involved in deconvoluting the components of molecular interactions are both complex and complicated but it is apparent that some of these components will be environmentally mediated. This is an important point and can be illustrated by for example the tendency that some polypeptides have to fold spontaneously into relatively well defined structures only under certain circumstances. In the early stages of protein biosynthesis interactions between amino acid subunits will be occurring in what is essentially an aqueous medium and so the nucleation steps of protein folding will be determined to some extent by this medium and information such as some of that presented here on intermolecular effects will be

TABLE 1.21

Enthalpic virial coefficients for the interaction of some N–actyl aminoacid amides in N,N dimethylformamide and in water at 25°C.

Solute	h_{AA} (J kg mol^{-2})	
	Solvent	
	DMF	Water
GLY	−609	−220
ALA	−886	268
VAL	−1432	1259
LEU	−1149	1714
PHE	−982	1049

relevant to the intramolecular subunit interactions. However as the folding proceeds water will be essentially excluded from the protein interior and the subunits will then be interacting in an environment which is largely if not entirely non-aqueous. Now for any particular 'buried' subunit its immediate environment will be complex but it is apparent from the large body of information which is available from x-ray structural studies that broadly speaking the interactions between subunits will be occurring in a molecular environment which consists of amidic and apolar regions. In view of this we recently suggested that useful and relevant information on interactions in protein interiors could be obtained from studies involving amidic solvents and a preliminary investigation has been published (ref.30) in which the amidic-apolar environment of protein interiors was represented by N,N dimethylformamide (DMF). The enthalpic virial coefficients for homotactic interactions obtained for some N-acetylaminoacid amides containing apolar side-chains are given in Table 1.21 along with those obtained in water. It is obvious from these that the overall trends in the results are quite different in the two solvents. Qualitatively, as the side-chain becomes more apolar, in water, the enthalpic virial coefficient becomes more positive (because of hydrophobic effects) whereas in DMF broadly speaking the virial coefficient becomes more negative. The marked differences stress the futility of attempting to rationalise solute-solute interactive behaviour without reference to the environment. We will not pursue the information in DMF since our feeling at the outset was that as an amidic environment to model protein interiors it is less appropriate than some others since it contains the tertiary amide group whereas of course it is the secondary amidic group which is most common in proteins. Rather we will consider briefly some recent results (ref.35) obtained using N-methylacetamide (NMA), which as a 'model' protein interior solvent seems to be much more satisfactory.

Table 1.22 gives the enthalpic virial coefficients found for some amidic solutes interacting both homotactically and heterotactically. As for DMF as a solvent the trends observed are quite different to those found in aqueous systems and unlike the latter the variation of the virial coefficients with side chain apolarity is by no means monotonic. This is clearly evident from the fact that the LEU–LEU coefficient is less negative than the VAL–VAL coefficient and that there are marked differences between the LEU–LEU and the isomeric ILEU–ILEU coefficients. Even a preliminary perusal of the information given in Table 1.22 shows that the systems cannot be treated using a SWAG approach and that in some ways NMA is a much

TABLE 1.22

Enthalpic virial coefficients of some amidic solutes in N-methylacetamide at 32°C.

Solute		h_{AB} (J kg mol^{-2})
A	B	
A	A	30
DMA	DMA	94
F	F	96
NMF	NMF	−10
DMF	DMF	−70
GLY	GLY	−100
ALA	ALA	−186
VAL	VAL	−543
LEU	LEU	−415
PRO	PRO	−80
PHE	PHE	−1045
ILEU	ILEU	−821
GLY	ALA	−120
GLY	VAL	−191
GLY	LEU	−114
ALA	VAL	−261
ALA	LEU	−339
VAL	LEU	−392

more structurally discriminating solvent than is water.

As mentioned earlier the native structure of a globular protein is determined by a balance of the interactions occurring between the constituent amino acid residues with each other and with the aqueous environment. Something which has been commented upon is that many studies which have used aqueous environments with model solutes have produced experimental information which has relatively weak correlations with features of protein folding and conformation. There are many reasons why this should be so including the invalidity of the chosen model solutes and the possible or indeed probable over-emphasis which has been made of the role of hydrophobic effects. For the present systems these latter are necessarily absent and there is one piece of evidence which indicated that the model systems (solute and solvent) chosen have some utility and link with events in protein chemistry.

The crystallographic coordinates of 36 proteins have been surveyed (ref.93) and using a statistical analysis a normalised measure of the frequencies of amino acid side-chain occurrences were deduced. Now if one makes the presumably realistic assumptions that enhanced frequency of contact between side chains arises because of relatively favourable energetics and that the enthalpic measures obtained here also report on such interactions one might expect to observe a correlation between enthalpic virial coefficients (h_{AB}) and measures of the intrinsic mutual affinity between the side chains on A and B. That such a correlation exists is shown in Fig.1.29 for eight out of the nine systems for which both virial coefficients and probability measures (P_{AB}) exist the correlation is very good indeed and the one system for which it is poor is for prolyl-prolyl interactions. We do not really know why this residue in particular should be so different but it might be relevant that of all the amino acid residues considered the prolyl is the one which is most likely to occur in chain reversals at the surface

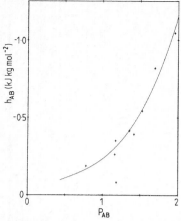

Fig.1.29 Correlation between enthalpic virial coefficients of some N–acetylaminoacid amides in the solvent N–methylacetamide with the probability of occurrence (P_{AB}) of side chain interactions in proteins.

of proteins (ref.94) and hence will be in an essentially aqueous environment. Of particular note are the numerically large values for both P_{AB} and h_{AB} for ILEU compared to its isomer LEU and generally it would seem that the experimental virial coefficients obtained in NMA as a solvent correlate rather well with at least one feature of protein architecture. A considerable amount of work is required to confirm these findings and some of this is the object of a current investigation in our laboratory.

1.4.9 Interactions in mixed solvents

In view of the fact that the interactions occurring between solutes in water and amidic solvents are so different and that, in principle at least, in a globular protein, water can permeate from the exterior into the interior thereby giving rise to mixed–solvent regions, we felt it would be worthwhile to explore the effect of changing solvent medium on molecular interactions (ref.35). A considerable amount of information has been obtained and only a few examples will be given.

Fig.1.30 presents the information on homotactic enthalpic virial coefficients for N–acetylamino acid amides in NMA–water mixtures as a function of the mole fraction of NMA. It is quite clear from this that very marked effects indeed are observed as the solvent composition changes and considerable amounts of the variation probably come from solvational changes. The significant contributions which come from such sources for amidic solutes in mixed solvents is illustrated in Fig.1.31 where the variations of both the enthalpy of solution (ref.95) and the enthalpic virial coefficient (ref.35) for NMF in DMF–water mixtures are presented. The striking 'mirror–image' behaviour of the two curves confirms the interplay between solvation and interaction. The fact that the amino acid amide interactions depend markedly on the solvent medium suggests that in protein interiors the same sort of effect might will prevail and that within the limits imposed by the polymer chain a given amino acid residue will attempt to get into an environment in which it is as energetically 'comfortable' as possible. An observation (ref.96) from Scheraga's group indicates that this is indeed the case.

The radial distributions of clusters of hydrophobic, hydrophilic and neutral (ref.96) amino acid residues in some 19 proteins were assessed by Meirovitch and Sheraga and they defined

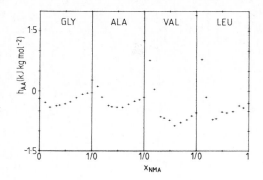

Fig.1.30 Homotactic enthalpic virial coefficients of some N–acetylaminoacid amides in mixtures of N–methylacetamide and water (x_{NMA} is the mole fraction of the former in the solvent mixture).

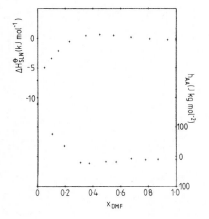

Fig.1.31 Variation of the enthalpy of solution (upper) and the homotactic enthalpic virial coefficient (lower) of the solute N–methylformamide in mixtures of N,N–dimethylformamide and water (x_{DMF} is the mole fraction of the former in the solvent mixture).

four spherical layers about the centres of mass of each protein. They then for each residue in each layer calculated the average fractions of hydrophobic, hydrophilic and neutral residues in an 8Å radius about the central residue. Superficially one might guess that if one considered the amino acid residues glycyl, alanyl, valyl and leucyl that the first would be in a predominantly hydrophilic environment and the last would be in a predominantly hydrophobic environment and the others would be in intermediate. This is however not always found to be so and in Table 1.23 we have correlated the ranking orders of the hydrophobicity of amino acid environments with the average location of the residues in the proteins. It is clear from this that although the above order does prevail near the centres of proteins it certainly does not apply elsewhere and this is particularly so near protein surfaces where for example alanyl residues are situated in more hydrophobic regions than glycyl residues and leucyl residues are in less hydrophobic environments that valyl residues. The changes in ranking order which are observed have some similarities to the variations in the virial coefficients in the NMA–water solvent systems and it appears that such systems may be useful to model some aspects at least of the molecular

TABLE 1.23

The ranking orders of some amino acid residues with the hydro-phobicity of their environments in different protein regions. The higher the numerical value the more hydrophobic is the residue's environment.

Amino acid residue	Hydrophobicity of environment			
	core	→		surface
glycyl	1	1	1	2
alanyl	2	2	1	1
valyl	3	4	3	4
leucyl	4	3	4	3

environments found within protein systems.

A final point which comes from this investigation and which has implications in several areas of protein chemistry bears on the establishment of hydrophobicity scales for amino acid residues. This has been a fairly active and contentious area over the years and in Table 1.24 the ranking orders from several sources of some amino acid residues are presented.

TABLE 1.24

Comparison of relative magnitudes[a] of h_{AA} and various sets of hydrophobicity constants.

Residue	h_{REL}	T	J	C	E	P	V
GLY	1	1	2	2	2	1	2
ALA	2	2	2	3	3	2	3
PRO	3	3	1	1	1	3	1
PHE	4	6	4	5	6	6	6
VAL	5	4	6	6	5	4	4
LEU	6	5	4	4	4	5	5

(a) Hydrophobicity increases with scale number.
(T) ref.97; (J) ref.98; (C) ref.99; (E) ref.100; (P) ref.101; (V) ref.102.

It is clear that no real consensus is obtained except in a very broad sense and it is suggested that the reason for this is that the hydrophobic nature of a group or molecule finds expression only when it interacts with another group or molecule and this expression will depend markedly both on the nature of the second group or molecule and on the general environment in which the interaction is occurring.

REFERENCES.

1   See for example the articles in, R. Jaenicke (Ed.), Protein Folding, Elsevier, Amsterdam, 1980.
2   C.B. Anfinsen and H.A. Scheraga, Advances in Protein Chem., 29 (1975) 205.
3   J.T.Edsall and H.A. Mackenzie, Advances in Biophysics, 16 (1982) 53.
4   C.N. Hassall, W.H. Johnson and N.A. Roberts, Bioorganic Chem. 8 (1979) 299.
5   C. Jolicoeur, B. Reidl, D. Desrochers, L.L. Lemelin, R. Zamojska and O. Enea, J. Solution Chem., 15 (1986) 109.
6   T.H. Lilley in, G.C. Barrett (Ed), The Chemistry and Biochemistry of the Amino Acids, Chapman and Hall, London, 1985, Chapter 21.
7   T.H. Lilley and R.H. Wood, J. Chem. Soc. Faraday Trans I 76 (1980) 901.
8   C.C. Briggs, T.H. Lilley, J. Rutherford and S. Woodhead J. Solution Chem., 3 (1974) 649.
9   J.J. Savage and R.H. Wood, J. Solution Chem., 5(1976) 733.

10 T.H. Lilley in, G. W. Neilson and J.E. Enderby (Eds.), Water and Aqueous Solutions, Adam Hilger, Bristol, 1986, 265.
11 R.H. Stokes, Aust J. Chem., 20 (1967) 2087.
12 G.R. Hedwig and T.H. Lilley, to be published.
13 C. Jolicoeur and J. Boileau, Canad. J. Chem. 56 (1978) 2707.
14 G.R Hedwig, to be published.
15 G.R. Hedwig, personal communication.
16 O. Enea and C. Jolicoeur, J. Phys. Chem., 86 (1982) 3870.
17 B. Riedl and C. Jolicoeur, J. Phys. Chem., 88 (1984) 3348.
18 F. Franks and D. Eagland, CRC, Critical Rev. in Biochem., 3 (1975) 165.
19 G.M. Blackburn, T.H. Lilley and E. Walmsley, J. Chem. Soc. Faraday Trans. I, 76 (1980) 915.
20 G.M. Blackburn, T.H. Lilley and E. Walmsley, J. Chem. Soc. Chem. Commun., (1981) 1091.
21 G.M. Blackburn, T.H. Lilley and E. Walmsley, J. Chem. Soc. Faraday Trans. I, 78 (1982) 1641.
22 G.M. Blackburn, T.H. Lilley and P.J. Millburn, J. Solution Chem., 13 (1984) 789.
23 K. Nelander, G. Olofsson, G.M. Blackburn, H.E. Kent and T.H. Lilley, Thermochim. Acta, 78 (1984) 303.
24 G.M. Blackburn, T.H. Lilley and P.J. Millburn, J. Chem. Soc. Chem. Commun., (1985) 299.
25 G.M. Blackburn, T.H. Lilley and P.J. Millburn, Thermochim. Acta, 83 (1985) 289.
26 G.M. Blackburn, T.H. Lilley and P.J. Millburn, J. Chem. Soc. Faraday Trans I, 81 (1985) 2191.
27 G.M. Blackburn, H.E. Kent and T.H. Lilley, J. Chem. Soc. Faraday Trans. I, 81 (1985) 2207.
28 G.M. Blackburn, T.H. Lilley and P.J. Millburn, J. Solution Chem., 15 (1986) 99.
29 G.M. Blackburn, T.H. Lilley and P.J. Millburn, J. Chem. Soc. Faraday Trans. I, 82 (1986) 2965.
30 H.E. Kent, T.H. Lilley, P.J. Millburn, M. Bloemendal and G. Somsen, J. Solution Chem., 14 (1985) 101.
31 T.E. Leslie and T.H. Lilley, Biopolymers 24 (1985) 695.
32 F.J. Millero, A. Lo Surdo and C. Shin J. Phys. Chem., 82 (1978) 784.
33 J. Kirchnerova, P.G. Farrell and J.T. Edward, J. Phys. Chem., 80 (1976) 1974.
34 L. Lepori and V. Mollica, Z. Phys. Chem. (Neue Folge), 123 (1980) 51.
35 T.E. Leslie and T.H. Lilley, to be published.
36 C. de Visser, W.J.M. Heuvelsland, L.A. Dunn and G. Somsen, J. Chem. Soc. Faraday Trans. I, 74 (1978) 1159.
37 C. de Visser, P. Pel and G. Somsen, J. Solution Chem., 6 (1977) 571.
38 C. de Visser, G. Perron, J.E. Desnoyers, W.J.M. Heuvelsland and G. Somsen, J. Chem. Eng. Data, 22 (1977) 74.
39 A. Lo Surdo, C. Shin and F.J. Millero, J. Chem. Eng. Data, 23 (1978) 197.
40 J.G. Kirkwood and F.P. Buff, J. Chem. Phys., 19 (1951) 774.
41 J.E. Garrod and T.M. Herrington, J. Phys. Chem., 73 (1969) 1877.
42 A. Ben–Naim, Water and Aqueous Solutions, Plenum, New York and London, 1974, pp. 137–169.
43 T.M. Herrington, A.D. Pethybridge, B.A. Parkin and M. G. Roffey, J. Chem. Soc. Faraday Trans. I, 79 (1983) 845.
44 S. Cabani, G. Conti and E. Matteoli, J. Solution Chem., 5 (1976) 751.
45 C. de Visser, G. Perron and J. E. Desnoyers, Canad. J. Chem., 55 (1977) 856.
46 L. G. Hepler, Canad. J. Chem., 47 (1969) 4613.
47 J. L. Neal and D. A. I. Goring, J. Phys. Chem., 74 (1970) 658.
48 G. Scatchard, Chem. Rev., 8 (1931) 321 ; 44 (1949) 7.
49 G. Scatchard and C. Raymond, J. Amer. Chem. Soc., 60 (1938) 1278.
50 H. L. Friedman, J. Chem. Phys., 32 (1960) 1351.
51 See for example, R. A. Robinson and R. H. Stokes, Electrolyte Solutions (2nd Edn.) Butterwoths, London, 1965, Chapter 8.
52 B.Y. Okamoto, R.H. Wood and P.T. Thompson, J. Chem. Soc. Faraday Trans. I, 74 (1978) 1890.
53 B.Y. Okamoto, R.H. Wood, J.E. Desnoyers, G. Perron and L. Delorme, J. Solution Chem., 10 (1981) 139.
54 J.P.E. Grolier, J.J. Spitzer, R.H. Wood and I.R. Tasker, J. Solution Chem. 14 (1985) 393.
55 J.J. Spitzer, S.K. Suri and R.H. Wood, J. Solution Chem. 14 (1985) 561.
56 J.J. Spitzer, S.K. Suri and R.H. Wood, J. Solution Chem. 14 (1985) 571.
57 J.J. Spitzer, I.R. Tasker and R.H. Wood, J. Solution Chem. 13 (1984) 221.
58 S.K. Suri and R.H. Wood, J. Solution Chem. 15 (1986) 705.
59 B.T. Doherty and D.R.Kester, J. Marine Res. 3 (1974) 285.
60 R.H. Wood, T.H. Lilley and P.T. Thompson, J. Chem. Soc. Faraday Trans I, 74 (1978) 1301.
61 E.A. Guggenheim, J. Chem. Soc. Faraday Trans 56 (1960) 1159.
62 H.L. Friedman, J. Solution Chem. 1 (1972) 387.
63 H.L. Friedman, J. Solution Chem. 1 (1972) 413.
64 H.L. Friedman, J. Solution Chem. 1 (1972) 419.

65 W.G. McMillan and J.E. Mayer J. Chem. Physics, 13 (1945) 276.
66 T.H. Lilley, E. Moses, and I.R. Tasker, J. Chem. Soc. Faraday Trans I, 76 (1980) 906.
67 T.H. Lilley and R.P. Scott, J. Chem. Soc. Faraday Trans I, 72 (1976) 184.
68 T.H. Lilley and R.P. Scott, J. Chem. Soc. Faraday Trans I, 72 (1976) 197.
69 G.M. Blackburn, T.H. Lilley and P.J. Millburn, to be published.
70 G.M. Blackburn, H.E. Kent and T.H. Lilley, to be published.
71 E.E. Schrier and E.B. Schrier, J. Phys. Chem., 71 (1967) 1851.
72 M.A. Gallardo and T.H. Lilley, to be published.
73 K.D. Kopple and D.H. Marr, J. Amer. Chem. Soc., 89 (1967) 6193.
74 K.D. Kopple and M. Ohnishi, J. Amer. Chem. Soc., 91 (1969) 962.
75 R. Deslauriers, Z. Grzonka, K. Schaumberg, T. Shiba and R.Walters, J. Amer. Chem. Soc., 97 (1975) 5093.
76 P.J. Cheek and T.H. Lilley, to be published.
77 G. Barone, P. Cacace, V. Elia and A. Cesaro, J. Chem. Soc. Faraday Trans I, 80 (1984) 2073.
78 G. Barone, personal communication.
79 A.K. Covington, T.H. Lilley, K.E. Newman and G.A. Porthouse, J. Chem. Soc. Faraday Trans I, 69 (1973) 963.
80 A.K. Covington, K.E. Newman and T.H. Lilley, J. Chem. Soc. Faraday Trans. I, 69 (1973) 963.
81 L.R. Pratt and D.C. Chandler, J. Solution Chem. 9 (1980) 1.
82 A. Cesaro Chapter in, H.J. Hinz (Ed.), Thermodynamics for Biochemistry and Biotechnology, Springer–Verlag, Berlin, 1986.
83 S.H. Gaffney, E. Haslam, T.H. Lilley and T.R. Ward, to be published.
84 M. Kabayama and D. Patterson, Canad. J. Chem., 36 (1958) 563.
85 F. Franks, D.S. Reid and A. Suggett, J. Solution Chem. 2 (1973) 99.
86 S. Ablett, M.D. Barrett, F. Franks, M.D. Pedley and D.S. Reid in, A. Alfsen and A.J. Bertreaud (Eds.), L'eau et les Systems Biologiques, C.N.R.S., Paris, 1976, 105.
87 G. Barone, G. Castruonovo and V. Elia, J. Solution Chem., 9 (1980) 607.
88 M.K. Kumaran, G.R. Hedwig and I.D. Watson, J. Chem. Thermodynamics, 14 (1982) 93.
89 R.D. Sheardy and E.J. Gabbay, Biochemistry 22 (1983) 2061.
90 T. Endo, Y. Hayashi and M. Okawara, Chem. Letters, (1977) 391.
91 T. Endo, H. Kawasaki and M. Okawara, Tetrahedron Letters, (1979) 23.
92 M.T. Cung, M. Marraud and J. Neel, Biopolymers 15 (1976) 2081.
93 D.C. Roberts and R.S. Bohacek, Int. J. Peptide Protein Res., 21 (1983) 491.
94 P.N. Lewis, F.A. Monamy and H.A. Scheraga, Biochim. Biophys. Acta, 303 (1973) 211.
95 A. Rouw and G. Somsen, J. Chem. Soc. Faraday Trans. I, 78 (1982) 3397.
96 H. Meirovitch and H.A. Scheraga, Macromolecules, 14 (1981) 340.
97 C. Tanford, The Hydrophobic Effect, Wiley, New York, 1973.
98 J. Janin, Nature 277 (1979) 491.
99 C. Chothia, J. Mol. Biol. 105 (1976) 1.
100 D. Eisenberg, R.M. Weiss, T.C. Terwilliger and W. Wilcox, Faraday Symp. Chem. Soc., 17 (1982) 109.
101 V. Pliska and J.L. Fauchere in, E. Cross and J. Meinhoffer (Eds.), Peptides : Structure and Biological Function, Pierce Chemical Co., Rockford, 1979, 249.
102 G. Von Heijne and C. Blomberg, Eur. J. Biochem., 97 (1979) 175.

CHAPTER 2

PROTEIN UNFOLDING

W. PFEIL

2.1   INTRODUCTION

The chapter concerns protein unfolding studies performed mainly during the past decade. In the first edition of
this book, first of all unfolding studies of small globular proteins were reviewed (ref. 1). In the meantime much more thermodynamic data related to protein unfolding became available. For
several reasons, there exists growing interest in protein stability. Therefore, data for Gibbs energy change at protein unfolding will be compiled in section 2.5, and an attempt will be
made to search for relations with structural and functional properties. The main interest during the past decade, however, was
focussed on investigations of proteins undergoing multi-state
transitions. Two cases will be considered in more detail in
sections 2.2 and 2.4: (i) three-state unfolding of α-lactalbumin
which leads to a reconsideration of the unfolded state, and (ii)
segmental unfolding of proteins consisting of domains. The corresponding types of transition profiles are schematically shown in
Fig. 2.1.

Case (a) in Fig. 2.1: For most small globular proteins
(M$<$20,000), such as lysozyme for example, a highly cooperative
transition can be observed both in GuHCl induced unfolding (ref.
2,3) and scanning calorimetry (ref. 4,5).

Case (b) in Fig. 2.1: Unfolding can include stable intermediates as found by the example of α-lactalbumin. In GuHCl induced

---

Abbreviations and symbols

DSC   differential scanning calorimetry
CD   circular dichroism
GuHCl   guanidine hydrochloride
T   temperature
ΔH^{cal} enthalpy change, calorimetrically determined
$\Delta H^{v.H.}$ enthalpy change, determined by van't Hoff treatment
$\varkappa$ ratio $\Delta H^{cal}/\Delta H^{v.H.}$
$\propto$ degree of conversion
C_p heat capacity
$(\theta)_{222}$ molar ellipticity at the given wavelength (nm)
M.W.   molecular weight

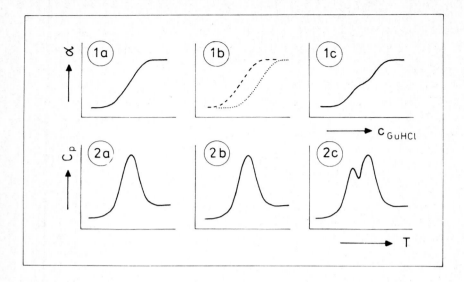

Fig. 2.1. The transition profiles of optically monitored GuHCl
induced transitions (1a-c) and scanning calorimetric recordings
(2a-c). See text.

unfolding followed by circular dichroism at different wavelengths
(222 nm and 270 nm) two consecutive transitions were found (ref.
6) whereas a single cooperative transition was observed in scanning calorimetry (ref. 7).

Case (c) in Fig. 2.1: Proteins consisting of domains show two
or more subsequent transitions in optically and calorimetrically
monitored experiments. An instructive example is pepsinogen (ref.
8,9).

More detailed knowledge about rather complicated transitions
was achieved by improvement of existing and by development of
new experimental techniques. The most elegant procedure is the
direct visualization of intermediates by urea gradient electrophoresis (ref. 10-12). Furthermore, optical measurements of transition profiles have gained importance by simultaneous recording
of absorbance, circular dichroism and fluorescence at various
wavelengths as well as by Raleigh scattering (ref. 13-15). On the
same sample about 30 data points can be simultaneously accumulated
reflecting local and global changes, such as changes of structural order of backbone and side chains, changes of environment
of various chromophores, exposure of side chains to solvent and,
finally, changes in molecular size.

Information about the nature of a transition, i.e. two-state or multi-state behaviour, can be obtained from scanning calorimetry comparing calorimetric and van't Hoff heat (ref. 5,16-19). An important step, however, was the introduction of deconvolution (ref. 20-22) which allows overlapping transitions to be resolved. Besides, a combination of various approaches is advantageous for checking the multiplicity of states involved in protein unfolding. Thus, scanning calorimetric studies of multidomain proteins have been successfully combined with investigations of proteolytic fragments for identifying the subtransitions (ref. 19,23). Recently, GuHCl-induced protein unfolding was followed by optical measurements combined with chromatographic runs (ref. 15). Similarly, GuHCl-induced transitions can be monitored by isothermal calorimetric titrations (ref. 7,24,25).

The thermodynamic criterion for two-state transitions is equality of calorimetric (ΔH^{cal}) and van't Hoff ($\Delta H^{v.H.}$) enthalpy (ref. 16). For a multi-state transition the ratio $x = \Delta H^{cal}/\Delta H^{v.H.}$ is greater 1. In 1974, careful scanning calorimetric studies of small globular proteins became available (ref. 5). For ribonuclease, lysozyme, cytochrome c, metmyoglobin and chymotrypsinogen $x = 1.05^{\pm}0.03$ was found, i.e. a value close to unity as expected for a two-state transition. Some more recent investigations including various proteins, denaturants and tests other than the thermodynamic criterion are listed in Table 2.1. For ribonuclease, lysozyme and chymotrypsinogen, two-state behaviour is now well substantiated by complete agreement of the tests and independence of the nature of the denaturant. For other proteins, such as pepsinogen and serum albumin, multi-state behaviour can be shown, again in agreement between various criteria and different ways of unfolding. More complicated, however, are cytochrome c, carbonic anhydrase and α-lactalbumin as examples. Using various denaturants, either two-state or multi-state behaviour can be observed. Various explanations are conceivable: (i) insufficient sensitivity of some of the methods in detecting the intermediates, (ii) existence of several native states of the proteins (the proteins are either metalloproteins or/and sensitive to salt content), (iii) different effectiveness of the various ways of unfolding, (iv) occurrence of apparent intermediates due to specific experimental conditions (e.g. solvation phenomena or biochemical and biophysical reasons for irreversible protein in-

TABLE 2.1

Unfolding transition characteristics of selected proteins determined by various approaches.

Protein	M.W.	Denaturant	Characteristics (a) two-state	multi-state	References
ribonuclease	13,600	GuHCl	ms (b), eq	—	15,26
		urea	uge, ms	—	10,14
		temperature	th, d (c), ms	—	5,14,20
lysozyme (HEW)	14,300	GuHCl	eq, ms	—	2,3,14
		urea	uge	—	10
		temperature	th, ms	—	5,14
chymotrypsinogen	25,200	urea	uge	—	10
		temperature	th	—	5
		pressure	el	—	27
cytochrome c	12,400	GuHCl	eq	—	28
		GuHCl	—	ms (b)	26
		urea	uge	—	10
		temperature	th	—	5
		pH	—	ms	14
α-lactalbumin	14,020	GuHCl	—	cd	6
		urea	uge (d)	—	10
		temperature	th	—	7
carbonic anhydrase	29,300	GuHCl	—	ms	14
		temperature	th	—	18
pepsinogen	39,400	GuHCl	—	ms (b)	14,15
		temperature	—	th, d, ms	8,9,14
serum albumin	66,200	GuHCl	—	ms	14
		urea	—	uge	10

Footnotes to Table 2.1

(a) Abbreviations
    cd - circular dichroism.
    d   - deconvolution (ref. 20-22).
    el - electrophoresis under pressure (ref. 27).
    eq - equilibrium treatment.
    ms - multidimensional spectroscopic analysis (ref. 13-15).
    th - thermodynamic criterion, i.e. ratio between calori-
          metric and van't Hoff heat (ref. 16).
    uge - urea gradient electrophoresis (ref. 10-12).
(b) Spectroscopic and chromatographic transition profiles (ref.
    15).
(c) Less than 5% intermediate over the transition range (ref. 20).
(d) No distinct intermediates were observed. At pH 8.6 a gradual
    transition exists between 3 M and 7 M urea which could be
    consistent with ref. 6.

---

activation in unfolding studies according to ref. 29). On the
other hand, it remains to be checked whether all the proposed
states really represent macroscopic thermodynamic states. Just
the α-lactalbumin example lead to controversial interpretations.
Thus, the paper will attempt to clarify the questions since the
proposed intermediates are important in view of current concepts
on protein folding (ref. 30-33).

2.2   THE STATES INVOLVED IN α-LACTALBUMIN UNFOLDING

α-lactalbumin (M.W. = 14,020) is evolutionally homologous to
lysozyme (ref. 34). The two proteins are of common ancestral
origin and characterized by identity of structurally important
amino acid residues, in particular, by identical position of the
four disulfide bridges. These properties are now well substantiated by sequences of eight α-lactalbumins from different
species and by nucleotide sequences (ref. 35,36). Based on the
similarity of lysozyme with α-lactalbumin, the three-dimensional
structure of the latter was proposed by a model building study
(ref. 37). The structure of α-lactalbumin in more detail was computed by a conformation refinement (ref. 38).
    The two proteins, however, show remarkable differences. Lyso-
zyme is an enzyme. α-lactalbumin serves as a modifier protein,
i.e. as non-catalytic subunit, in the galactosyltransferase
system (ref. 39). The protein seems to be more flexible than
lysozyme as indicated by hydrogen exchange studies (ref. 40) and
by reducibility of disulfide bonds (ref. 41). The most interesting
difference, however, concerns α-lactalbumin unfolding under the

influence of GuHCl. In 1976, in careful studies, a three-state
unfolding pathway was shown for bovine α-lactalbumin (ref. 6,42)
and later for human α-lactalbumin (ref. 43). In these studies
non-coincidence of the apparent transition curves was observed
which were obtained from the ellipticity changes at far (222 nm)
and at near (270 nm and 296 nm) ultraviolet wavelengths as schematically shown in Fig. 2.1 (case b). The properties of the
states involved in α-lactalbumin unfolding will be discussed
below. However, another important difference between lysozyme
and α-lactalbumin remains to be mentioned. In 1980 it was found
by thermal unfolding studies of α-lactalbumin in the presence of
excess calcium and EDTA, respectively, that α-lactalbumin is a
calcium binding metalloprotein (ref. 44).

Important unfolding studies of α-lactalbumin were made before
calcium binding was known. Nevertheless, the former results
remain valid. The protein can be prepared by the usual procedures
in a calcium containing form due to the presence of a high affinity calcium binding site (ref. 45-47). Thus, almost the same half
transition temperature in thermal unfolding of α-lactalbumin was
reported in independent studies although the transition temperature is sensitive to calcium content of α-lactalbumin (ref. 44).
Of more importance for a critical evaluation of former studies is
the low thermostability of the apo-form. The transition temperature is dependent on ionic strength and might be below 30°C
(ref. 48-50). Thus, in studies of apo α-lactalbumin at unfavourable conditions, a mixture of various states has to be taken into
account. Furthermore, complications may arise since the calcium
free form of α-lactalbumin is able to bind chelating agents (ref.
51) and monovalent cations (ref. 52). Another problem in handling
α-lactalbumin is its tendency to aggregate, in particular, under
conditions where intermediates were found. This renders investigations more difficult, which require higher protein concentration
than optical and calorimetric studies. Thus, a discrepancy was
found in α-lactalbumin studies using sedimentation, viscosity,
light scattering, X-ray and neutron scattering (ref. 53-57).

Measurements of physical properties of α-lactalbumin under
various external conditions using different approaches have shown
that a variety of conformations with different amount of secondary and tertiary structure exists. The following states were proposed:

N-state = native protein. The calcium containing protein at neu-
          tral pH is characterized by typical properties of small
          globular proteins, such as corresponding values of
          intrinsic viscosity (ref. 55), characteristic circular
          dichroism spectra in the aromatic and peptide region
          (ref. 6), fluorescence properties (ref. 58) etc. The
          native protein is able to undergo a cooperative unfol-
          ding transition with significant enthalpy and heat
          capacity changes (ref. 6,7,59,61).

Apo-state = calcium free protein. The protein exhibits native-
          like properties but reduced stability (ref. 48,49). In
          contradiction to former findings (ref. 55), both bovine
          and human apo α-lactalbumin are able to undergo a co-
          operative thermal transition (ref. 48,49).

A-state = acid protein. The protein at about pH 2 has an unfolded-
          like circular dichroism spectrum in the aromatic region
          and a native-like spectrum in the peptide region (ref.
          6,43,55,60). The A-state shows, compared with the
          native protein, a slightly increased size (ref. 55,57)
          and enhanced heat capacity (ref. 7). The A-state does
          not show a cooperative thermal transition (ref. 25,55,
          61; Fig. 2.2).

P-state = partly unfolded protein. The protein at neutral pH and
          moderate GuHCl concentration shows properties similar
          to those of the A-state. The P-state is best accessible
          in human α-lactalbumin since the two subtransitions
          are more apart from each other than in bovine α-lact-
          albumin (ref. 43).

T-state = thermally unfolded protein. Descriptions of the T-state
          so far given are still controversial and will be dis-
          cussed below in more detail.

U-state = GuHCl unfolded protein with intact S-S bridges. The
          protein in high molar urea or guanidinium salt solu-
          tions is devoid of the main elements of native struc-
          ture (ref. 6,55). Compared with the native protein the
          U-state has increased molecular size (ref. 55,57) as
          usually found for cross-linked random coil (ref. 62-64).

Further intermediates obtained in the presence of LiCl and $NaClO_4$
are similar to the A-state (ref. 65,66). These states are not
considered here since the thermodynamic approach for the deter-

mination of ΔG seems to fail. As recently shown by the ribonuclease example, the transitional Gibbs energy change exceeds even the value of GuHCl induced unfolding (ref. 67).

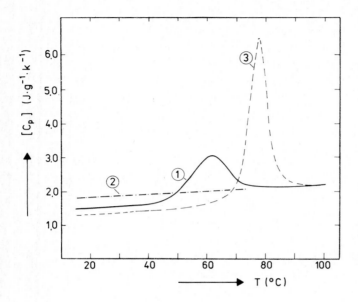

Fig. 2.2. Temperature dependence of partial heat capacity of bovine ∝-lactalbumin and lysozyme. (1) bovine ∝-lactalbumin, N-state at pH 6.3, (2) bovine ∝-lactalbumin, A-state at pH 2, (3) lysozyme, pH 4.5. From ref. (61).

In general, circular dichroism measurements in the peptide region are of basic importance in unfolding studies. Since it is known that the GuHCl induced transitions of ∝-lactalbumin in the near and far ultraviolet region do not proceed simultaneously, the optical properties represent a main criterion in definition of the various states. Corresponding data referring to circular dichroism band in the peptide region are summarized in Table 2.2 and Fig. 2.3. As it can be shown in Fig. 2.3, $(\theta)_{222}$ values of N, apo and **A** states from different references fall in a rather narrow range. The unfolded state, however, cannot be characterized by a single value due to the intriguing solvation phenomena. The optical properties of the U-state, i.e. $(\theta)_{222}$ below $40^{\circ}C$, are linearly dependent on both GuHCl concentration and temperature (ref. 6,42). If one would formally extrapolate the CD signal at 222 nm ($25^{\circ}C$) to zero GuHCl concentration by the equation

TABLE 2.2

Molar ellipticity of various states of bovine and human α-lact-albumin at 222 nm $(-(\theta)_{222}\cdot 10^{-3}/\text{deg cm}^2 \text{ dmol}^{-1})$ (a)

State	$-(\theta)_{222}\cdot 10^{-3}$	T($^{\circ}$C)	pH	Conditions	Ref.
Bovine α-lactalbumin					
native (N)	9.35	25	6.5	0.1M KCl	6
	8.95	25	6.0		(b)
	9.65	20	7		55
	12.3	25	5.5	0.15M KCl	66
	9.9	25	8	20mM Tris + CaCl$_2$	52
	9.55	25	7		60
calcium free (apo)	9.6	11.4	8	10mM Borat + NaCl	48
	10.5	25	8	20mM Tris +0.2M NaCl	52
acid (A)	9.5	25	2.5	0.1M KCl	6
	9.2	20	2		55
	12.2	25	2.2	0.15M KCl	66
thermally unfolded (T)	6.84	90	7	holo form	55
	6.67	58	8	apo form	48
	6.4-6.5	80	7	holo form	60
	6.4-6.5	80	8	apo form	60
GuHCl unfolded (U)	1.3	25	6.5	6.9M GuHCl	6
	2.16	25	2.5	6.3M GuHCl	6
	2.14	25	6	5M GuHCl	(b)
	1.0	25	6	8M GuHCl	(b)
	2.5	25	5.5	6M GuHCl	66
	2.5	25	2.2	6M GuHCl	66
	3.8±0.2	80	6	6M GuHCl	(c)
Human α-lactalbumin					
native (N)	9.1	25	7	1M GuHCl	43
	8.69	20	7		55
calcium free (apo)	9.24	20	7	10mM EDTA	55
partly unfolded (P)	8.52			2M GuHCl	55
acid (A)	12.9	25	2.2	1M GuHCl	43
	13.8	20	2		55
GuHCl unfolded (U)	2.1	20	7	6M GuHCl	55

(a) Part of the values was taken from figures in the given refs.
(b) Calculated based on the equations given in the legend to
    Fig. 1 in ref. 42.
(c) Extrapolated value based on the data from ref. 42 and ref. 60.

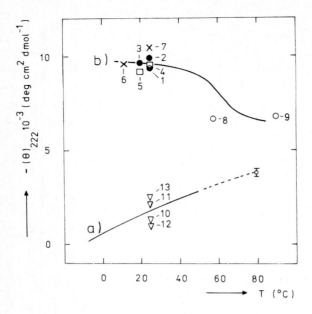

Fig. 2.3. Temperature dependence of circular dichroism band at 222 nm for various states of bovine α-lactalbumin. Solid line a) unfolded protein in 6 molar GuHCl (ref. 42). Solid line b) thermal transition of native α-lactalbumin (ref. 60).
● 1-3: native protein, ref. 6,52,55.
□ 4-5: acid form, ref. 6,55.
× 6-7: apo form, ref. 48,52.
o 8-9: thermally unfolded protein, ref. 48,55.
∇ 10-13: GuHCl unfolded protein, ref. 6,42,66.
See also Table 2.2.

given in ref. 42, a value of $-(\theta)_{222} \approx 4 \cdot 10^{-3}$ deg cm^2 dmol^{-1} for the U-state would be obtained. The value is suited to illustrate the considerable influences of GuHCl on the spectral properties in the far ultraviolet region.

The thermally unfolded state is similar to, but not identical with, the typical intermediates, A and P states (ref. 59). The similarity is based on complete absence of specific tertiary structure in all these forms as indicated by circular dichroism in the near ultraviolet region (ref. 6,42,55,59). Whereas in the A and P states the CD band in the far ultraviolet region is completely retained, the thermal transition includes additionally to the changes in the near ultraviolet region reduction of CD amplitude in the far ultraviolet (Fig. 2.3, ref. 60). Further differences were shown in fluorescence (ref. 58) and calorimetric investigations (ref. 7). Comparing $(\theta)_{222}$ of the T state at 80°C

TABLE 2.3

Gibbs energy change at various transitions of bovine and human
α-lactalbumin. ΔG (kJ mol^{-1}) at 25°C.

Transition	ΔG	pH	Approach	Ref.
Bovine α-lactalbumin				
N $\longrightarrow$ U	15.9	5.5	GuHCl, pH	68
	17.6	5.5	GuHCl	66
	27.2	6.65	GuHCl	6
	28.5	7	GuHCl	69
	17.6	7.0	GuHCl	70
	18.4	7.0	urea	70
	17.6	7.0	GuHSCN	70
	26.4	7.3	GuHSCN	71
	18.0	7.3	GuHSCN	71
	23.8	7.3	GuHSCN	71
	17.6	7.3	GuHSCN	71
	22.2	7.3	GuHCl	71
N $\longrightarrow$ T	21.9	6.3	DSC	7
N $\longrightarrow$ A	17.2		GuHCl	6
	18.4		spectra, pH	6
	26.8		pH	66
apo $\rightarrow$ T	2.5	8.0	DSC, I=0.01	49
Human α-lactalbumin				
N $\longrightarrow$ U	26.4	6.0	GuHCl	43
N $\longrightarrow$ A	9.2	6.0	GuHCl	43
apo $\rightarrow$ T	5.1	8.0	DSC, I=0.01	49

with the GuHCl unfolded protein at 25°C, it remains a high content of "residual structure" which exceeds that of other small
globular proteins. On the other hand, relative to the CD amplitude of the "maximally unfolded" state (U) in high molar GuHCl
solution at 80°C, the effect is much less pronounced, even if the
dependence of $(\theta)_{222}$ on GuHCl concentration is disregarded. This
implies the question whether it is reasonable to regard the U
state as reference state (see section 2.3).

   Now the question arises whether the changes in optical properties correspond to changes in thermodynamic parameters. In Tables
2.3 and 2.4 Gibbs energy and enthalpy changes are collected for
the various α-lactalbumin transitions. The mean values are com-

TABLE 2.4

Enthalpy changes at various transitions of bovine and human
α-lactalbumin. ΔH in kJ mol^{-1}.

Transition	ΔH	T($^\circ$C)	Method	Ref.
Bovine α-lactalbumin				
N $\longrightarrow$ U	120	25	circular dichroism	42
	106	25	isothermal calorimetry	72
	122	25	isothermal calorimetry	7
N $\longrightarrow$ T	130	25	scanning calorimetry	7
N $\longrightarrow$ A	75	25	circular dichroism	42
	99	25	pH jump	68,73
	100	25	isothermal calorimetry	72
apo $\longrightarrow$ T	187	25	circular dichroism	48
	175	25	scanning calorimetry	49
	155	31	fluorescence	50
Human α-lactalbumin				
N $\longrightarrow$ P	100	40	isothermal calorimetry	25
apo $\longrightarrow$ T	254	40	scanning calorimetry	49

pared with the optical changes in the schemes given in Fig. 2.4
and 2.5 for bovine and human α-lactalbumin, respectively. Furthermore, the distribution coefficient K_d, a chromatographically
determined relative measure of molecular size, is added (W. Pfeil
and M.L. Sadowski, unpublished results). K_d is defined by

$$K_d = (v_e - v_o)/(v_i - v_o) \tag{1}$$

according to ref. (74,75), where v_e is the elution volume of the
protein, v_o the void volume (Blue Dextran) and v_i is the elution
volume of a low molecular internal standard (DNP glycin). The
main advantage of chromatography on an analytical column is the
possibility to use a low protein concentration, thus avoiding
uncertain corrections for protein association which might be the
source of disagreement between former findings (see above).

From Fig. 2.4 it follows for the acid state of bovine α-lact-
albumin that not only CD band at 222 nm but also the size are
close to the native state. This result is in agreement with
former findings (ref. 55,57). On the other hand, enthalpy and

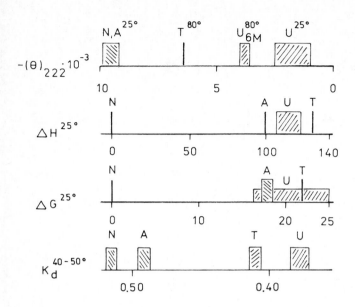

Fig. 2.4. Comparison of the changes of molar ellipticity, enthalpy, Gibbs energy and molecular size at various transitions of
bovine α-lactalbumin.

Gibbs energy change at the transfer N ⟶ A are closer to the
GuHCl unfolded state than to the native one. Practically indistinguishable from the U state, however, is the thermally unfolded
protein with respect to ΔH and ΔG of the N ⟶ T transition.
Molecular size of the T state measured chromatographically on
thermally unfolded apo α-lactalbumin at 40-50°C appears in difference to earlier findings (ref. 55) much larger than that of the
A state. As a further example for the noncoincidence of optical
and thermodynamical changes at the N ⟶ A transition may serve
human α-lactalbumin (Fig. 2.5). Here, $(\theta)_{222}$ of the A form is
much larger than that of the native protein, whereas ΔG amounts
to about one-half of the N ⟶ U transition.
    Thus, the magnitude of circular dichroism band in the far
ultraviolet region does not tell much about thermodynamic properties, at least in the case of α-lactalbumin. When at the A ⟶ U
transition the whole secondary structure of bovine α-lactalbumin
is disrupted, the same transition is accompanied by a hardly
detectable enthalpy change which does not exceed 6 kJ mol^{-1}
according to calorimetric measurements (ref. 72).

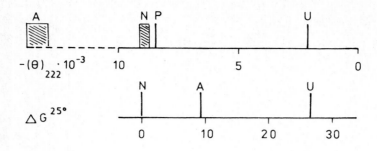

Fig. 2.5. Comparison of the changes of molar ellipticity and
Gibbs energy at various transitions of human α-lactalbumin.

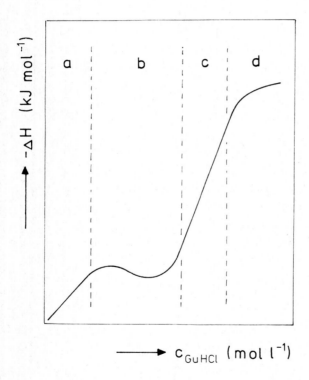

Fig. 2.6. Scheme of isothermal calorimetric GuHCl titration
curves of proteins. See text.

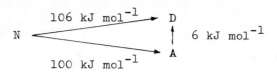

Since the acid state is characterized by nearly the same secondary structure content as the native protein (ref. 55,60), but its
disruption proceeds with marginal enthalpy change, there is no
reason to overestimate the thermodynamic importance of the (less
pronounced) residual structure in thermal unfolding. Already on
lysozyme (ref. 24) and later on bovine α-lactalbumin (ref. 7) it
was shown that thermally and GuHCl unfolded protein are thermodynamically indistinguishable.

In view of the three-state unfolding transition of α-lactalbu-
min in GuHCl, the origin of cooperativity of the structural
transition remains to be explained. The best way is to solve the
problem by isothermal calorimetric GuHCl titrations of human
α-lactalbumin since the consecutive transitions are more apart
from each other than those of bovine α-lactalbumin (ref. 43).
Isothermal calorimetric titration curves of proteins with GuHCl
can be divided into four parts (ref. 1,24), Fig. 2.6:
(a) the pretransitional region: almost linear increase of $(-\Delta H)$
    starting from zero denaturant concentration,
(b) the transition region: sigmoidal part exhibiting reduced
    slope or sigmoidal curvature at appropriate GuHCl concen-
    tration,
(c) the postdenaturational region: almost linear increase of
    $(-\Delta H)$,
(d) the saturation region: deflection, i.e. no further increase
    of the $(-\Delta H)$ function above 6-7 mol 1^{-1} GuHCl.
In case of human α-lactalbumin, calorimetric titrations were
performed at various conditions (ref. 25). Since enthalpy change
at protein unfolding is usually temperature dependent, the
transition can be more convincingly demonstrated at elevated
temperature. From Fig. 2.7 there follows an onset at about
0.25 mol 1^{-1} GuHCl and a transition midpoint at about 0.9-1.0
mol 1^{-1} GuHCl. The enthalpy change amounts to about 100 kJ mol^{-1}
and the transition is accompanied by a considerable heat capacity
change (ref. 25). The calorimetrically observed transition coincides with the first of the two consecutive transitions monitored

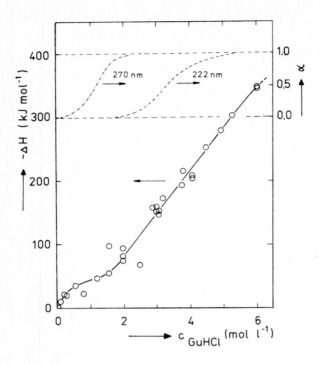

Fig. 2.7. Isothermal calorimetric titration of human α-lactalbu-min with GuHCl at 40°C. Inset: the transitions monitored by circular dichroism at 270 nm and 222 nm.

by circular dichroism at 270 nm, i.e. it corresponds to the destruction of the specific environment of the aromatic side chains. A heat effect of the second transition at 2-5 mol l^{-1} GuHCl corresponding to the destruction of the secondary structure and to unfolding of the chain cannot be found by means of the calorimetric titration. This means that the effect is (i) either absent, or (ii) is rather small and therefore compensated by the increase of the enthalpy of the additional GuHCl binding upon this transition. However, some details discussed in ref. (25) speak in favour of the second explanation (ii). This concerns, in particular, the observed posttransitional slope in Fig. 2.7 which does not, as usually (ref. 1.24), exceed the pretransitional slope. It is noteworthy that the transition heat N⟶P as determined here to about 100 kJ mol^{-1} is about 2.5 times smaller than that of the thermal transition of the apo state in scanning calorimetry (ref. 49, see also Table 2.4). Thus, the intermediate is unlike the thermally unfolded protein. The calorimetric titra-

tion of the A state (not shown here) proceeds without any transition at moderate GuHCl concentration. The curve, however, has an
increased slope compared with the native form which bears resemblance to thermally unfolded lysozyme (ref. 24).

Summarizing, the first of the two consecutive transitions
monitored by circular dichroism in the aromatic region is really
a cooperative transition which is characterized by the mean heat
effect and heat capacity change despite of the small changes in
molecular size. The second transition monitored by circular
dichroism in the peptide region appears as a gradual transition
in calorimetry without large heat effect. The T state (thermally
unfolded protein) is according to the thermodynamic criteria
unfolded, i.e. similar to the GuHCl unfolded protein.

The intermediate state of α-lactalbumin is really of interest
since it is characterized by secondary structure, native-like
molecular size but marginal thermodynamic stability. The intermediate for which later the term "molten globule" was introduced
(ref. 76), had been originally defined by PTITSYN and coworkers
as follows (ref. 55):
"... all properties of the A, P, Apo and T forms of α-lactalbu-
mins are consistent with the model of a compact globule with
native-like secondary structure and with slowly fluctuating
tertiary structure." [*]

The proposal could have importance in protein folding, even if
apo and T states claimed for the "molten globule" are now better
defined as native and unfolded, respectively. Nevertheless, in
the present section the question arose how to define thermodyna-
nic states of proteins when calorimetric and optical criteria
are not in agreement.

2.3   EVALUATION OF THE THERMODYNAMIC STATES

Description of protein states in precise terms of thermodynamics is a prerequisite for their understanding. Assuming state X
to be defined by a set of external variables such as temperature,
pH, ionic strength etc. (T, pH, I, ...), the state is characterized by heat capacity, specific volume, Gibbs energy, enthalpy
etc. (C_p, V, G, H, ...). Changing from one set of external
variables to another, a cooperative transition to state Y might

[*]After finishing this paper, further studies on α-lactalbumin
became available (ref. 183,184).

occur accompanied by changes of the thermodynamic characteristics (C_p, V, G, H, ...). The macroscopic description is an average over the properties of microscopic states.

The main states in protein unfolding are the so-called native (N) and unfolded (U) ones. The denotation 'native' and 'unfolded' implies that additional, i.e. nonthermodynamic terms are introduced in order to achieve a more concrete description of the rather complex properties of biopolymers. So the term 'native' stands for a variety of properties of a protein, such as enzymic or other biological activity, spectral properties, sequence, three-dimensional structure etc. Knowledge of some of these properties is necessary to bring thermodynamic quantities in context with the corresponding state, e.g. C_p^N, the heat capacity of the native state. Moreover, nonthermodynamic characteristics sometimes open up attractive ways for indirect determination of thermodynamic quantities (see, e.g., ref. 18). On the other hand, problems can arise from mixing of thermodynamic and nonthermodynamic properties. Thus, it is well known that gross conformation of most enzymes is maintained over rather extended ranges of temperature and pH, sometimes even in the presence of moderate denaturant concentration. Nevertheless, enzymic activity or specific optical properties may disappear due to minor or local structural changes in a rather narrow range. At the same time, thermodynamic functions need not show discontinuities, i.e. they confirm the properties of still existing 'native' state in terms of thermodynamics and structure.

About thermodynamic states there exist misleading interpretations in the biochemical literature when they are solely based on indirect observations. As an example, recently the case of D-amino acid oxidase was analyzed in more detail (ref. 78). For the protein, a 30% decrease of tryptophan fluorescence was found as the temperature was raised from 8 to 22°C (ref. 79). The reversible transition with an apparent enthalpy change of 326 kJ mol^{-1} was interpreted as a conformational transition. Based on these findings the heat capacity function of the protein was constructed (ref. 78, see also Fig. 2.8). However, no indication of any transition was found by means of highly sensitive scanning calorimetry (the proposed transition heat is beyond all sensitivity problems of modern instruments). The obvious discrepancy between nonthermodynamic and calorimetric findings in the present

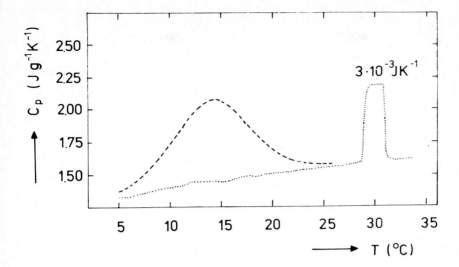

Fig. 2.8. D-amino acid oxidase: expected (broken line) and calorimetrically determined (dotted line) heat capacity function. See text. Redrawn from ref. 78.

case can now be explained by dynamic quenching processes (ref. 80,81). Such non-coincidence of indirect conclusions about thermodynamic states with calorimetric results can often be observed, in particular in more complex systems. Actually, proper-ties such fluorescence depolarization, NMR, EPR, enzyme kinetics and others whose magnitude depends on stochastic relaxation processes could lead to similar discrepancy (ref. 80,81).

Biochemical and thermodynamic properties of the native state are well characterized on numerous proteins. The native state is a naturally occurring one, and the experimental investigations are feasible. On the contrary, the unfolded state is more contro-versial. The state must be produced artificially and the most effective way of protein unfolding and data treatment are still disputed.

Protein unfolding can be defined according to SCHERAGA (ref. 82): "We shall use the term 'denaturation' ([*]) to describe the

---

[*]In the present paper the term 'unfolding' is used instead of 'denaturation'. Denaturation is primarily related to biological activity. However, loss of activity need not imply gross confor-mation changes. Unfolding is per se a disruption of protein structure. However, random coil need not be achieved as it will be shown below.

process in which a protein or polypeptide is transformed from an ordered to disordered state without rupture of covalent bonds."

TANFORD's work (ref. 62,63) makes a distinction between the random coil and "incompletely disordered state, with some residual cooperative structure". According to TANFORD "the completely disordered state is achieved by all proteins examined so far in concentrated GuHCl solutions" (ref. 62). Accordingly, GuHCl unfolded protein is widely accepted as the reference state.

An enormous amount of observations on physicochemical properties of unfolded proteins was collected by TANFORD (ref. 62,63) and LAPANJE (ref. 64). Even if restricting to one single example, the distinction between GuHCl and thermally unfolded protein seems to be justified. So the intrinsic viscosity of lysozyme increases from $\eta = 3.0$ cm^3 g^{-1} for the native protein at 25°C to only $\eta = 4.7$ cm^3 g^{-1} for the thermally unfolded one at 60-75°C (ref. 62,83,84). The corresponding values for GuHCl unfolded lysozyme with intact or broken disulfide bridges are 6.5 cm^3 g^{-1} and 17 cm^3 g^{-1} at 25°C (ref. 62), respectively. However, it remains open whether the incompletely disordered state is really a kind of "cooperative structure" in the sense of the above definition. Thus, on heating, the viscosity of GuHCl unfolded lysozyme drops from 6.5 cm^3 g^{-1} to about 4.8 cm^3 g^{-1} at 55°C and achieves just the value of the thermally unfolded protein (ref. 85). The temperature dependent changes of intrinsic viscosity resemble the temperature dependent changes of optical properties $(\Theta)_{222}$ discussed in the preceeding section 2.2 (Fig. 2.3) by the example of the homologous α-lactalbumin. Careful studies of temperature dependencies of optical rotatory dispersion were recently reported for both α-helices and random coil polypeptides (ref. 86).

Upon heating the properties of GuHCl unfolded proteins change continously and approach the properties of thermally unfolded protein (viscosity, $(\Theta)_{222}$). By means of calorimetric investigations it was shown that GuHCl and thermally unfolded lysozyme (ref. 1,24) and α-lactalbumin (ref. 7) are indistinguishable with respect to the enthalpy and heat capacity changes. Furthermore, no cooperative transition was found calorimetrically at titration of thermally unfolded lysozyme with GuHCl (ref. 24). This finding was unexpected since optically active structure in thermally unfolded protein was detected by changes in ORD or CD that occur when GuHCl is added to thermally unfolded protein (ref. 77). The structure has not been characterized and its significance is uncertain. No changes were found by means of NMR when urea is added to thermally unfolded lysozyme (ref. 177). On the other hand, structure that differs from the classical secondary structure was identified on predominantly unfolded polypeptides by means of NMR (ref. 178,179) and on thermally unfolded protein by means of IR spectroscopy (ref. 180). In more recent studies only gradual changes of $(\Theta)_{222}$ were found at titration of thermally unfolded α-subunit of tryptophan synthase with urea and GuHCl (ref. 87). All these findings are consistent with the existence of a unique unfolded state in GuHCl and thermal unfolding. However, they suggest also non-ideality of both GuHCl and thermally unfolded protein which becomes more pronounced at elevated temperature. It is worth mentioning that even for GuHCl unfolded proteins at $25^{\circ}C$ deviations from random coil properties exist. Evidence suggesting that GuHCl unfolded protein still contains regions of ordered structure after treatment with GuHCl was given originally by TANFORD (ref. 62, p. 171 ff). Further observations based on fluorescence studies, electrostatic interactions, optical changes at reductive cleavage of disulfide bridges etc. were mentioned in ref. (64).

It seems to be an advantage of optical methods to compare directly proteins at quite different conditions, e.g. thermally unfolded protein at elevated temperature with GuHCl unfolded protein at room temperature. However, even if GuHCl is "invisible" in the used spectral region, the presence of the denaturant cannot be disregarded for thermodynamic reasons: a two-component system consisting of protein and water as in the case of thermal unfolding must be regarded as qualitatively different from a three-component system containing additionally GuHCl. Some of the changes produced by GuHCl are:

(a) high molar GuHCl solution (6 molar) contains no more than about 5 moles water per mol GuHCl,

(b) GuHCl is able to interact with protein. The number of GuHCl molecules bound by lysozyme is 67 (ref. 88), i.e. a formal calculation for the complex gives an increase in mol mass of the protein by about 45% and an increase of volume by about 20%,

(c) transfer of protein in high-molar GuHCl is accompanied by

considerable transfer free energies (ref. 89) and by large
heat effects, reaching e.g. for lysozyme more than -500 kJ
per mol protein (ref. 24),

(d) GuHCl induced structural transitions in proteins are primarily related to the change in the number of denaturant binding
sites (Δn). There is no obvious relationship between Δn and
ΔH of the structural transition. GuHCl induced transitions
with zero enthalpy change are conceivable,

(e) preferential binding heat and temperature dependence of denaturant binding, which were originally assumed to be negligible (ref. 90), play an important role in GuHCl induced
transitions (ref. 24) and render van't Hoff treatment more
difficult.

Thus, the problem to define thermodynamic states is far from
being trivial. Observations made by means of nonthermodynamic
methods need not be conclusive, as it was shown by the example
of D-amino acid oxidase (ref. 78,81). On the other hand, considerable thermodynamic effects due to solvation phenomena produced
by denaturants remain to be taken into account.

Based on direct, i.e. calorimetric measurements, the macroscopic states 'native' and 'unfolded' are well substantiated. The
two states are characterized by individual heat capacity functions, and they are separated from each other by cooperative
transitions representing discontinuities of the heat capacity
functions. Thermodynamically the unfolded state appears as unique
for thermally and GuHCl unfolded protein. However, non-ideality
of the unfolded state (compared with random coil) is obvious and
becomes more pronounced at elevated temperature. In contrast to
the ideal properties of the random coil, the unfolded protein
must be defined as a real state characterized by the degree of
disorder which can be achieved under specific experimental conditions.

The macroscopic states considered here are composed of numerous microscopic states. Protein is a relatively small system
showing large statistical fluctuations (ref. 91). Since the first
considerations on the distribution of the microscopic states
(ref. 16,92), evidence supporting the concept was obtained (ref.
181). According to hydrogen exchange studies the activation
parameters represent a rather extended probability density function (ref. 93,94,182). Further evidence showing remarkable width

of probability density was given by a statistical mechanical treatment of proteins (ref. 95-97). The treatment is closely re- lated to calorimetrically determinable heat capacity functions (ref. 96-99). Special attention was paid to the possibility that 'apparent intermediate' states of the protein occur even when the macroscopic state does not show discontinuities (ref. 95, p. 123 ff.). Further experimental findings, in particular changes of properties of unfolded protein were discussed in ref. (100,101) and found to be consistent with the concept of distribution in the unfolded state. In view of the width of probability density of native and unfolded states of bovine α-lactalbumin in the transition region (Fig. 2.9), both the real properties and the proposed 'apparent intermediates' are not surprising.

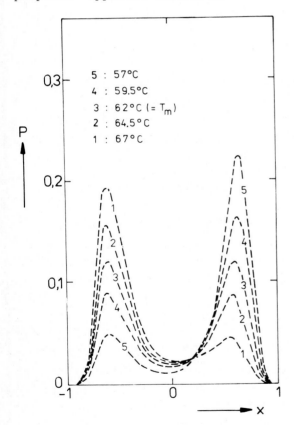

Fig. 2.9. Probability density versus structural state parameter x for α-lactalbumin obtained by IKEGAMI treatment of scanning calorimetric curves. x corresponds to the fraction of noncova- lent bonds being in bonded state. See also ref. (7,98,99).

On the α-lactalbumin example, additionally to the native (N) and unfolded (T or U) states, the acid and partly unfolded forms (A and P states) were proposed. These states are according to the properties summarized in section 2.2 (Fig. 2.4 and 2.5) thermodynamically distinguishable from N and U states. A and P states can be considered as thermodynamically stable intermediates in α-lactalbumin unfolding which occur under specific conditions. As it was shown in section 2.2, the enthalpy change A $\longrightarrow$ U, corresponding to rupture of secondary structure, is rather small. The problem which contributions stabilize the intermediate to what extent is still open. However, the intermediates are separated from the native state by a cooperative transition accompanied by heat capacity and enthalpy changes. A and P states are devoid of the native-like properties of aromatic side chains. The aromatic-aromatic interactions analyzed in detail recently (ref. 102) could well explain cooperativity and thermodynamic properties of the transition. On the other hand, the changes of properties of the aromatic side chains at the N $\longrightarrow$ P and N $\longrightarrow$ A transition are hardly understandable without assuming water penetration into the interior of α-lactalbumin. Since enthalpy of formation of H-bonds is much larger in nonpolar than in aqueous surrounding, the cooperativity of the transition could be, besides the aromatic-aromatic interaction, also due to changes in enthalpy state of H-bonds.

2.4   UNFOLDING OF PROTEINS WHICH DO NOT REPRESENT SINGLE COOPERATIVE SYSTEMS

Development of the thermodynamic approach in protein unfolding parallels investigations in other fields, in particular studies aimed to elucidate the biopolymer structure. Accordingly, calorimetric studies turned towards more complex proteins during the past decade. A prerequisite, however, is analysis of more complex transitions in such a way that subtransitions can be identified. This became possible by the statistical mechanical deconvolution which is based on the average excess enthalpy function corresponding to eq. (2):

$$\langle \Delta H(T) \rangle \; = \; \int_{T_o}^{T} \langle \Delta C_p \rangle \; dT \tag{2}$$

$\langle \Delta H(T) \rangle$ can be obtained from scanning calorimetric data by direct integration of the excess apparent molar heat capacity functions. Thus, the partition function can be calculated:

$$Q(T) = \exp \int_{T_o}^{T} \left[\frac{<\Delta H(T)>}{RT^2} \right] dT \qquad (3)$$

From the partition function all the thermodynamic quantities of the system can be obtained.

Both deconvolution (ref. 20-22,103) and most recent achievments in studying proteins which do not represent single cooperative systems were recently reviewed (ref. 19). Therefore, this section will be restricted to only a few examples.

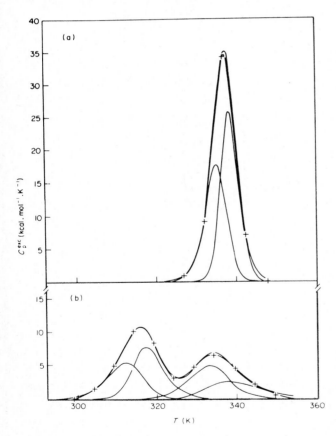

Fig. 2.10. Deconvolution of the excess heat capacity function of pepsinogen (a) and pepsin (b) at the same solvent condition: pH 6.5, 100 mM NaCl. Crosses indicate calculated heat capacity function, which almost coincides with the experimental one. From ref. (9), with permission.

The advantage of deconvolution were first shown in a fascina-
ting manner by the example of pepsin and pepsinogen (ref. 9).
For pepsinogen a single heat absorption peak was found showing
significant deviation of the ratio $\varkappa = \Delta H^{cal}/\Delta H^{v.H.}$ from unity
even when varying the experimental conditions (pH, NaCl, urea)
over a wide range. The transition represented in Fig. 2.10a)
is characterized by $\varkappa = 2.0$ and can be resolved in two subtransitions differing in transition temperature and heat ($T_1=62.1^{\circ}C$,
$\Delta H_1=456$ kJ mol^{-1}; $T_2=65.5^{\circ}C$, $\Delta H_2=632$ kJ mol^{-1}).

Thermal unfolding of pepsin is even more complex. In an extended range of experimental conditions (pH, ionic strength, inhibitor) two consecutive transitions are visible. The two heat
absorption peaks, each representing multi-state transitions cen-
tered around $40^{\circ}C$ ($\varkappa_1 = 1.6$) and $65^{\circ}C$ ($\varkappa_2 = 2.0$) can be resolved
in four single two-state transitions as shown in Fig. 2.10b).
The findings are in fair agreement with the molecular structure
of pepsin according to ref. (104). All the four cooperative units
of pepsin marked by dotted lines in Fig. 2.11 have a compact
structure with a well developed hydrophobic core.

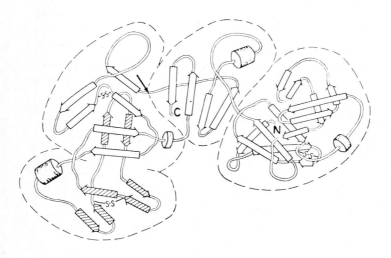

Fig. 2.11. Scheme of pepsin structure according to ANDREEVA and
GUSTCHINA. Broken lines encircle the compact regions with hydrophobic cores. The arrow indicates the site of cleavage of the
polypeptide chain at fragmentation (residues 179 to 180).
From ref. (9), with permission.

The current state-of-the-art in deconvolution can best be shown by the plasminogen example (ref. 23). Plasminogen sequence consists of 790 amino acid residues and is schematicall represented in Fig. 2.12.

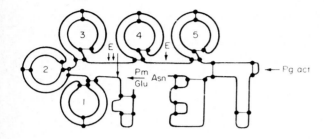

Fig. 2.12. Scheme of primary structure of plasminogen. Arrows indicate sites of cleavage of the chain by elastase (E), plasminogen activator (Pg. act.) and plasmin (Pm). -•-•- Disulfide crosslinks. From ref. (23), with permission.

Making use of the sites for proteolytic cleavage shown in Fig. 2.12, the melting pattern of the corresponding fragments can be obtained. The calorimetric recordings of the fragments (Fig. 2.13) indicate the existence of seven independent cooperative structures within the plasminogen molecule. Five of them are formed by the homologous regions, the characteristic kringles. Two cooperative parts can be attributed to the C-terminal part that contains the cleavage site activated at interaction with the plasminogen activator. From more detailed studies reported in ref. (23) it follows that plasminogen activation (cleavage Arg 560-Val 561) is accompanied by marked thermodynamic changes. They are due to increase of interdomain motility. The situation bears similarity with the deco-operation found by the pepsinogen-pepsin example (ref. 9).

Deconvolution of excess enthalpy functions can gain importance when complementary investigations are possible for identifying the cooperative units. One of the most conclusive approaches is puzzling out proteolytic fragments as shown by the plasminogen example. In studies of calcium binding proteins destabilization of domains upon removal of the bivalent cations was applied (ref. 105). Domains of lactoperoxidase were identified combining calorimetry, absorption spectrometry and circular di-

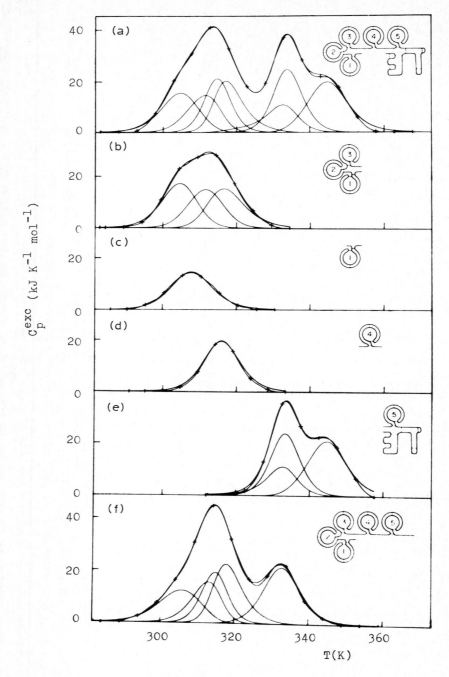

Fig. 2.13. Deconvolution of the excess heat capacity functions of (a) Lys-plasminogen, (b) fragment K 1-3, (c) fragment K 1, (d) fragment K 4, (e) miniplasminogen, and (f) plasminogen heavy chain at pH 3.4. Crosses indicate the experimentally determined

Legend to Fig. 2.13, continued
function to distinguish it from the calculated one using parameters of transitions estimated by deconvolution analysis. The
structure of the studied molecules is given schematically in the
upper right corner of each panel. From ref. (23), with permission

chroism, since the haem group contained in the less thermostable
domain could be used as an internal label (ref. 106). The capa-
bilities of deconvolution of calorimetric curves in combination
with independent techniques are far from being exhausted as it
was recently shown by prediction of structural pecularities of
troponin c (ref. 105, see also ref. 107 for a comment on thermodynamic properties and X-ray structure of troponin c).

2.5   THE STABILITY OF GLOBULAR PROTEINS

2.5.1   General aspects

There exists a growing demand in data characterizing protein
stability. In fact, protein stability is one of the most important aspects in the understanding of protein structure and dynamics, folding, functioning etc. Recently protein stability has
gained importance in view of industrial applications of proteins
as well as attempts to influence protein stability in a favourable way for practical reasons.

The term stability is ambigous in so far as it can be related
to functional properties under more or less extreme conditions
(denaturation). Thermodynamics, however, provides a precise definition. The negative Gibbs energy change for protein unfolding
can be considered as a measure of the net stabilization energy
of the protein structure. Various approaches for the determination of ΔG_{unf} are available, and their implications have been
considered in detail (ref. 1,18,62-64,92,99,108,109,115). Most
convenient are, at present, scanning calorimetry and denaturant
induced unfolding for the determination of ΔG_{unf}.

From scanning calorimetry the temperature function of Gibbs
energy change at constant pH can be obtained by means of equation (4):

$$\Delta G_{unf}(T) = \Delta H(T_{1/2}-T)/T_{1/2} + \Delta C_p(T-T_{1/2}-T\ln(T/T_{1/2})) \qquad (4)$$

where ΔH is the enthalpy change at transition temperature $T_{1/2}$
(half conversion) and ΔC_p the heat capacity change (temperature
independent, see ref. 115). The approach does not involve any

assumptions, and the desired data can be determined with high accuracy by modern scanning calorimetric devices (ref. 103).

Denaturant induced unfolding can be performed at the desired temperature and pH. However, extrapolation to zero denaturant concentration is problematic. Most frequently, equation (5) derived from denaturant binding models (ref. 3,62,63) or linear extrapolation according to equation (6) (ref. 109-111) are used:

$$K_{unf} = K^o_{unf}(1 + ka_{\pm})^{\Delta n} \qquad (5)$$

$$\ln K^o_{unf} = \ln K_{unf} + m_{(denaturant)} \qquad (6)$$

Here K^o_{unf} is the equilibrium constant for protein unfolding at zero denaturant concentration, k the denaturant binding constant, $a_{\pm}$ mean ion activity of GuHCl, and Δn the preferential interaction parameter. The results of the two extrapolation procedures may differ. A systematic analysis of the problem including proper choice of the denaturant binding constant was given in ref. 112. Comparison of Gibbs energy change in protein unfolding obtained by various approaches is contained in ref. (99,109,115). It turns out that the differences between the considered approaches must not be overestimated. In particular, there is no indication of any diminishing Gibbs energy change at thermal unfolding due to the 'residual structure' (ref. 99). When more subtle differences are to be considered, such as influence of point mutations on stabilization energy, it is advisable to compare data obtained by the same approach.

In Table 2.5 selected values of Gibbs energy changes for unfolding of various proteins are summarized (for a more comprehensive data collection see ref. 115). The data refer, where possible, to standard conditions, i.e. 25°C, pH 7 and the absence of denaturants. The proteins listed in Table 2.5 differ greatly with respect to their functional properties and, so far known, with respect to their teriary structure. Nevertheless, their stability falls into a rather narrow range of about -20 to -60 kJ mol^{-1}, thus roughly confirming the predictions made on this account already in TANFORD's early studies (ref. 113,114).

The data collected in Table 2.5 enable reconsideration of former proposals. Thus, the theory of helix-coil transitions implies a relationship between chain length and thermodynamic quantities (ref. 157). Recently, a theoretical prediction was

TABLE 2.5

Gibbs energy change at protein unfolding. Selected values.

Protein	pH	T (°C)	ΔG_{unf} (kJ mol^{-1})	Approach	M.W. (kD)	Type [a]	Ref.
G-actin	7.5	25	27	heat	41.7	II	116
alkaline phosphatase, apomonomer	7.5	30	31.8	DSC	47	I	117
apolipoprotein AI	8	23	15.5	GuHCl	28.3	IV	118
apolipoprotein CII	7.4	25	11.7	GuHCl	8.8	IV	119
arabinose binding protein	7.4	25	38.5*	DSC	33.2	II	120
aspartate aminotransferase, apo form, α-subunit	8.2	30	126	DSC	44.6	I	121
carbonic anhydrase	-	-	45	H/D exchange	28.7	I	122
α-chymotrypsin	4	25	48.5	DSC	25.6	III	5
chymotrypsinogen	4	25	49	GuHCl	25.6	III	123
cytochrome b5 (fragment 1-90)	7	25	25	DSC	10.6	II	124
cytochrome c, bovine, ferric	4.8	25	37.7	DSC	12.4	II	5
iso-2 cytochrome c	7.2	20	13	GuHCl	12.5	II	125
iso-1 cytochrome c	7	20	8.8	GuHCl	12	II	126
DNAse I	7	25	39	GuHCl	28.7	III	127
flagellin	7	25	72*	DSC	32.6	II	128
galactose binding protein (+Ca^{2+})	7.8	25	11*	DSC	40+48	II	129

	pH	T (°C)					
glucose dehydrogenase	-	-	6.9	GuHCl	115	I	130
glycinin, monomer (without NaCl)	7.6	25	64*	DSC	26	II	131
growth factor, epidermal	7.5	25	67	GuHCl	6.2	IV	132
growth hormone, bovine	8	25	58.6	GuHCl	21.5	IV	133
growth hormone, ovine	8	25	40.6	GuHCl	21.5	IV	133
growth hormone, rat	8	25	33.5	GuHCl	21.5	IV	133
hemolysin	8.15	25	32	DSC	20	IV	134
immunoglobulin, various domains	7.5	25	24	GuHCl	22.3	IV	135
immunoglobulin, fragment c_L	7.5	25	24	GuHCl	22.1	IV	136
insulin	3	20	27	H/D exchange	6	IV	137
lac repressor, headpiece	8	25	10.8	DSC	5.7	II	138
α-lactalbumin, bovine (+Ca^{2+}) b)	6.3	25	21.9	DSC	14	IV	7
α-lactalbumin, human (+Ca^{2+})	6-7	25	26.4	GuHCl	14	IV	43
α-lactalbumin, bovine (-Ca^{2+})	8	25	12.5	DSC, I=0.1	14	IV	49
α-lactalbumin, human (-Ca^{2+})	8	25	6.1	DSC, I=0.1	14	IV	49
β-lactamase	7	25	25	GuHCl	28.8	III	139
β-lactoglobulin	3.15	25	42.7	urea	18.4	IV	109
lipase, aspergillus	7	-	46	heat	2x25	III	140
lipase, rhizopus	7	-	16.8	heat	48	III	140
lysozyme (HEW)	7	25	60.7	DSC	14.3	III	5
lysozyme, phage T4, wild type	-	25	25	heat	18.6	III	141
myoglobin	6.6	25	39.7	GuHCl	17.2	II	112
ovalbumin	7	25	24.8	GuHCl	42.7	IV	142
papain	3.8	25	76*	DSC	23.4	III	143
parvalbumin III (0.1 mM CaCl2)	7	25	46	DSC	11.4	II	144

	pH	(°C)		Method		Type[a]	Ref.
pepsin	6.5	25	45.2	DSC	36.6	III	9
pepsinogen	6.5	25	65.7	DSC	36.7	III	9
phosphoglycerate kinase, horse	7.4	25	26	GuHCl	44.4	I	146
phosphoglycerate kinase, yeast	7.5	25	22.3	GuHCl	45.8	I	145
phosphoglycerate kinase, Thermus therm.	7.5	25	49.7	GuHCl		I	145
phycocyanin	6	25	18.4	urea	17.8	II	147
ribonuclease	7	25	58.6*	DSC	13.7	III	148
serum albumin	-	-	32.6	H/D exchange	66.5	IV	149
subtilisin inhibitor	7	25	59*	DSC	11.5	IV	150
thermolysin (fragment 206-316)	7.5	25	31	GuHCl	11.8	III	151
thioredoxin	7	25	30.5	GuHCl	11.7	II	152
toxins: cytotoxin	4.5	25	24*	DSC	6.8	IV	153
toxins: neurotoxin I	5.7	25	24*	DSC	8.0	IV	153
toxins: neurotoxin II	5.0	25	36*	DSC	6.9	IV	153
troponin c	-	25	19.7*	DSC	18	II	154
trypsin	-	25	67	DSC	22.3	III	18
trypsin inhibitor	3	25	51.7	DSC	6.5	IV	155
tryptophan synthase [c], α-subunit, wild type	7	25	35.1	DSC	28.6	I	156

* Data were calculated based on other thermodynamic values given in the Ref.

a) Type: I - intracellular enzyme
     II - intracellular nonenzyme
    III - extracellular enzyme
    IV - extracellular nonenzyme

b) Further data see Table 2.3.
c) Further data see Table 2.8.

made showing that spherical molecules with fewer than 70 and more
than 2000 residues are not stable. According to the prediction,
a stability maximum should exist around 200 residues (ref. 158).
The experimental data for Gibbs energy change at protein unfolding plotted against molecular weight (Fig. 2.14) seem not to
support the idea. On the other hand, part of the proteins contains disulfide bridges and prosthetic groups which have considerable influence on ΔG_{unf}.

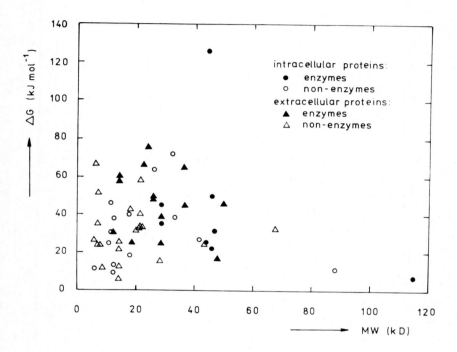

Fig. 2.14. Gibbs energy change at protein unfolding (data from
Table 2.5) plotted versus molecular weight.

Protein stability meets both structural and functional requirements (ref. 99). For further discussion of the problem of protein
stability, it is advantageous to follow a classification into
four protein types according to biological distinction: intracellular enzymes, intracellular nonenzymes, extracellular enzymes and extracellular nonenzymes. The classification is well
substantiated by an analysis of more than 350 proteins with respect to amino acid composition, molecular weight, folding type,
amount of disulfide bridges, etc. (ref. 159,160). Some of the

TABLE 2.6

Typical characters of the four groups of proteins according to NISHIKAWA et al. (ref. 159,160).

Group	I	II	III	IV
localization	intracellular		extracellular	
function	enzyme	nonenzyme	enzyme	nonenzyme
dominant molecular weight	20-60,000	8-30,000	15-60,000	8-30,000
variation in amino acid composition	small	large	small	large
content of nonpolar amino acids a)	38.7	36.0	36.0	33.2
relatively high content of specific amino acids	Ala, Leu	Lys, Ala, Leu	Ser, Tyr	Cys, Pro
dominant folding type	α/β	all types, more α	β or $\alpha+\beta$	α or β
arrangement of β sheet	parallel		antiparallel	
disulfide bonds	no		yes	

a) Percent of total amino acids of Ala, Val, Leu, Ile, Phe, Tyr and Trp.

a)

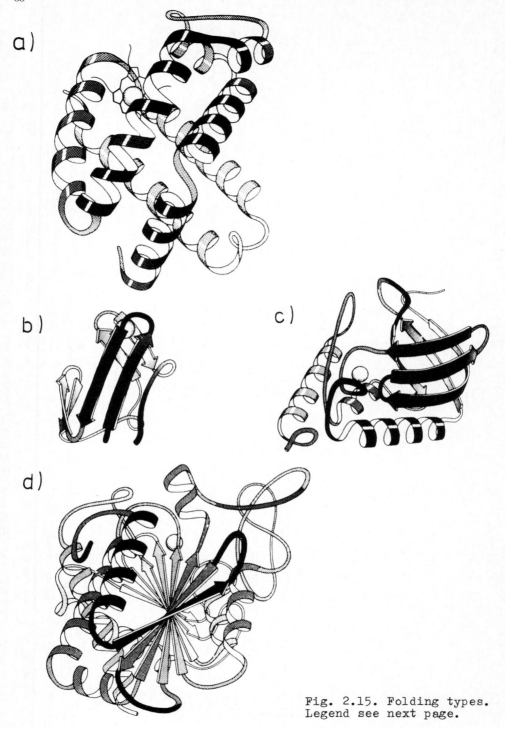

b)

c)

d)

Fig. 2.15. Folding types.
Legend see next page.

Fig. 2.15. The different folding types shown on examples.
a) α protein: hemoglobin β subunit.
b) β protein: rubredoxin.
c) $\alpha + \beta$ protein: staphylococcal nuclease.
d) α/β protein: carboxypeptidase.
Form ref. (162), with permission.

characteristics are summarized in Table 2.6. The main differences
between intra- and extracellular proteins which could have importance for the stability problem are the higher content of nonpolar amino acid residues and the absence of disulfide bridges in
intracellular proteins. On the other hand, most enzymes are
larger than nonenzymes, and do not show the same variability in
amino acid composition. Practically, all structure types can be
found, i.e. α, β, mixed types $\alpha + \beta$ and alternating α/β (Fig.
2.15). Mixed types occur more frequently among enzymes.

In Table 2.7 average values for ΔG_{unf} of globular proteins are
given in the four categories of intra- and extracellular enzymes
and nonenzymes. The four types show significant differences
between enzymes and nonenzymes ($p > 0.99$ except for the intracellular enzymes). At the same time, both intra- and extracellular nonenzymes have the same stability notwithstanding the presence of disulfide bridges of the the latter.

TABLE 2.7
Comparison of the types of folding and ΔG_{unf} for the four biological groups according to NISHIKAWA et al. (ref. 159,160).

Biological group	folding type	S-S bridges	no. of examples	ΔG_{unf} [a]
intracellular proteins:				
enzymes	α/β	no	8	(43 ± 33)
nonenzymes	all types, more α	no	14	32 ± 19
extracellular proteins:				
enzymes	β or $\alpha + \beta$	yes	14	47 ± 17
nonenzymes	α or β	yes	21	30 ± 15

[a] ΔG_{unf} average values, in kJ mol^{-1}.

These findings seem to suggest the importance of functional
requirements in evolution of optimal protein structure. So, the

question is still under discussion why are enzymes so large (ref. 163-165). Possibly, the problem of size can be seen in connection with protein stability, although the findings summarized in Fig. 2.14 did not support the idea of a correlation between protein stability and molecular weight. On the other hand, among the nonenzymes are much proteins which are involved in forming bio-polymer complexes during functioning. It is conceivable that lower stabilization energy could facilitate conformative adapti-bility of the proteins (ref. 99,166,167).

2.5.2 Influence of amino acid replacements on protein stability

Considerable interest is now being focussed on stability dif-ferences caused by amino acid replacements. Corresponding inves-tigations were originally started on protein families (e.g. cyto-chrome c and myoglobin etc.), on homologous proteins from meso-philic and thermophilic organisms, and on specific protein frag-ments characterized by lack of terminal residues (review in ref. 99). These investigations include, as a rule, several amino acid replacements occurring simultaneously. At the same time, so far X-ray data are available, main chain fold seems to remain un-altered. For more detailed considerations, recently systematic studies of single amino acid replacements became available. Of

TABLE 2.8
Gibbs energy change (kJ mol^{-1}) at unfolding of mutants of α sub-unit of tryptophan synthase at 25°C.

Type	Position	ΔG_{unf}	pH	Approach	Ref.
wild type	Gly 211	24.1	7.8	DSC	168
mutant	Arg 211	22.2	7.8	DSC	168
mutant	Gln 211	26.4	7.8	DSC	168
wild type	Glu 49	36.8	7.0	GuHCl	169
wild type	Glu 49	35.1	7.0	DSC	156
mutant	Gln 49	26.4	7.0	GuHCl	169
mutant	Gln 49	31.0	7.0	DSC	156
mutant	Ser 49	31.0	7.0	GuHCl	169
mutant	Ser 49	32.2	7.0	DSC	156
mutant	Met 49	55.6	7.0	GuHCl	169
mutant	Val 49	50.2	7.0	GuHCl	169
mutant	Tyr 49	36.8	7.0	GuHCl	169
mutant	Leu 49	63.0	7.0	GuHCl	170

particular interest are Gibbs energy changes for unfolding of
α subunit of tryptophan synthase (ref. 156,168-170). As it can
be seen from Table 2.8, point mutations in the two positions
(Glu 49 and Gly 211) have rather different influence on ΔG_{unf}.
Replacement of Gly 211 by Arg and Gln, i.e. rather drastic
changes with respect to the side chains, cause only slight
changes of ΔG_{unf}. On the other hand, replacement of Glu 49 by
Gln leads to a more pronounced effect. By introduction of nonpolar groups instead of Glu 49, nearly twofold increase of protein stability can be found. In fact, location of the replaced
residues is of grat importance with respect to the stability of
a protein. So from X-ray study of wild type and temperature-
sensitive lysozyme from bacteriophage T4 it follows that "extra
energy of thermal stabilization can be provided without disturbance of the tertiary structure of the protein" (ref. 171). From
thermodynamic studies of phage T4 lysozyme and four of its
mutants (ref. 172) it can be concluded that "modification of
buried residues invariably leads to altered stability" (ref. 173).
The same concerns partly buried side chains, especially those of
some hydrophobic character. Only conservative changes of side
chains located at the surface seem not to have a major influence
on protein stability. However, generally the absence of alterations of protein structure upon amino acid replacement cannot be
presumed. As it was instructively shown by the example of bovine
and porcine phospholipase A_2, only one substitution (Val 63 $\rightarrow$ Phe)
might produce considerable changes of a 12 residues loop (ref.
174-176). The structural differences are given in Table 2.9. At
the same time, the two structures can well be superimposed for
the other part of the molecule (rms = 0.47 $\overset{\circ}{A}$, ref. 175), although
15% of the 124 amino acid residues are different.

TABLE 2.9
Structural differences between bovine and porcine phospholipase A_2
after optimal superposition of the complete molecules (ref. 176).

position	58		60			63		65				70		
sequence bovine p.	Leu	Asp	Ser	Cys	Lys	Val	Leu	Val	Asp	Asn	Pro	Tyr	Thr	
sequence porcine p.	Leu	Asp	Ser	Cys	Lys	Phe	Leu	Val	Asp	Asn	Pro	Tyr	Thr	
distance C_α atoms, $\overset{\circ}{A}$	0.6	1.8	1.9	1.0	2.8	8.3	8.2	7.9	6.1	3.7	0.8	1.1	0.5	

2.5.3 Ligand equilibria influencing apparent protein stability

Various proteins require specific ions in order to attain
their biologically active conformation. Consequently, both the
additon of excess ion concentration and the removal of the ions
by chelating agents can considerably influence thermostability.
The effect was originally analyzed by the example of calcium
binding proteins (ref. 99,144). In unfolding of the native
protein (N) carrying n moles Ca^{2+}, release of the ions must be
taken into account:

$$N \rightleftharpoons U + nCa^{2+} \tag{7}$$

Since usually the equilibrium constant is formulated solely for
the protein by $K = (N)/(U)$, it follows

$$\Delta G^{app} = - RT \ln K + n RT \ln a_{Ca^{2+}} \tag{8}$$

The presence of excess calcium leads to a considerable influence
on ΔG^{app} as it was shown by the phase diagram of parvalbumin
(ref. 99). From the mentioned phase diagram it becomes also visible that the thermal transition temperature can be shifted towards higher values at increasing calcium ion concentration
solely by shifting the calcium binding equilibrium.

A more general treatment of protein unfolding at varying ligand
concentrations was recently analyzed by the example of arabinose-
binding protein (ref. 120). Assuming an oligomeric protein which
binds n ligand molecules (L), for reversible two-state unfolding
with simultaneous protein and ligand dissociation the following
equation can be formulated (ref. 120):

$$N_m L_n \rightleftharpoons mU + nL \tag{9}$$

$$K = (U)^m (L)^n / (N_m L_n) \tag{10}$$

If total ligand and protein subunit concentration $(L)_o \gg (N)_o$
are introduced and the extent of conversion is defined as
$\alpha = (U)/(N)_o$, one obtains:

$$K = m \frac{\alpha^m}{1-\alpha} (N)_o^{m-1} (L)_o^n \tag{11}$$

The temperature dependence of K according to the van't Hoff equation leads to

$$\frac{\Delta H^{v.H.}}{RT_m} + n \ln (L)_o + (m-1) \ln (N)_o = \text{constant} \qquad (12)$$

Accordingly, a plot of $\ln (L)_o$ versus $1/T_m$ at constant $(N)_o$ has a slope of $-\Delta H^{v.H.}/nR$ (see Fig. 2.16).

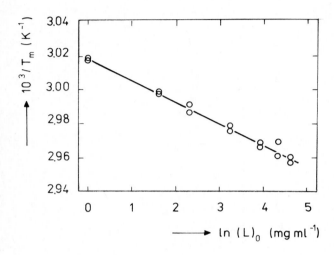

Fig. 2.16. Plot of $1/T_m$ versus $\ln(L)_o$ according to eq. (12) L-arabinose-binding protein in the presence of various concentrations of arabinose. Redrawn from ref. (120).

Equation (12) provides a thermodynamic basis for understanding the dependence of transition temperature on ligand concentration. That means, conformation changes need not be assumed for explanation of the often mentioned stabilizing effect of substrates and ions in protein unfolding. In the careful study of arabinose-binding protein (ref. 120) unfolding temperature was increased by the presence of excess arabinose or galactose, "an effect which is due solely to displacement by the added ligand of the unfolding-dissociation equilibrium".

REFERENCES

1   W. Pfeil and P.L. Privalov, in M.N. Jones (Ed.), Biochemical
     Thermodynamics, Elsevier Sci. Publ. Co., Amsterdam, Oxford,
     New York, 1979, pp. 75-115.
2   K.C. Aune and C. Tanford, Biochemistry 8 (1969) 4579-4585.
3   K.C. Aune and C. Tanford, Biochemistry 8 (1969) 4586-4590.
4   N.N. Khechinashvili, P.L. Privalov and E.I. Tiktopulo,
     FEBS Lett. 30 (1973) 57-60.
5   P.L. Privalov and N.N. Khechinashvili, J. Mol. Biol. 86
     (1974) 665-684.
6   K. Kuwajima, K. Nitta, M. Yoneyama and S. Sugai, J. Mol. Biol.
     106 (1976) 359-373.
7   W. Pfeil, Biophys. Chem. 13 (1981) 181-186.
8   P.L. Mateo and P.L. Privalov, FEBS Lett. 123 (1981) 189-192.
9   P.L. Privalov, P.L. Mateo, N.N. Khechinashvili, V.M. Stepa-
     nov and L.P. Revina, J. Mol. Biol. 152 (1981) 445-464.
10   T.E. Creighton, J. Mol. Biol. 129 (1979) 235-264.
11   T.E. Creighton, J. Mol. Biol. 137 (1980) 61-80.
12   T.E. Creighton, in R. Jaenicke (Ed.), Protein Folding,
     Elsevier/North Holland Biomed. Press, Amsterdam, New York,
     1980, pp. 427-445.
13   A. Wada, H. Tachibana, K. Hayashi and Y. Saito, J. Biochem.
     Biophys. Methods 2 (1980) 257-269.
14   A. Wada, Y. Saito and M. Ohogushi, Biopolymers 22 (1983)
     93-99.
15   Y. Saito and A. Wada, Biopolymers 22 (1983) 2105-2122.
16   R. Lumry, R.L. Biltonen and J.F. Brandts, Biopolymers 4
     (1966) 917-944.
17   T.Y. Tsong, R.P. Hearn, D.P. Wrathall and J.M. Sturtevant,
     Biochemistry 9 (1970) 2666-2677.
18   P.L. Privalov, Adv. Protein Chem. 33 (1979) 167-241.
19   P.L. Privalov, Adv. Protein Chem. 35 (1982) 1-104.
20   R.L. Biltonen and E. Freire, CRC Crit. Rev. Biochem. 5 (1978)
     85-124.
21   E. Freire and R.L. Biltonen, Biopolymers 17 (1978) 463-479.
22   E. Freire and R.L. Biltonen, Biopolymers 17 (1978) 481-496.
23   V.V. Novokhatny, S.A. Kudinov and P.L. Privalov, J. Mol.
     Biol. 179 (1984) 215-232.
24   W. Pfeil and P.L. Privalov, Biophys. Chem. 4 (1976) 33-40.
25   W. Pfeil, V.E. Bychkova and O.B. Ptitsyn, FEBS Lett. (1986),
     in press.
26   A. Salahuddin and C. Tanford, Biochemistry 9 (1970)
     1342-1347.
27   S.A. Hawley and R.M. Mitchell, Biochemistry 14 (1975)
     3257-3264.
28   D. Puett, J. Biol. Chem. 248 (1973) 4623-4634.
29   T.J. Ahern and A.M. Klibanov, Science 228 (1985) 1280-1284.
30   G.E. Schulz and R.H. Schirmer, Principles of Protein Struc-
     ture, Springer-Verlag, New York, Heidelberg, Berlin, 1979.
31   R.L. Baldwin and T.E. Creighton, in R. Jaenicke (Ed.)
     Protein Folding, Elsevier/North Holland Biomed. Press,
     Amsterdam, New York, 1980, pp. 217-260.
32   T.E. Creighton, Proteins - Structures and Molecular Princip-
     les, Freeman, New York, 1983.
33   R. Jaenicke, Angew. Chem. 96 (1984) 385-402.
34   K. Brew, F.J. Castellino, T.C. Vanaman and R.L. Hill, J. Biol.
     Chem. 245 (1970) 4570-4582.

35 O.U. Beg, H. Von Bahr-Lindström, Z.H. Zaidi and H. Jörnvall, Eur. J. Biochem. 147 (1985) 233-239.
36 J.G. Shewale, S.K. Sinha and K. Brew, J. Biol. Chem. 259 (1984) 4947-4956.
37 W.J. Browne, A.C.T. North, D.C. Phillips, K. Brew, T.C. Vanaman and R.L. Hill, J. Mol. Biol. 42 (1969) 65-86.
38 P.K. Warme, F.A. Momany, S.V. Rumball, R.W. Tuttle and H.A. Scheraga, Biochemistry 13 (1974) 768-782.
39 R.L. Hill and K. Brew, Adv. Enzymol. 43 (1975) 411-490.
40 H. Takesada, M. Nakanishi and M. Tsuboi, J. Mol. Biol. 77 (1973) 605-614.
41 K.S. Iyer and W.A. Klee, J. Biol. Chem. 248 (1973) 707-710.
42 K. Kuwajima, J. Mol. Biol. 114 (1977) 241-258.
43 M. Nozaka, K. Kuwajima, K. Nitta and S. Sugai, Biochemistry 17 (1978) 3753-3758.
44 Y. Hiraoka, T. Segawa, K. Kuwajima, S. Sugai and N. Murai, Biochem. Biophys. Res. Commun. 95 (1980) 1098-1104.
45 E.A. Permyakov, V.V. Yarmolenko, L.P. Kalinichenko, L.A. Morozova and E.A. Burstein, Biochem. Biophys. Res. Commun. 100 (1981) 191-197.
46 K. Murakami, P.J. Andree and L.J. Berliner, Biochemistry 21 (1982) 5488-5494.
47 S.C. Bratcher and M.J. Kronman, J. Biol. Chem. 259 (1984) 10875-10886.
48 Y. Hiraoka and S. Sugai, Int. J. Peptide Protein Res. 23 (1984) 535-542.
49 W. Pfeil and M.L. Sadowski, studia biophys. 109 (1985) 163-170.
50 E.A. Permyakov, L.A. Morozova and E.A. Burstein, Biophys. Chem. 21 (1985) 21-31.
51 M.J. Kronman and S.C. Bratcher, J. Biol. Chem. 258 (1983) 5707-5709.
52 Y. Hiraoka and S. Sugai, Int. J. Peptide Protein Res. 26 (1985) 252-261.
53 M.J. Kronman, R. Andreotti and R. Vitols, Biochemistry 3 (1964) 1152-1160.
54 W.R. Krigbaum and F.R. Kügler, Biochemistry 9 (1970) 1216-1223.
55 D.A. Dolgikh, R.I. Gilmanshin, E.V. Brazhnikov, V.E. Bychkova, G.V. Semisotnov, S.Yu. Venyaminov and O.B. Ptitsyn, FEBS Lett. 136 (1981) 311-315.
56 Y. Iizumi, I. Miyaka, K. Kuwajima, S. Sugai, K. Inoue, M. Iizumi and S. Katano, Physica 120 B (1983) 444-448.
57 K. Gast, D. Zirwer, H. Welfle, V.E. Bychkova and O.B. Ptitsyn, Int. J. Biol. Macromol. (1986), in press.
58 P.B. Sommers and M.J. Kronman, Biophys. Chem. 11 (1980) 217-232.
59 K. Kuwajima and S. Sugai, Biophys. Chem. 8 (1978) 247-254.
60 K. Kuwajima, Paper presented at the EMBO Workshop on Protein Folding, Cambridge, 1984.
61 W. Pfeil, Paper presented at the 5th International Conference on Chemical Thermodynamics, Ronneby, Sweden, 1977.
62 C. Tanford, Adv. Protein Chem. 23 (1968) 121-282.
63 C. Tanford, Adv. Protein Chem. 24 (1970) 1-95.
64 S. Lapanje, Physicochemical Aspects of Protein Denaturation, John Wiley and Sons, New York, 1978.
65 S. Maruyama, K. Kuwajima, K. Nitta and S. Sugai, Biochim. Biophys. Acta 494 (1977) 343-353.
66 C.C. Contaxis and C.C. Bigelow, Biochemistry 20 (1981) 1618-1622.

67 F. Ahmad, Canad. J. Biochem. and Cell Biol. 63 (1985)
    1058-1063.
68 N. Kita, K. Kuwajima, N. Nitta and S. Sugai, Biochim. Bio-
    phys. Acta 427 (1976) 350-358.
69 S. Sugai, H. Yashiro and K. Nitta, Biochim. Biophys. Acta
    328 (1973) 35-41.
70 F. Ahmad and C.C. Bigelow, J. Biol. Chem. 257 (1982)
    12935-12938.
71 K. Takase, K. Nitta and S. Sugai, Biochim. Biophys. Acta
    371 (1974) 352-359.
72 T. Okuda and S. Sugai, J. Biochem. 81 (1977) 1051-1056.
73 K. Nitta, N. Kita, K. Kuwajima and S. Sugai, Biochim. Biophys.
    Acta 490 (1977) 200-208.
74 K.G. Mann and W.W. Fish, Methods in Enzymol. 26 (1972) 28-42.
75 G.K. Ackers, Adv. Protein Chem. 24 (1970) 343-446.
76 M. Ohgushi and A. Wada, FEBS Lett 164 (1983) 21-24.
77 K.C. Aune, A. Salahuddin, M.H. Zarlengo and C. Tanford,
    J. Biol. Chem. 242 (1967) 4486-4489.
78 J.M. Sturtevant and P.L. Mateo, Proc. Natl. Acad. Sci. USA
    75 (1978) 2584-2587.
79 V. Massay, B. Curti and H. Ganther, J. Biol. Chem. 241 (1966)
    2347-2357.
80 A. Cooper, Proc. Natl. Acad. Sci. USA 78 (1981) 3551-3553.
81 P.L. Mateo, in M.A.V. Ribeiro da Silva (Ed.), Thermochemistry
    and its Applications to Chemical and Biochemical Systems,
    D. Reidel Publ. Co., Dodrecht, 1984, pp. 541-568.
82 H.A. Scheraga, Protein Structure, Academic Press, New York
    and London, 1961, pp. 81-128.
83 K. Hamaguchi and H. Sakai, J. Biochem. 57 (1965) 721-732.
84 M. Kugimiya and C.C. Bigelow, Canad. J. Biochem. 51 (1973)
    581-585.
85 F. Ahmad and A. Salahuddin, Biochemistry 13 (1974) 245-249.
86 S. Hvidt, M.E. Rodgers and W.F. Harrington, Biopolymers 24
    (1985) 1647-1662.
87 K. Ogasahara, K. Yutani, M. Suzuki and Y. Sugino, Int. J.
    Peptide Protein Res. 24 (1984) 147-154.
88 J.C. Lee and S.N. Timasheff, Biochemistry 13 (1974) 257-265.
89 M.Y. Schrier and E.E. Schrier, Biochemistry 15 (1976)
    2607-2612.
90 C. Tanford and K.C. Aune, Biochemistry 9 (1970) 206-211.
91 A. Cooper, Proc. Natl. Acad. Sci. USA 73 (1976) 2740-2741.
92 J.F. Brandts, in S.N. Timasheff and G.D. Fasman (Eds.),
    Structure and Stability of Biological Macromolecules,
    Marcel Dekker Inc., New York, 1969, pp. 213-290.
93 R.B. Gregory, D.G. Knox, A.J. Percy and A. Rosenberg,
    Biochemistry 24 (1982) 6523-6530.
94 S.W. Englander and N.R. Kallenbach, Quart. Rev. Biophys.
    16 (1984) 521-655.
95 A. Ikegami, Biophys. Chem. 6 (1977) 117-130.
96 M.I. Kanehisa and A. Ikegami, Biophys. Chem. 6 (1977)
    131-149.
97 A. Ikegami, Adv. Chem. Physics 46 (1981) 363-413.
98 W. Pfeil, Biomed. Biochem. Acta 42 (1983) 849-854.
99 W. Pfeil, Mol. Cell. Biochem. 40 (1981) 3-28.
100 M.I. Kanehisa and T.Y. Tsong, J. Mol. Biol. 124 (1978)
    177-194.
101 A. Galat, Int. J. Biochem. 15 (1983) 715-719.
102 S.K. Burley and G.A. Petsko, Science 229 (1985) 23-28.
103 P.L. Privalov and S.A. Potekhin, Methods in Enzymol., (1986),
    in press.

104 N.S. Andreeva and A.E. Gustchina, Biochem. Biophys. Res.
    Commun. 87 (1979) 32-42.
105 T.N. Tsalkova and P.L. Privalov, J. Mol. Biol. 181 (1985)
    533-544.
106 W. Pfeil and P.I. Ohlson, Biochem. Biophys. Res. Commun.
    (1986), in press.
107 C. Shutt, Nature 315 (1985) 15.
108 R. Lumry and R.L. Biltonen, in S.N. Timasheff and G.D. Fas-
    man (Eds.), Structure and Stability of Biological Macromole-
    cules, Marcel Dekker Inc., New York 1969, pp. 65-212.
109 C.N. Pace, CRC Crit. Rev. Biochem. 3 (1975) 1-43.
110 R.F. Greene Jr. and C.N. Pace, J. Biol. Chem. 249 (1974)
    5388-5393.
111 J.A. Schellman, Biopolymers 17 (1978) 1305-1322.
112 C.N. Pace and K.E. Vanderburg, Biochemistry 18 (1979) 288-292.
113 C. Tanford, J. Amer. Chem. Soc. 84 (1962) 4240-4247.
114 C. Tanford, J. Amer. Chem. Soc. 86 (1964) 2050-2059.
115 W. Pfeil, in H.-J. Hinz (Ed.), Thermodynamic Data for Bio-
    chemistry and Biotechnology, Springer-Verlag, Heidelberg,
    1986, in press.
116 C.C. Contaxis, C.C. Bigelow and C.G. Zarkadas, Canad. J.
    Biochem. 55 (1977) 325-331.
117 J.F. Chlebowski, S. Mabrey and M.C. Falk, J. Biol. Chem. 254
    (1979) 5745-5753.
118 C. Edelstein and A.M. Scann, J. Biol. Chem. 255 (1980)
    5747-5754.
119 W.W. Mantulin, M.F. Rohde, A.M. Gotto Jr. and H.J. Pownall,
    J. Biol. Chem. 255 (1980) 8185-8191.
120 H. Fukada, J.M. Sturtevant and F.A. Quiocho, J. Biol. Chem.
    258 (1983) 13193-13198.
121 A. Relimpio, A. Iriarte, J.F. Chlebowski and M. Martinez-
    Carrion, J. Biol. Chem. 256 (1981) 4478-4488.
122 P. Cupo, W. El-Deiry, P.L. Whitney and W.M. Awad Jr., J. Biol.
    Chem. 255 (1980) 10828-10833.
123 M.M. Santoro, J.F. Miller and D.W. Bolen, ASBC/AAI Annual
    Meeting, see Fed. Proc. 43 (1984) 1838.
124 P. Bendzko and W. Pfeil, Biochim. Biophys. Acta 742 (1980)
    669-676.
125 B.T. Nall and T.A. Landers, Biochemistry 20 (1981) 5403-5411.
126 E.H. Zuniga and B.T. Nall, Biochemistry 22 (1983) 1430-1437.
127 N. Okabe, E. Fujita and K.I. Tomita, Biochim. Biophys. Acta
    700 (1982) 165-170.
128 O.V. Fedorov, N.N. Khechinashvili, R. Kamiya and S. Asakura,
    J. Mol. Biol. 175 (1984) 83-87.
129 D.K. Strickland, T.L. Andersen, R.L. Hill and F.J. Castellino,
    Biochemistry 20 (1981) 5294-5297.
130 R.E. Thompson, S.W. Morrical, D.P. Campell and W.R. Carper,
    Biochim. Biophys. Acta 745 (1983) 279-284.
131 A.N. Danilenko, E.K. Grozav, T.M. Bibkov, V.Ya. Grinberg and
    V.B. Tolstoguzov, Int. J. Biol. Macromol. 7 (1985) 109-112.
132 L.A. Holladay, C.R. Savage Jr., S. Cohen and D. Puett,
    Biochemistry 15 (1976) 2624-2633.
133 L.A. Holladay, R.G. Hammonds Jr. and D. Puett, Biochemistry
    13 (1974) 1653-1661.
134 H. Uedaira, T. Honda, Y. Takeda, T. Miwatani, H. Uedaira and
    A. Ohsaka, Int. Sympos. Thermodynamics of Proteins and Bio-
    logical Membranes, Granada, Spain, 1983, Abstract P-II-20.
135 A. Sumi and K. Hamaguchi, J. Biochem. 92 (1982) 823-833.
136 Y. Goto and K. Hamaguchi, J. Biochem. 86 (1979) 1433-1441.

137 A. Hvidt and S.O. Nielsen, Adv. Protein Chem. 21 (1966)
    287-386.
138 H.-J. Hinz, M. Cossmann and K. Beyreuther, FEBS Lett. 129
    (1981) 246-248.
139 B. Robson and R.H. Pain, Biochem. J. 155 (1976) 331-344.
140 M.P. Tombs and G. Blake, Biochim. Biophys. Acta 700 (1982)
    81-89.
141 J.A. Schellman and R.B. Hawkes, in R. Jaenicke (Ed.),
    Protein Folding, Elsevier/North Holland Biomed. Press,
    Amsterdam, New York, 1980, pp. 331-343.
142 F. Ahmad and A. Salahuddin, Biochemistry 15 (1976) 5168-5175.
143 E.I. Tiktopulo and P.L. Privalov, FEBS Lett. 91 (1978) 57-58.
144 V.V. Filimonov, W. Pfeil, T.N. Tsalkova and P.L. Privalov,
    Biophys. Chem. 8 (1978) 117-122.
145 H. Nojima, A. Ikai, T. Oshima and H. Noda, J. Mol. Biol. 116
    (1977) 429-442.
146 M. Desmadril, A. Mitraki, J.M. Betton and J.M. Yon, Biochem.
    Biophys. Res. Commun. 118 (1984) 416-422.
147 C.-H. Chen, O.H.W. Kao and D.S. Berns, Biophys. Chem. 7 (1977)
    81-86.
148 T.Y. Tsong, R.P. Hearn, D.P. Wrathall and J.M. Sturtevant,
    Biochemistry 9 (1970) 2666-2677.
149 M. Ottesen, Methods Biochem. Anal. 20 (1971) 135-168.
150 K. Takahashi and J.M. Sturtevant, Biochemistry 20 (1981)
    6185-6190.
151 C. Vita and A. Fontana, Biochemistry 21 (1982) 5196-5202.
152 R.F. Kelley and E. Stellwagen, Biochemistry 23 (1984)
    5095-5102.
153 N.N. Khechinashvili and V.I. Tsetlin, Molekularnaya Biol. 18
    (1984) 786-791.
154 T.N. Tsalkova and P.L. Privalov, Biochim. Biophys. Acta 624
    (1980) 196-204.
155 E. Moses and H.-J. Hinz, J. Mol. Biol. 170 (1983) 765-776.
156 K. Yutani, N.N. Khechinashvili, E.A. Lapshina, P.L. Privalov
    and Y. Sugino, Int. J. Peptide Protein Res. 20 (1982)
    331-336.
157 D. Poland and H.A. Scheraga, Theory of Helix-Coil Transitions
    in Biopolymers, Academic Press, New York and London, 1970.
158 K.A. Dill, Biochemistry 24 (1985) 1501-1509.
159 K. Nishikawa, Y. Kubota and T. Ooi, J. Biochem. 94 (1983)
    981-995.
160 K. Nishikawa, Y. Kubota and T. Ooi, J. Biochem. 94 (1983)
    997-1007.
161 J.S. Richardson, Adv. Protein Chem. 34 (1981) 167-339.
162 J.S. Richardson, in R. Jaenicke (Ed.), Protein Folding,
    Elsevier/North Holland Biomed. Press, Amsterdam, New York,
    1980, pp. 41-52.
163 D.G. Kell, Trends Biochem. Sci. 7 (1982) 349.
164 T.A.J. Payens, Trends Biochem. Sci. 8 (1983) 46.
165 P.A. Srere, Trends Biochem. Sci. 9 (1984) 387-390.
166 K. Wüthrich, G. Wagner, R. Richarz and W. Braun, Biophys. J.
    32 (1980) 549-560.
167 W.S. Bennett and R. Huber, CRC Crit. Rev. Biochem. 15 (1984)
    291-384.
168 C.R. Matthews, M.M. Crisanti, G.L. Gepner, G. Velicelebi and
    J.M. Sturtevant, Biochemistry 19 (1980) 1290-1293.
169 K. Yutani, K. Ogasahara, A. Kimura and Y. Sugino, J. Mol.
    Biol. 160 (1982) 387-390.
170 K. Yutani, K. Ogasahara, K. Aoki, T. Kakuno and Y. Sugino,
    J. Biol. Chem. 259 (1984) 14076-14081.

171 M.G. Grütter, R.B. Hawkes and B.W. Matthews, Nature 277 (1979) 667-669.
172 R. Hawkes, M.G. Grütter and J. Schellman, J. Mol. Biol. 175 (1984) 195-212.
173 M.G. Grütter and R.B. Hawkes, Naturwissenschaften 70 (1983) 434-438.
174 B.W. Dijkstra, K.H. Kalk, W.G.J. Hol and J. Drenth, J. Mol. Biol. 147 (1981) 97-123.
175 B.W. Dijkstra, R. Renetseder, K.H. Kalk, W.G.J. Hol and J. Drenth, J. Mol. Biol. 168 (1983) 163-179.
176 B.W. Dijkstra, W.J. Weijer and R.K. Wierenga, FEBS Lett. 164 (1983) 25-27.
177 C.M. Dobson, P.A. Evans and K.L. Williamson, FEBS Lett. 168 (1984) 331-334.
178 A. Bundi, R.H. Andretta and K. Wüthrich, Eur. J. Biochem. 91 (1978) 201-208.
179 C. Boesch, A. Bundi, M. Opplinger and K. Wüthrich, Eur. J. Biochem. 91 (1978) 209-214.
180 O.B. Fedorov and N.N. Khechinashvili, Doklady Akad. Nauk USSR 226 (1976) 1207-1210.
181 A. Ansari, J. Berendzen, S.F. Browne, H. Frauenfelder, I.E.T. Iben, T.B. Sauke, E. Shyasmunder and R.D. Young, Proc. Natl. Acad. Sci. USA 82 (1985) 5000-5004.
182 R.B. Gregory and R. Lumry, Biopolymers 24 (1985) 301-326.
183 D.A. Dolgikh, L.V. Abaturov, I.A. Bolotina, E.V. Brazhnikov, V.N. Bushuev, V.E. Bychkova, R.I. Gilmanshin, Yu.O. Lebedev, G.V. Semisotnov, E.I. Tiktopulo and O.B. Ptitsyn, Eur. Biophys. J. 13 (1985) 109-122.
184 K. Kuwajima, Y. Harushima and S. Sugai, Int. J. Peptide Protein Res. 27 (1986) 18-28.

CHAPTER 3

CONFORMATIONAL TRANSITIONS IN NUCLEIC ACIDS

H.H. KLUMP

3.1   Introduction

Since the first edition of Biochemical Thermodynamics in 1979 the
investigations on conformational changes of nucleic acids and
their constituents have flourished due to two new developments.The
first   and   formost   improvement   concerns   the   biochemical
preparation techniques of native DNAs (Hirose et al 1978) and DNA-
fragments (Patel,1981) which can be   produced   by restriction
enzymes (Arber,1965) and be prepared in   great   quantities (Cech,
1981).The   second improvement is based on the advancement of
the   experimental   techniques   to   determine   the   energetics   of
conformational transitions by the help of highly   sensitive
scanning microcalorimeters (Privalov,1974), which are now commercially available and are used in many laboratories.The focus of
interest has changed from small cellular RNAs such as tRNAs
(Hinz,1977) back to DNA (Klump,1984) and DNA-like polynucleotides
(Klump,1985) . In addition to the established conformers (B-DNA,A-
RNA)   two new conformation families have been added and   attracted
much attention namely the structure-family of the closed circular
plasmids (Seidel,1985) and the cosmids (Klump,1987), the family of
the natural DNA vectors, used in genetic engineering (Winnacker,
1985), and the family of left-handed DNA structures. (Pohl,1972;
Klump,1986,1987).
In the mean time numerous experimental studies have confirmed that
the isomerization equilibrium between the right-handed B- and the
left-handed Z-conformations of DNA is determined by a well known
set of external parameters(Soumpasis,1986).As extensively reviewed
(Rich,1982) the conformational choice of the DNA present is determined
   1.) by the precise chemical structure of the polynucleotide,
(Behe,1985) i.e. by its base sequence in the first place. Only
alternating purine/pyrimidine sequences are capable to invert
their handedness. The equilibrium is shifted towards the left-handed conformer (Klump,1985) by chemical modifications at the level
of the bases , either by methylation or bromination of the 5'
position of the pyrimidine moiety (Klump,1987). The preferred conformation is
2.)determined by the set of environmental conditions in the close
vicinity of the double helix.(Saenger,1986) The choice of the type
and the concentration of the ionic species (Pohl,1976) determines
whether the left-handed or the right-handed conformer is favoured.
Special DNA binding ligands such as monoclonal antibodies against
one conformer are also capable of stabilising the handedness of the
selected   species   (Rich,1984).   The   introduction   of   certain

* present address:Department of Biochemistry, University of Cape
Town, Rondebosch,R.S.A.

cosolvents into the aqueous environment of the polynucleotides
(Azorin,1983)), which tend to lower the water activity and the
dielectric coefficient , the temperature , and the pH can select
for a special conformer.(Haniford,1983)
3.) in the case of closed circular molecules the degree of
topological stress , i.e. supercoiling , and intervention of
competitive relaxation processes such as denaturation or cruci-
formation will determine which conformer will be the
thermodynamically most stable under the set of conditions present.
Singleton,1984)The quantitative analysis of the complex equilibria
involved requires knowledge of central thermodynamic quantities
such as the stability constant of base pairs in the two
conformations , the enthalpy of the B to Z transition, the entropy
of the inversion of the handedness to calculate the transition
free enthalpy for this process (Szu,1985), the free enthalpy
involved in creating B/Z junctions or the equivalent nucleation
probability (Irikura,1985), and the dependence of these parameters
on the intrinsic and extrinsic factors identified above.(Frank-
Kamenetskii,1985) It is certainly a long way to go until we have
accomplished this distant goal.
The prime aim of this survey is to give a progress report of the
scientific projects in the field of nucleic acid energetics. Only
those formalisms, which were not dealt with in the forgoing rewiew
in this series are extensively outlined here. For the established
formalisms the reader is kindly referred to the chapter on
conformation changes in nucleic acids by H.-J. Hinz (Hinz,1979) in
the forgoing edition of Biochemical Thermodynamics.

3.2   Conformational Transitions in Helical Polynucleotides

3.2.1 Energetics of Helix Coil Transitions in B-DNAs

It is well established that under a set of given environmental
conditions such as pH, the total activity of monovalent or
divalent cations and/or added cosolvents the relative stability
of the B-DNA duplex structure depends on its primary structure,
(Breslauer,1986),i.e. its base sequence. More precisely, the
detailled conformation and hence the local stability depends
primarily on the character of the next nearest neighbour base
pair (Klump,1987). Ten different nearest neighbour interactions
have to be considered in any Watson-Crick type DNA
structure.(Watson,1953) These pairwise interactions are AA/TT,
AT/TA, TA/AT, CA/GT, GT/CA, CT/GA, GA/CT, CG/GC, GC/CG,and
GG/CC.(Nelson,1981)The overall stability and the melting behavior
of any given B-DNA duplex structure can be predicted from its
known primary sequence if this set of relative stabilities (ΔG)
and the temperature dependence (ΔH, ΔCp) of the next-nearest
neighbour interactions is known.(Marky,1983)
Unfortunately , comparatively few such studies on DNA oligomers
have been performed so that the relevant thermodynamic data
required to predict the conformational stability of the DNA
secondary structure are rather sparse. The seriousness of this
deficiency is dramatized by the fact that investigators attempting

to evaluate sequence dependent structural preferences in the DNA
strands have resorted to the use of the available such as will be
shown subsequently inadequate thermodynamic data , derived for RNA
structures (Tinoco,1973). In fact, comparison of available data
suggest that serious errors may be introduced by applying RNA data
to the analysis of the sequence -dependent structural preferences
in DNA molecules. Consequently, a meaningful evaluation of

sequence-dependent structural preferences within the DNA double-helix indeed will require an independently evaluated DNA data set.
      Several years ago this program was initiated with the long term goal to obtain the required thermodynamic data characteristic for B-DNA structures . To this aim mainly two independent physical methods have been applied and improved, namely high performance differential scanning microcalorimetry (DSC) (Privalov,1975) and UV spectroscopy (Crothers,1964), to characterize thermally induced conformational transitions , i.e. in the first place the helix to coil transitions of especially designed and selected synthetic polynucleotides and/or of natural DNA sequences.
By combining the results from these studies , as will be extensively outlined below, it is now possible to resolve and to assign thermodynamic profiles for all given DNA sequences. It can readily be demonstrate that DNA duplex stability is preferentially governed by the thermodynamic situation and thus is exclusively dependent on the next nearest neighbour base pair frequencies. Consequently , using the new B-DNA thermodynamic data library it will be possible to predict the stability and calculate the melting behavior of any B-DNA double helix from its primary sequence.
This predictive ability should prove its value in a number of important biochemical applications, such as
(i) the calculation of the minimal lenth required of an oligomer as a gene-probe to form a stable duplex with a target gene at a given set of hybridization conditions,
(ii) the estimation of the melting temperature Tm of a duplex structure formed between an oligomeric probe and its complementary gene segment,
(iii) the identification of potential sites of local melting within a polymeric duplex by prediction of the sequence-dependent melting temperatures of local characteristic DNA domains (e.g. promoter sequences),
(iv) the evaluation of the stability of highly repetitive short DNA stretches such as satellite DNAs ,
(v) the prediction of the impact of a specific transition or transversion on the gross stability of a certain DNA sequence , and finally
(vi) the calculation and comparison of the stability of a B-DNA duplex with the stability of the identical sequence in the alternative conformational states such as Z-,A- or C-DNA structures once these non-B-DNA conformations are thermodynamically characterized.
Table I gives a complete listing of the calorimetrically obtained thermodynamic parameters Tm, ΔH, ΔS, and ΔG° for native double helical DNAs.

Table I

Thermodynamic Parameters for the Thermal Denaturation   of   Native
DNAs

	Tm °C	ΔH* kJ/mbp	ΔS J/mbp K	ΔG kJ/mbp	Lit.
Salmon DNA, pH 6	25	34.8	116.8	–	ab
Herring spermatozoa DNA	25	21	70.5	–	c
Ps.fluorescens DNA, pH 6	25	32.8	110.1	–	b
S marcessens DNA, pH	25	32.8	110.1	–	b
Sea urchin DNA, pH 6	25	33.6	112.8	–	b
Calf thymus DNA	72	29.3	84.9	4.0	d
T2 phage DNA	78	35.6	1o1.4	5.2	e
T2 phage DNA	75	38.8	111.5	5.6	f
Cl.perfringens	55	32.3	98.5	2.9	g
M.lysodeikticus	79	35.6	1o1.2	3.3	g
Salmon sperm DNA, pH 7	61	32.8	97.9	3.7	g
Calf thymus	77	3o.1	88.o	4.5	h
Calf thymus pH 10.3	68.8	43.5	126	3.3	h
Salmon DNA	67	33.5	83.7	4.4	k
Calf thymus DNA, pH 6.8	72.6	33.8	98	4.6	i
Calf thymus DNA, pH 7.5	66	31.4	92	3.8	j
Salmon sperm DNA, pH 7.5		31.4	92	4.0	j
Herring sperm DNA	68	33.1	81	5.o	k
Human placenta DNA	68	33.5	1o2	5.2	k
Chicken blood DNA	67	33.5	113	5.2	k
Calf thymus DNA	66	33.5	99	5.2	k
Cl perfringens DNA	61	32.2	96	3.5	k
E coli DNA	73	34.3	99.4	5.3	k
M lysodeikticus DNA	81	36.0	1o1.6	6.1	k
T7 phage DNA	70	34.7	1o1.2	5.9	k
PM 2 phage DNA lin.	72	33.o	95.8	5.4	k
Plasmid pzmc 134 DNA lin.	72	33.1	97.2	5.4	k
Plasmid ColE 1 amp-DNA oc	68	36.4	1o6.7	4.6	n
Plasmid ColE 1 amp-DNA lin.	68	37.2	1o9	4.7	n
PM 2 phage DNA oc	67	45.2	132.6	5.7	n
PM 2 phage DNA oc	45	25.5	80.2	1.6	n
Plasmid ColE 1 amp-DNA ccc	81	26.8	75.7	4.2	l
Plasmid ColE 1 amp-DNA lin	51	22.1	68.2	1.8	l
Plasmid ColE 1 amp-DNA oc	51	21.8	67.3	1.8	l
Plasmid ColE 1 amp-DNA rel	82	32.4	91.3	5.2	l
Calf thymus DNA	45	22.6	71.1	1.4	l
Calf thymus DNA bulk	70	33.9	98.8	4.5	m
Calf thymus DNA sat.IIIa	7o.5	34	99.2	4.5	m
Calf thymus DNA sat III	72	34.5	1oo	4.5	m
Calf thymus DNA sat IV	74	35.5	1o2.3	5.0	m
Calf thymus DNA sat I	77.5	36.8	1o5.1	5.5	m
Calf thymus DNA sat II	81	38.o	1o7.3	6.0	m
Human DNA sat I	69	33.5	102	5.2	m
Human DNA sat II	72	33.5	102	5.3	m
Human DNA sat III	75	33.5	102	5.3	m
Irido Virus VI	78	34.o	96.9	5.1	o

a)Sturtevant,J.M.,and Geidoschek,E.P.(1969) J.Am.Chem.Soc.80,2911
b)Bunville,L.G.,Geiduschek,E.P.,Rawitscher,M.H.,and Sturtevant J.
(1965)Biopolymers 3,213-240
c)Ackermann,Th.,and Rüterjans,H.(1964)Ber.Bunsenges.Phys.Chem.68,
850-856
d)Rüterjans,H.(1965)Thesis,Westf.Wilhelms-Uni.,Münster,W.-Germany

104

e)Privalov,P.,Kafiani,K.,and Monaselidze,D.P.(1965)Biopizika,10.
393-398
f)Privalov,P.,Ptitsyn,O.,Birshtein,T.,(1967)Biopolymers 8,559-71
g)Klump,H.,and Ackermann,Th. (1971)Biopolymers 10.513-522
h) Shiao,D.,and Sturtevant,J.(1973) Biopolymers 13,1829-36
i)Gruenwedel.D.,(1974)Biochim.Biophys.Acta 340.16-30
j)Klump,H.,and Burkart,W.,(1977)Biochim.Biophys.Acta 475,601-604
k)Klump,H.,and Herzog,K.(1984)Ber.Bunsenges.Phys.Chem.88,20-24
l)Seidel,A.(1985)Thesis,Uni.Regensburg,W.-Germany
m)Klump,H.(1987)Ber.Bunsenges.Phys.Chem.91,206-212
n)Klump,H.,and Niermann,Th.manuscript in preparation
o)Klump,H.,Beaumais,J.,and Devachelle,M.(1983)Arch.Virology,75,269
*) From the observed difference of the experimental base line below and
above the transition ΔCp is assumed to be zero (cf. p. 107).

3.2.2 Energetics of Helix Coil Transitions in Synthetic Polynucleotides

In the mean time a large body of thermodynamic data has been
acquired.Table II shows the thermodynamic data for the helix coil
transition of synthetic polynucleotides:

Table II

(i) The A/U System

Polynucleotide	Tm °C	[Na+] mol/ltr	ΔH KJ/mbp	ΔS J/mbp K	dTm/dlgNa+ °C	Lit
poly rA poly rU	55.5	0.100	32.2	96.7	20.0	a
poly rA poly dU	44.0	0.1oo	28.6	88.5	28.5	a
poly dA poly rU	46.0	0.100	30.6	92.6	15.5	a
poly dA poly dU	50.0	o.100	31,6	95.6	17.0	a
poly d(A-U)	50.0	0.100	31.5	95.4	17.0	a
poly r(A-U)	37.2	0.100	26.6	85.8	16.5	a

(ii) the A/T System

Polynucleotide	Tm °C	[Na+] mol/ltr	ΔH KJ/mbp	ΔS J/mbp K	dTm/dlgNa+ °C	Lit
poly rA poly rT	78.0	0.100	34.8	99.2	18.0	a
poly rA poly dT	59.0	0.100	33.3	100.9	11.0	a
poly dA poly rT	69.0	0.100	32.8	95.9	16.5	a
poly dA poly dT	65.0	0.100	32.3	95.6	16.5	a
poly d(A-T)	54.5	0.100	31.2	95.3	16.0	a
poly r(A-T)	76.0	0.100	-	-	17.0	a

(iii) the I/C System

poly rI poly rC	60.0	0.100	33.0	99.1	19.0	a
poly dI poly rC	35.0	0.100	25.2	81.8	25.0	a
poly dI poly dC	45.0	0.100	29.6	93.1	18.0	a
poly rI poly dC	53.0	0.100	31.0	95.1	19.0	a
poly d(I-C)	56.5	0.100	31.6	95.9	18.0	a
poly r(I-C)	-	-	--	--	--	

(iv) the G/C System

poly dG poly dC	82.0	0.010	35.1	98.9	19.0	a
poly rG poly dC	103.0	0.010	41.0	109.0	14.0	a
poly rG poly rC	106.0	0.010	41.2	108.7	13.0	a
poly d(G-C)	96.5	0.010	39.8	107.7	19.0	a
poly r(G-C)	109.0	0.010	42.3	110.7	13.0	a
poly d(G-m5C)	101.0	0.010	40.5	108.3	15.0	b

(v) the Mixed Alternating Sequences

poly d(A-C)poly d(G-T)	83	0.100	35.8	100.4	a
poly d(A-G)poly d(C-T)	82	0.100	36.0	102.4	a

(vi) the Strictly Alternating Triplet Sequences

poly d(AAC) poly d(GTT)	78	0.100	--	--	--	c
poly d(ATC) poly d(GAT)	73	0.100	--	--	--	c
poly d(AGC) poly d(GTC)	80	0.100	--	--	--	c
poly d(AGG) poly d(CCT)	75	0.100	--	--	--	c
poly d(AAT) poly d(ATT)	55	0.100	31.5	96.0	16	d
poly d(TAC) poly d(GTA)	65	0.010	34.1	102.1	--	d
poly d(TTC) poly d(GAA)	60	0.010	32.4	97.3	--	d
poly d(TTG) poly d(CAA)	65	0.010	33.9	100.3	--	d
poly d(ATC) poly d(GAT)	68	0.010	--	--	--	d

a)Klump,H.,(1987)Thermochem Acta.in press
b)Klump,H.,(1986)FEBS Letters,196,175-179
c)Breslauer et al (1986)Proc.Natl Acad.83,3746-40
d)Klump,H.,Siepmann,I.,and Jovin,Th.(1987)manuscr.in prep.

3.2.3 Influence of Nearest Neighbours on Energetics of Base Pairing in Polynucleotides

Out of these double helical polynucleotides six sequences can be selected to account for all possible next nearest neighbour inter- action, present in long double helical complexes. Poly dA poly dT and poly d(A-T) will account for the two alternative A/T neighbourhoods, poly dG poly dC and poly d(G-C) will account for possible GC neighbourhoods , and poly d(A-G) poly d(C-T) and poly d(A-C) poly d(G-T) will cover the possible mixed nearest neighbour influences.

Table III: Thermodynamic Data of Six Selected Deoxy-polynucleotides

Polymer	Tm oC	ΔH KJ/mbp	ΔS J/ mbp K	[Na+] m/l	ΔG° KJ/mbp	
1. poly dA	74	33.8	97.3	0.200	4.77	a)
poly dT	75	34.3	98.6	0.200	4.89	
2. poly d(A-T)	60	31.1	93.4	0.100	3.27	
3. poly d(A-G) poly d(C-T)	82	36.0	101.2	0.100	5.78	
4. poly d(A-C) poly d(G-T)	83	35.3	99.5	0.050	5.75	
	93	36.1	98.6	0.2oo	6.79	
5. poly dG poly dC	83	38.0	106.7	0.002	6.19	
6. poly d(G-C)	91	39.5	108.5	0.0.02	6.98	

a) Klump,H.,(1985)Thermochem.Acta 85,457-563

The ΔH values in Table III represent a mean value of at least three different measurements. $\Delta S = \Delta H/Tm$, and $\Delta G_T = \Delta H (1- T/Tm)$.

Since the unanimous observation of all workers in the field reveals that no significant heat capacity changes are detectable for the dissolved polymers before and after the helix coil transi- tion, it is acceptable to assume that the transition enthalpies, listed in Table III , are indeed temperature independent. (Bres- lauer,1986)The free enthalpy data listed in Table III are calculated for 25° C.Following standard thermodynamic procedures it is possible to calculate the free enthalpy value for any given temperature ($\Delta G° = \Delta H° -T \Delta S°$). It is possible to predict the stability of any B-DNA duplex structure from its primary sequence.

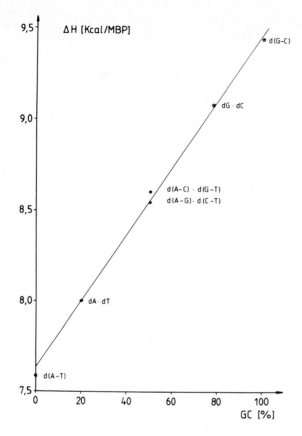

Fig.1: Transition enthalpy per mole base pairs vs. gross sequence composition (% GC).The enthalpy difference $\Delta\Delta H$ = ΔH_{GC} - ΔH_{AT} equals 1.79 Kcal (7.48 KJ) , the corresponding entropy difference results to 1.3 cal (5.43 J K⁻¹) The slope corresponds to the equation ΔH_{GC} = 7.63 (1 + 0.235 X$_{GC}$) Kcal /mbp. ΔS_{GC} = 23.62 (1 + 0.o55 X$_{GC}$) cal/mbp K, and Tm = 50 + 55 X$_{GC}$ °C. X$_{GC}$ is the percentage of GC base pairs within the sequence.

3.2.4 Prediction of Thermodynamic Properties of Polynucleotides from Primary Sequence Data

It is also possible to predict the melting temperature of the given sequences . The validity of this data set will be tested with sequences taken from the literature.(Aboul-ela,1985)We will also assume that the transition entropy is independent of the sequence and we will show ,that this assumption is strongly supported by the experimental results.
Aboul-ela et al have prepared a series of oligonucleotides ,which seem exceptionally suitable to test this data set, for these sequences are only varied in a single position in the centre of the sequence. Thus end effects which are always difficult to account for,can be neglected in this case,i.e. they will

contribute in the same way to the observed enthalpies and transition temperatures. As a first aproximation we will assume that the outer base pairs will add only half of the normal contribution of the stabilizing enthalpy as compared to the same base pair in the middle of a sequence , due to the absence of stacking forces on one side of the outer base pairs. It is also assumed that there is no nucleation enthalpy to be considered in the computation.
The model sequences consist basically of the following base pairs:

$$5'CAAAXAAAG\ 3'$$
$$3'GTTTYTTTC\ 5'$$

where X and Y are complementary bases and X respectively Y represents any of the four canonical bases.The prediction of the transition enthalpy is based on the assumption that there are only next nearest neighbour interactions , i.e. that the sum of the individual interactions amounts to the transition enthalpy of the complete sequence.The transition enthalpy (ΔHvH)is independently accessable for oligomers from the UV melting curves,if we view the transition as an all-or-none process.

Table IV: Comparison of Calculated and Observed ΔH_T and ΔG_T

Table IV

Oligomer duplex	Tm obs oC	ΔHpred KJ/mole	ΔHobs KJ/mole	ΔGpred KJ/mole	ΔGobs KJ/mole
CAAATAAAG GTTTATTTC	41	269.9	244.9	37.42	35.53
CAAAAAAAG GTTTTTTTC	45	275.7	273.2	40.57	40.13
CAAACAAAG GTTTGTTTC	48	279.1	269.6	42.38	42.22
CAAAGAAAG GTTTCTTTC	50	279.7	262.5	42.41	39.71

The good agreement between the predicted and the observed values shows that the different assumptions are valid and the data set given here is suitable to be applied to oligomers as well as to a given sequence within a long duplex structure.
There is another independent backup for this data set namely the correlation between the ΔH_{pred} and the observed Tm values. Since ΔS is sequence-independent changes of Tm and of ΔH should go in the same direction.

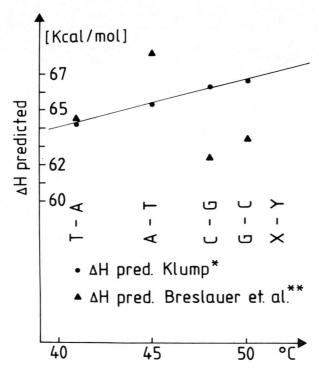

Fig.2: Correlation between the calculated ΔH per mole oligomers and the experimentally observed transition temperature Tm ,taken from the literature (Tinoco et al 1985) .X represents the varied base in the 5'strand and Y the complementary base in the 3'strand.

There is an other data set in the literature , compiled by Breslauer and coworkers(1982), which contains the thermodynamic data for the disruption of interactions in an existing duplex at 1 M NaCl, 25°C ,and pH 7.The units are Kcal/mol of interaction for ΔH and ΔG, and cal/K per mol of interactions for ΔS.The two data sets are comparable but not identical, mainly due to the fact that the second set ,despite the observed high Tm values for the mixed sequences , which contain as many AT base pairs as GC base pairs, lists an interaction enthalpy for the GA/TC neighbours as well as for the CA/GT neighbours ,which amounts only to half the interaction enthalpy of the GC/CG neighbours. The interaction enthalpy is even significantly lower than the interaction enthalpy of AT/TA neighbours.This discrepancy has to be reinvestigated .
In a similar manner as outlined for the prediction of the transition enthalpies the corresponding values for the free enthalpy change can be evaluated and utilized to compile the free transition enthalpy of any given sequence from its primary structure.This calculation is based on the same assumptions as the calculation of the ΔH values.Again there are some discrepancies between the two data sets There is a so-called symmetry term included in the set of Breslauer and his coworkers to account for the palindromic properties of the short sequences as compared to directional sequences of the same length.And there is an additional helix initiation free energy of 5 Kcal for duplexes,

which exclusively contain GC base pairs and of 6 Kcal for those
containing only AT sequences. These are contributions which are
not derived from the experimental (calorimetric) Tm and ΔH
values.Even if there is a good agreement between the "observed"
free enthalpy change and the computed value,it needs a more
comprehensive justification to include the two extra contributions
into the calculation.
Anyway, both data sets can be utilized to predict duplex free
energies with a sufficiently high level of accuracy This also
holds for the prediction of ΔH and of Tm from the primary
structure.

3.3   Property Diagrams of Helical Polynucleotides

3.3.1   Property Diagrams for all Synthetic Polynucleotides with
B-DNA Structure

Based on the large body of data from numerous thermal transition
measurements , monitored by a variety of physical methods, which
are capableof followingthe transition from helix to random coil
conformation and mirror the degree of transition, in the
temperature interval applied ,of a large variety of synthetic
polynucleotide sequences over a wide range of counter ion
concentrations we can design a "property diagram".(Klump,1987)We
obtain a two dimensional" thermodynamic phase surface" for the Tm
values as a function of the two most important system variables,
namely the net base composition of a given sequence (percent GC-
base pairs) and the counter ion concentration in theenvironment ,
given on a logarithmic scale. The obtained phase surface can
best be described as a tilted plain, with the lowest corner at

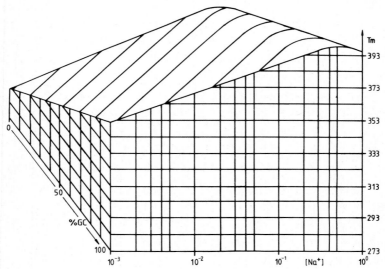

Fig.3:The transition temperature Tm of synthetic polynucleotide
sequences (calculated per mole base pairs) as a function of the
sodium concentration and the sequence composition (% GC)

pure AT-sequences and lowest sodium concentration.The Tm changes
linearly as a function of log Na+ for a given GC content and
linearly with the change in the percentage of GC base pairs at a
fixed sodium concentration.The two edges of the plain are
determined exclusively by the alternating A-T and G-C
sequences.This will be discussed in the following section.Around
physiological sodium concentrations the plain starts to level off,
forming a ridge at highest thermal stability , and starts to bend
downwards at still higher cation concentrations.This diagram
enables us to predict the transition temperature for any given
regular synthetic polynucleotide in the range of counter ion concentration between 1 mM and 1 M per liter.

It is obvious from this diagram that the upper limit of thermal
stability for any given sequence coincides with physiological
salt conditions, and it throws some light on the established habit
of linear extrapolation of low-salt data into the range of molar
cation concentrations(Tinoco,1971).The range above 100°C is
explored here for the first time , due to an improvement of the
experimental technique (Klump,1985). This temperature range

becomes accessible through the shift of the boiling temperature of
the solvent water by a moderate outside pressure applied onto the
surface of the sample. It can be demonstrated that there is no
detectable pressure dependence of the transition temperature by
comparison of two independent measurements of the same sample with
and without outside pressure. It is also clear from this approach
that the polynucleotide backbone is quite insensitive to
hydrolysis at high temperatures, otherwise the transition would be
irreversible after exposure of the sample to high temperatures,
which is in fact not observed.The strictly alternating poly d(G-C)

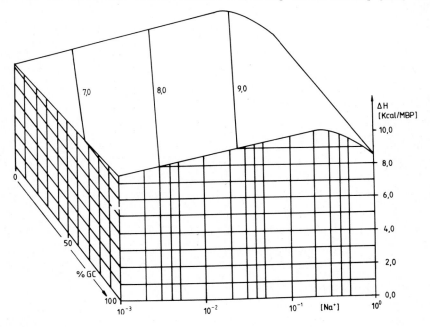

Fig.4: The transition enthalpy of synthetic polynucleotide sequences as a function of the sodium concentration and the mean value
of the sequence composition.

at 0.3 M NaCl turns out to be the most stable polynucleotide sequence of all unmodified sequences.Only the left handed poly d(G-m5C) has shown a higher maximal transition temperature of 132.5 °C.

From a series of corresponding calorimetric experiments over the same range of counter ion concentration , and for the same set of polynucleotide sequences it is possible to draw a phase diagram, which gives the transition enthalpy ΔH per base pair for all possible polynucleotide sequences.The two dimensional phase surface represents all experimentally obtained ΔH values as function of the two variables, the net GC content , and the concentration of the sodium ions on a logarithmic scale.

The upper edge represents the course of the transition enthalpy of poly d(A-T) between 1 mM and 1 M Na+ , the lower edge of the property plane refers to the course of the transition enthalpy of poly d(G-C) in the same range of counter ion concentration. The plane that stretches between these two margins represents the corresponding transition enthalpy for any mixed polynucleotide sequence , calculated per mol of base pairs. It is satisfactory that the experimentally obtained values which refer to the two possible sequences, which both contain exactly 50% CG base pairs, namely poly d(G-A) poly d(C-T) and poly d(A-C) poly d(G-T), fit right onto the plain and support the assumption of a linear dependence of the transition enthalpy with respect to base composition of a given sequence, when the sodium ion concentration is kept constant.As will be demonstrated for the natural B-DNA sequences their corresponding ΔH values will also fit into the property surface. The corner of lowest transition enthalpy is again determined by the low GC content of the sequence and the low sodium concentration.The plain rises towards the region of physiological Na+-concentration and highest GC content of the sequences, and starts to decline towards the region of even higher sodium concentration.

One of the basic assumptions for the calculation of the thermodynamic parameters is that the helix coil transition can be strictly treated as an equilibrium reaction. From each calorimetric experiment we obtain two parameters , namely the transition enthalpy ΔH per base pair when the analytical concentration of subunits is determined independently,(Vogel,1975) and the transition temperature Tm,i.e. the temperature, at which half the sequences are in the helical state and the other half of the sequences are already randomly coiled.

The free enthalpy at T=Tm is zero by definition of a true equilibrium, and hence, the enthalpy divided by the Kelvin temperature (Tm) gives the corresponding transition entropy for the experimental conditions applied.Thus we have a full set of transition entropies as function of the two variables ,stated above,which enables us to construct the property diagram, which represents the transition-entropy-surface for all possible polynucleotide sequences.

This diagram shows some remarkable features . If we connect the points of identical transition entropy on the surface we get a series of straight lines, which run parallel to the axis , which represents the variation of the sequence composition.This particular feature holds for all counter ion concentrations.It is obvious from this finding that the transition entropy is independent of the composition of the sequence, i.e. it is exclusively a conformational entropy contribution from the gain of

rotational freedom along the backbone, when the ordered double helical structure is abandoned.
Tinoco(1962) has outlined a procedure to evaluate the upper limit for the transition entropy per base pair, following the Boltzmann formalism, which relates the probability of states to the entropy. De Voe and Tinoco have assumed that the helix coil transition will affect six single bonds per nucleotide to rotate free around , with three prefered positions of the residues along the bonds .If we compute the transition entropy per base pair we have to account for two backbones in a double helical structure. Thus we get the following formula

$$\Delta S = 2 (6 R \ln 3)$$

The numerical result yields 108.7 KJ/K mbp (26 Kcal/K mbp).A comparison of the published ΔS values for a large variety of poly-nucleotides and the calculated upper limit for the conformational entropy change due to the helix coil transition shows a remarkable coincidence.

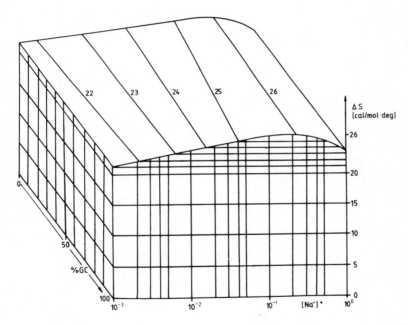

Fig.5: The transition entropy of synthetic polynucleotide sequences as a function of the sodium concentration and the bulk sequence composition.

The resulting threshold value from the phase diagram fit also extremely well into this picture . It is not accidental that the model-free experimentally obtained ΔS value and the model calculation of De Voe et al , based exclusively on the change in conformation entropy, lead to the same result. This nummerical value, now backed by the experimental approach, justifies the frequently made assumption of a sequence independent entropy change in the literature, but it surely contradicts the assumption

114

that ΔS is also temperature independent. (Record,1978) The
experimentally based results show clearly that the entropy rises
with increasing counter ion concentration.
Following this line of arguments we have a quantitative approach
to the most important thermodynamic state function ΔG and its
dependence of two variables, the sodium concentration and the
sequence composition.The free enthalpy change can be calculated
from the experimental values for any given temperature , only
assuming that the transition enthalpy is virtually temperature
independent. This can be justified from the observation of the
absence of a sizable increase of the heat capacity during the
conformational change of the polymers.

$$\Delta G_T = \Delta H_{o.t} (1 - T / T_m).$$

The surface, which represents the free enthalpy change associated
with the helix coil transition , calculated for $25^{O}C$ and per base
pair, rises from the corner of the plane at low cation concentration and low GC content of the sequences considered,to the range
of physiological salt concentration and high GC content.

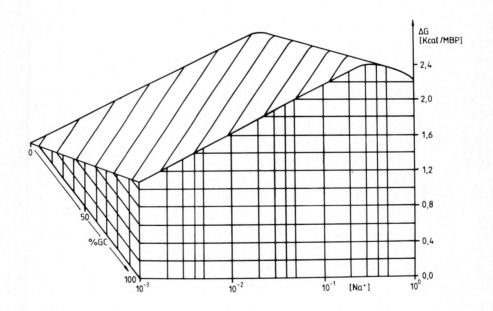

Fig.6: The Gibbs energy of the helix coil transition of synthetic
polynucleotide sequences as a function of the sodium concentration
and the bulk sequence composition .

Both variables contribute to the increase of ΔG. As can be viewed
from the diagram immediately, the pure AT-sequence is not stable
at 1 mM NaCl and at room temperature.This finding , derived from
the property diagram, can immediately be checked by experiments to
be correct. In fact Gruenwedel had reported this fact, based on
his careful investigation on the counter ion dependence of the
transition enthalpy of poly d(A-T). (Gruenwedel,1974)The property

plane which represents the ΔG function is remarkably more tilted
than the other three planes shown here before. For the first time
the full set of thermodynamic state functions is shown . The
complete representation of the ΔG function e.g. allowsevaluationof
the difference of ΔG₀c and ΔGₐτ for a given set of environmental
conditions. For 10 mM NaCl and 298 K this difference amounts to
1.3o Kcal/mbp (5.34 KJ/mbp) , indeed very close to the value of
1.25 Kcal/mbp (5,23 KJ/mbp) ,reported by Kallenbach(1968) for the
analogous RNA sequence. The property diagram allows evaluation of
the most stable sequences and the environmental conditions under
which they are most stable.
The complete set of experimentally based thermodynamic state
functions is now at hand . This was the goal of numerous
thermodynamic studies .
We can use this set to predict the thermal stability of any
synthetic polynucleotide sequence, and we can discriminate
between solvent effects and sequence effects on the thermal
behavior, and can settle the discussion on the influence of the
nearest neighbours on the stability of a given base pair.

3.3.2   Property Diagrams of all Native B-DNA Sequences

To quantify the influence of the nearest neighbours within a DNA
sequence we will present a second line of evidence ,derived from
an independent set of experiments, to support the validity of the
results obtained for the selected polynucleotide sequences. This
alternative approach is based on the same experimental techniques
but it relies on selected native DNA samples as the basis for a
pronounced variation in base composition.
To fulfil the requirement of a large variability of base
compositions not only eukaryotic DNAs such as from calfs, humans,
or fish are investigated -these show only a very limited variation
in the gross GC content, namely around 42% GC- but viral DNAs as
well as bacterial DNAs , both from eu- and archae-bacteria, are
included in this investigation.This selection enlarges the span of
observable sequence composition from 27% GC to 72% GC. The
variation of the sodium concentration is varied in the same range
as in the polynucleotide research.
In the course of this investigation it became apparent that
microcalorimetry in its present state is capable of distinguishing
between individual DNAs .

3.3.3   Synopsis of the two Sets of Property Diagrams

As outlined above for all polynucleotides either from the increase
of the absorbance, recorded at 260 nm as function of the
temperature, or from the maximum of the heat capacity vs.

temperature curves,or from any suitable physical method the Tm
values can be obtained , and can be presented as function of the
two variables, namely the counter ion concentration, plotted on a
logarithmic scale, and the sequence composition of the DNAs under
study, i.e. the gross GC content of the bulk DNA. From all these
Tm values a surface area in the three dimensional property diagram
is constructed.The "Tm plane" is extrapolated to the limits of
0.0% GC and 100.0% GC base pairs which are inaccessible for
native DNA samples. Within the plane the point of identical Tm
values can be connected to give "isotherms", i.e. lines of equal
thermal stability. The representation of the transition
temperatures , derived from melting curves of native DNAs, on a
surface, is remarkably similar to the same plot of Tm values for
the sythetic polynucleotides.

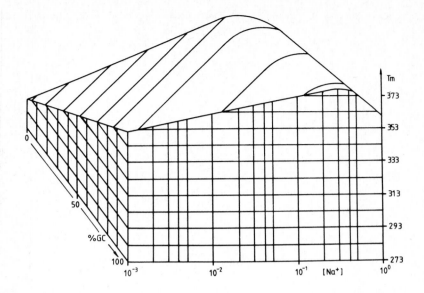

Fig.7: Transition temperature of native DNA sequences as function of the bulk sequence composition , and the sodium concentration.

A superposition of both diagrams reveals the coincidence of the upper and the left-hand edge . The "Tm plain" for the DNAs is only slightly less tilted towards the ridge of highest thermal stability. The maximal thermal stability for any sequence , derived from the DNA data (104oC) is about 20 degrees lower than the corresponding value derived from the polynucleotide data. The discrepancy of the two data sets goes when one would take the Tm values of poly dG poly dC as the reference state of highest stability instead of the Tm values of poly d(G-C).It is remarkable that the extrapolated Tm values for pure poly d(A-T) from the DNA data and the experimentally obtained Tm values obtained for this polynucleotide are identical. The same holds also for the Tm values of the mixed alternating sequences (50% GC) and the natural 50% mixture (E.coli DNA) at low counter ion concentration. The Tm difference betweem the two extremes , i.e. 0% and 100% GC, is 55°C for DNAs and 58°C for polynucleotides.
The corresponding ΔH plane , derived from a large number of measurements of different DNAs , again is almost identical to the same plane derived from the polynucleotide measurements. Within the margin of experimental error the "DNA plot" and the "polynucleotide plot" are virtually identical.

The maximal stability of any GC base pair is about 41.8 KJ per mol base pairs (10 Kcal/mbp). It holds also for the natural DNA sequences that the range of maximal thermal stability falls into the range of physiological counter ion concentration. The three dimensional plot which shows the ΔS function dependence of the two system variables, the net GC content and the sodium ion concentra- tion, derived from the DNA measurements has the same general features as the corresponding plot derived from the corresponding synthetic polymers. The connection of equal ΔS values results in straight lines which all run parallel to the (% GC)-axis , as was already shown for the polynucleotides. It is now established that

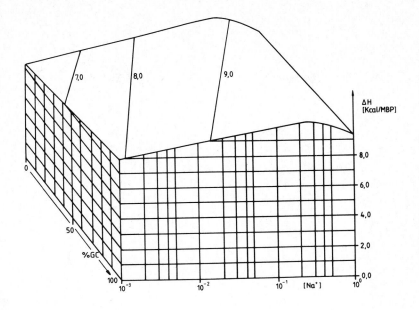

Fig.8: Transition enthalpy of native DNA sequences as function of the sodium concentration and the bulk sequence composition.

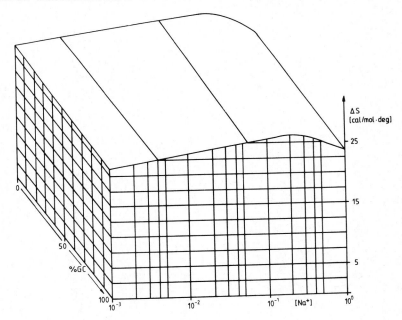

Fig.9: The transition entropy as function of the bulk sequence composition and the net sodium concentration , calculated per mole base pairs.

also for the helix coil transition of DNAs there is no detectable
dependence of the transition entropy from the primary structure.
The theoretical considerations about the origin of the entropy
change discussed above for the polynucleotides and the computa-
tion of the upper limit of this value is strongly supported by the
DNA data.
The maximal entropy change due to the helix coil transition of a
DNA sequence is about 108.7 J/K mbp (26 Kcal/K mbp).

As the final example of the phase diagrams we present the ΔG
function, calculated for 25°C, derived from ΔH and Tm values of
DNA sequences, as function of the net GC content of a sequence and
of the NaCl concentration (on a logarithmic scale).The property
plane looks very similar to the corresponding plane derived from
the polynucleotide data. The least stable sequences are the pure
AT sequences at low counter ion concentration, the maximal
stability is reached at physiological conditions for high GC

content, and the further increase of cation concentration leads to
a decrease of the stability.(Gruenwedel,1974)
From this plot a number of interesting conclusions can be drawn.
The difference in transition free enthalpy between a pure AT
sequence and a pure GC sequence, extrapolated from the
intermediate mixed sequences of native DNAs at a cation
concentration of 1 mM yields 1.2 Kcal per mol base pairs (5.0
KJ/mbp), which is close to the experimentally obtained difference
for the polynucleotide sequences (5.4 KJ/mbp). The maximal
variation of the transition free enthalpy across the property
plane amounts to 2.2 Kcal/mbp(9.2 KJ/mbp). This difference can be
converted into a Tm difference of 100°C.
Summing up the results from the large body of experimental data it
can be safely statet, that the polynucleotide data can be utilized
to complete the set of DNA data to yield the given property
diagrams, and vice versa. The DNA data correspond to the
polynucleotide data in the composition range where no experimental
polynucleotide data are available. Small differences between the
two data sets seem to be real,i.e. outside the margin of
experimental error of the methods applied.It is reasonable to

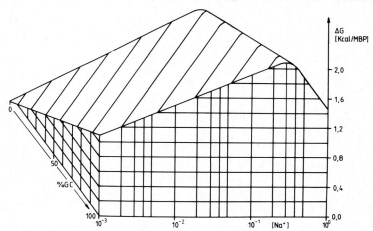

Fig.10:Gibbs energy for the helix coil transition of native DNA
sequences as function of the sodium concentration and the bulk sequence composition.

explain the difference by the assumption that not all the polynucleotides , used to evaluate this data set, exhibit strictly the canonical B-DNA structure.

3.4   Energetics of Helix Helix Transitions

3.4.1   Energetics of the B-Z DNA Transition

The Z structure is characterized by a dinucleotide repeat.Thus it is convenient to consider sequences of 2n basepairs (bp) as present in the canonical alternating polynucleotides poly d(G-C) e.g.to adopt the Z form under certain experimental conditions.This is demonstrated by crystallographic and solution studies.(Rich 1982) The state with all bp's in B form is taken as the reference state. The transition free energy difference equals

$$\Delta G(k,m,n)=-RT \ lnU(k,m,n) \qquad (1)$$

where $U(k,m,n)$ is the statistical weight (relative to the B state) of a microstate with k units (2 bp's each) in the Z form and m B/Z junctions.The associated free energy difference can be further decomposed into the form

$$\Delta G(k,m,n,)= 2k \quad g + 2m \ F_j + \quad \Delta G_s (k,m,n) \qquad (2)$$

where g (in kcal $mole^{-1}$) is the intrinsic free energy difference between the Z and the B forms for one mole bp's, F_j is an effective positive junction free energy for one mole B/Z junctions,(Soumpasis,1986) and ΔG_s is the change in total free energy of supercoiling due to the relaxation of the whole circular plasmid molecule accompanying the B-Z transition of k units in an inserted stretch of an alternating pu/py sequence.(We condider a particular sequence optionally embedded in a closed circular DNA carrier which can be placed under topological stress, because this is an experimentally observable situation). Studies of sufficiently short DNA oligomer transitions are very convenient for obtaining energetic parameters because of the validity of an all-or-none formalism to a first aproximation Breslauer,1986). One can define an apparent equilibrium constant K by the help of some spectroscopically observable "degree of transition" Θ ,i.e. the average fraction of units in the Z conformation .Thus

$$K= \Theta/(1-\Theta) \qquad and \qquad lnK = - \Delta G/RT \qquad (3),(4)$$
$$(3),(4)$$

Assuming that the solution is dilute with respect to the DNA component,

$$\Delta G= G_z^1 - G_B^1 = G_z^1(n,p,T) - G_B^1(n,p,T) \qquad (5)$$

where G_z^1 is the total Gibbs free energy of the system comprising just one mole of noninteracting DNA molecules all in conformation Z and $n_1....n_r$ moles of solvent components constituting the composition vector n with 1:water,2:cosolvent,3:first electrolyte;

p is the pressure ; and T is the temperature.G_B^1 is similarly defined.
With this formulation, all thermodynamic derivatives of lnK have a clear meaning.For example,

$$(d \ lnK/ \ dT) \ = \ \Delta H(n,p,T)/RT^2$$

$$(d \ lnK/ \ dn_j) = \ -\mu_j(n,p,T)/RT$$

$$(d \ lnK/ \ dp) \ = \ -\Delta V(n,p,T)/RT$$

where ΔH, μ_j, and ΔV are the differences in enthalpy, chemical potential of the j-th component and solution volume of the two systems. In practice the derivatives of lnK have a clearcut meaning and describe variations in lnK with respect to precisely those quantities that are experimentally varied, namely the amounts of the different components (salts, cosolvents) present in solution, temperature T and pressure p.

A logarithmic dependence of the apparent equilibrium constant K on NaCl concentration was initially observed in the study of poly d(G-C)(Pohl,1972). Cooperative transitions were also observed with poly d (G-C) undergoing B-Z transitions in the presence of organic solvents either alone or as a cosolvent at very low salt concentrations. It is noticable that the NaCl-induced B-Z transition of poly d(G-C) is essentially isoenergetic; i.e. the enthalpy change is less than 0.3 kcal/mole bp from calorimetric and from spectroscopic equilibrium-kinetic studies. The reaction has to be entropically driven and probably dominated by interionic interactions.(Klump,1986) In all other situations ,i.e. in mixed solvents, or under the influence of bound di- or oligovalent cations and other conformationally selective ligands, the B-Z transition of poly d(G-C) can also be induced by temperature variations at constant solvent composition. The latter is also true for DNA sequences methylated or halogenated at the 5'position of the cytosine moiety.(Chaires,1986)Such substitutions lead to a selective stabilization of the left-handed conformation of the two canonical alternating purine-pyrimidine families,d(G-C) and d(A-C)d(G-T).(Singleton,1984)With temperature dependent transitions, the apparent spectroscopically determined van't Hoff transition enthalpy change (for an unknown size of the cooperative unit)

$$\Delta H_{vH} = \ 4R \cdot T^2 (d \ \theta/dT)$$

is related to the intrinsic transition enthalpy per base pair (determined by calorimetry). In the case of long polymers described by means of the infinite lenghth Ising model (Roy,1983)

$$\Delta H_{vH} = \ l_o \ \Delta H_{cal}$$

where l_o is the calculated length of the cooperative unit .

The thermodynamic parameters for the temperature dependent B-Z transitions in DNA polymers obtained from van't Hoff analysis of spectroscopic data and of some directly obtained calorimetric data are presented in Table V.

Several conclusions can be drawn:

(i) DNAs with 5' cytidine substitutions (methyl,halogen) in both sequence families show a similar apparent transition enthalpy ΔH per base pair of ca 4.2 KJ per mole base pairs, derived from spectroscopic measurements, which is very close to the directly determined value from adiabatic scanning microcalorimetry.(4.2 KJ/mbp for poly d(G-m^{5}C))

(ii)the halogen substituents are more effective than a methyl group in stabilizing the Z form. Poly d(G-br^{5}C) and poly d(G-io^{5}C) are constitutively left-handed at all ionic strengths;

(iii) the cooperative lengths,i.e. junctional free energies, vary greatly from one polymer to another,i.e.are DNA-sequence specific as well as a function of environmental conditions. Thus reported values for the cooperative length l₀ for d(G-C) duplexes range from 25 bp's to 100-1000 bp's.Three recent estimates for the methylated polymer are 180,226,and 110 bp's. and finally
(iv)the van't Hoff plots for oligomers are linear.

Table V

Thermodynamic Parameters for B-Z Transitions in Linear DNA

Polynucleotid	T °C	ΔH_{v} KJ/mole	ΔS J K^{-1} mol	ΔH_{o} KJ/mbp	ΔS_{o} J/mbp K
poly d(G-C)	38	o	o	o	o
poly d(G-m³C)	40	-	-	4.2	13.4
poly d(G-m³C)	42	-	-	4.2	13.3
poly d(G-m³C)	43	-	-	2.1	6.7
poly d(A-m³C)d(G-T)	57	920	2800	-	-
poly d(A-brC)d(G-T)	40	815	2692	3.8	12.1
poly d(A-mC)d(G-brU)	40	2006	6370	4.6	14.7
poly d(A-brC) d(G-brU)]	40	1714	6700	5.0	16.0

Other data on B-Z transitions are available for more complex solution conditions.Szu and Charney (1985) have reported the thermodynamic parameters of the thermal transition of poly d(G-m³C) in 0.5M MgCl₂,50mM NaCl from least square fitting of experimental degrees of transition Θ vs. temperature curves to the Ising model expression.Using their data one obtains an apparent transition enthalpy of 5.85 KJ mole^{-1}, and an entropy value of 18.8 J K^{-1} mole^{-1} at Tm = 39.3 °C. Both values compare

favourable with those derived from a recent calorimetric study of the same DNA in 10mM MgCl₂, 50mM NaCl(Klump,1986). A smaller value was published in a very recent study of the same polymerunder similar conditions (1 mM MgCl₂,50 mM NaCl), namely 2.5 KJ for the transition enthalpy ΔH per base pair, and 8.4 J per degree and mole base pairs for the transition entropy.(Chaires,1986).
In mixed solvents under certain extreme conditions, the B-Z transition can be rendered endothermic, i.e. the B conformation is favoured upon heating.(Latha,1983)
Only a very short and incomplete basic formalism is presented here and the reader is kindly refered to the original literature,but it is worthwhile to understand the few pertinent thermodynamic data related to the B-Z transition.
It is apparent that one is in a position to make preliminary yet quantitative predictions and to separate the contributions of intrinsic sequence-dependent effects from more global ones. However, additional information is required in order to construct more precise and extensive phase diagrams e.g..

3.4.2   Energetics of the A-B DNA Transition

It is well known that the B-DNA double helix can be reversibly
converted into the A-DNA conformation by decreasing the water-
activity a. This decrease can be brought about in solutions by
addition of non-electrolytes that mix well with water,e.g.
alcohols or esters.(Ivanov,1985) To evaluate the energetics of the
A-DNA to B-DNA transition it would be possible to compare the
helix to random coil transition of the two conformers in seperate
experiments. Both helical conformations can be disordered by
increasing the temperature of the system. The energetics of the
order/order transition would simply be the difference of the
energetics of the two seperate helix to random coil transitions.
But there is a different approach to the A-DNA/B-DNA/coil equilibrium .Plotting the transition temperature for the transition from
the ordered to the disordered conformation as a function of the
water activity a point should exist in the (Tm,a)-plane ,at which
all three different conformations are at equilibrium .A note of
caution has to be added here again . Since the B to A , as well as
the B to coil and A to coil transitions are not first order phase
transitions, i.e. they have a finite width, a sizable area has to
exist around the tripple point where all three "phases" coexist
simultaneously.
Until recently the experimental approach to determine the
energetics of these transitions was limited to the regions in the
phase diagram where only two forms were at equilibrium.Unfortuna-
tely, owing to DNA aggregation, no true course was obtainable for
the branch of the phase diagram that describes the A/coil
equilibrium. As to the B/A equilibrium branch, this was shown to
be independent of temperature. Recently Ivanov et.al.(1985) have
used a different approach to gain information on the transition
energy for the A/B transition.
Two DNA samples with a different GC content were
used:Cl.perfringens (31% GC) and calf thymus (42% GC).To obtain
aqueous solutions of different water activity repeated additions

of small aliquotes of trifluoroethanol (TFE) to the original DNA
solutions were performed. The samples were heated at a rate of 0.7
deg/min throughout the transition interval. The main parameters of
the B/A transition ,the energy and the cooperative length,were
estimated from the slopes of the different branches A-B, A-Coil,
and B-Coil in the vicinity of the triple point. The cooperative
length varies between 10 to 20 base pairs, which corresponds to
the energy for the B/A junction of 5.0-7.5 KJ per mole base
pairs.The presence of A-DNA conformation was determined by
circular dichroism.

3.4.3   Energetics of A-Z RNA Transitions

Very recently it was successfully demonstrated for the first time
that corresponding inversions (B- to Z-DNA) of helical
handedness occur also in double stranded RNAs(Soumpasis,1986),
contradicting a statement of Dickerson (1981), which in an
apodictic way states that it is immediately clear that there is
nothing like a Z-RNA because of the 2'OH-group of the ribose
ring.This statement seems to be premature in the light of the
results presented here for a series of non-modified poly(G-C) and
the methylated and halogenated analogs.

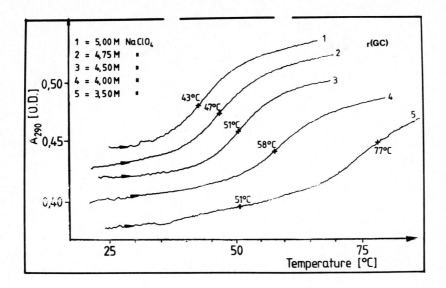

Fig.11 Absorption changes upon the order/order transition as function of the temperature for different perchlorate concentrations.
(1)5 M; (2) 4.75 M; (3) 4.5 M; (4)4.0 M; (5)3.5 M;monitored at A₂₉₀
Heating rate was 1°/min.

UV spectroscopic studies have shown that poly(G-C) in aqueous
solution containing 4 M NaClO₄ can reversibly be converted into a
left-handed Z-RNA by rising the temperature to 60°C. This conversion temperature will linearly decrease as function of the
logarithmic increase of the perchlorate concentration up to 110°C.
Above this temperature the order order transition merges with the
helix coil transition. The reversibility and cooperativity of the
helix/helix conversion facilitates the investigation and the
quantitative evaluation of the transition enthalpy per base pair
by the help of scanning microcalorimetry.As will be discussed in
more detail in the following paragraph in a 5 M NaClO₄ solution
the conversion temperature drops to 41°C ,the conversion enthalpy
per mole bp's yields 4.17 KJ, the corresponding entropy change
amounts to 13.4 J K⁻¹ .The van't Hoff enthalpy , which can be
derived from the temperature dependence of the degree of transition, i.e. the fraction of polymer strands which populates the
left-handed conformational state, is calculated to 820 KJ per mole
of the cooperative unit. (Klump,1987) From the ratio of the two
enthalpies the size of the cooperative unit (bps) can be determined to amount to about 200 bps. Very similar results are obtained
for the poly (G-m⁵C) and for poly (G-br⁵C).(cf. Table V).
For the first time the quantitative approach to investigate all
order/order transitions was made . Instead of looking for the
traditional order/disorder transitions in the form of the helix/
random coil transition the three possible order/order transitions
are viewed in a comprehensive manner(Reley,1966), namely

(i)the disproportionation,(Cantor,1980)i.e. the rearrangement of two identival double helices to a triple helix and a partly stacked polypurine single strand;and

(ii) the displacement reactions ,i.e. the exchange of one of the two strands within the double helix to a single strand from the solution to form a new double helix with an enhanced thermal stability (this displacement reaction can even lead to a crosswise exchange of one strand each between two double helical complexes).
One of the prerequisites, necessary to perform the desired experiments, was the improvement of the instrumentation for the UV and the calorimetric measurements to a temperature range above the normal boiling point of the solvent water.(Klump,1986)The possibility to apply an outside pressure of some 10 atm. to the surface of the solution under investigation brings the registration of the melting temperature of even the most stable polynucleotides into the range where it can be seperated from the boiling point of the solvent. The highest transition temperature observed so far is 140° C.In the mean time the sophistication of adiabatic scanning microcalorimeters has been improved to allow scans of very small samples(0.5 ml) up to 132° C.The accuracy and the precision of the instrumentation allowdeterminationofenthalpy changes as low as 1.0 KJ per mole subunits. As it turned out there is no fundamental difference in the occupation of conformational space of secondary structures between DNAs and RNAs.

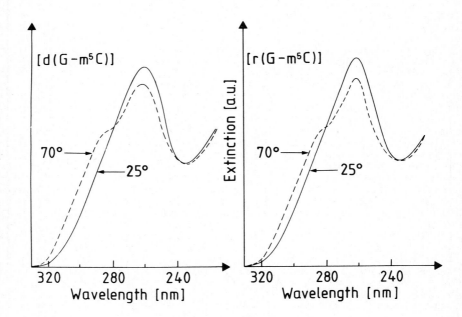

Fig.12a:Spectral changes accompanying the B-DNA to Z-DNA transition and the A-RNA to Z-RNA transition .

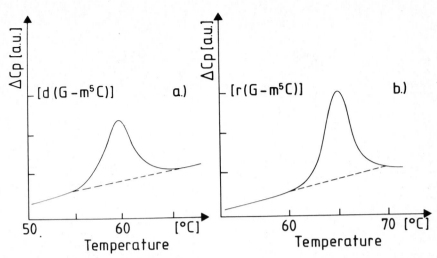

Fig.12b:Apparent heat capacity change as function of the
temperature for the thermally induced inversion of helical
handedness.

As a rule RNA helices are more stable than DNA helices.So at least
in principle it should be possible to induce the analogue to the
Z-DNA structure in a RNA double helix for strictly alternating
purine/pyrimidine sequences in a suitable environment. With the
help of ^{32}P-NMR and CD spectroscopy (Behe,1984)it was demonstrated
that poly (G-C) can in fact invert its handedness from the
canonical A-RNA conformation into the left-handed Z-RNA
conformation .This statement is based on the assumption that
spectroscopical similarities reflect a conformational similarity.
The experimental prove of a left-handed RNA structure by X-ray analysis is still lacking.A continuous scan of the inversion process
by NMR spectroscopy is not possible and the registration of the
change of a CD signal at a fixed wave length due to the
order/order transition allows only scanning of the course of the
reaction (degree of transition as function of the changing
environmental conditions) to yield the van't Hoff enthalpy.These
methods are short of giving quantitative results for the
conformational unit (e.g. bp).The registration of the CD signal
can be replaced by the registration of the UV absorbance at 290 nm
because the Z-RNA structure is characterized by a higher
extinction coefficient at this wave length as compared to the A-
RNA conformation .The recorded UV-melting curves are symmetrical
with respect to the transition temperature Tm ,i.e. the
temperature, at which the absortion increased to half of its
final value. This shape reflects the presence of only one
inversion center per polymer strand. The inversion of the handedness can best be described as an intramolecular process.This
observation is supported by a moderate inversion rate which seems
to be independent of the temperature.(Jovin,1987)
The heat capacity changes ,due to the structural changes within
the helical structure in the limited temperature range of the
thermal transition , is a measure of the transition enthalpy of
the polymer sample under investigation in the calorimeter vessel.
Knowing the actual polymer concentration through a colorimetric
phosphorus analysis (Vogel,1975) allows computation of the true

transiton enthalpy per base pair.(4.2 KJ/mole bps).It is remarkably close to the experimentally observed value for the B-DNA to Z-DNA transition of pol d(G-m^{5}C) in 10 mM MgCl$_2$,50 mM NaCl.(cf Table.V) The positive sign of ΔH is acceptable due to the fact that the right-handed conformation is stable at room temperature even at very high counter ion concentrations.This is different from the analogous deoxypolymers. From the experimental observations it is obviously justified to treat the inversion process as a true equilibrium. Consequently the standard thermodynamic equations hold to calculate the other thermodynamic parameters, such as the true transition enthalpy as well as the true free enthalpy change at any given temperature, taking the transition enthalpy as temperature-independent.From the Gibbs-Helmholtz equation at T=Tm follows that the transition entropy is the ratio of the enthalpy and the temperature, yielding an entropy value of 13.4 J K^{-1}for the inversion reaction. This value is very small as compared to the helix/coil transitions (ca. 90 J K^{-1}),but

it is not unreasonable for an order/order transition, for we start from a doublehelical conformation and we end up in a helical con-formation. Hence there is no contribution from conformational entropy from a backbone elements which are freely rotating around the single bonds along the chains, but only contributions from changes in solvation and from changes in surface charge density between the two conformers. It is impossible to give any detailed description of the various contributions to the entropy change. Since both the transition enthalpy and the transition temperature are small it is not surprising that the free enthalpy change due to the transition, calculated for standard conditions, is not bigger than 0.27 KJ per mole bps.
There is some caution necessary in the application of the standard formalism to calculate the cooperative length for the order/order process. There are always three distinct regions in each double strand, namely the left-handed helix, the right-handed helix , and the A-Z junction in between. There is only one center of inversion possible in each polymer molecule. It is formally acceptable to determine the cooperative length by dividing the van't Hoff enthalpy by the calorimetrically determined enthalpy of inversion to yield the "size of a mole" also for this process. However, this does not mean that the number of base pairs which undergo the conversion simultaneously are located in a loop of this calculated size (200 bps).

Table VI gives an compilation of the experimentally obtained thermodynamic data for the A-RNA to Z-RNA transition.

Table VI

Energetics of A-RNA to Z-RNA Transition

Polynucleotide	Tm °C	$\Delta H_{o\,a\,1}$ KJ mbp^{-1}	$\Delta S_{o\,a\,1}$ J K^{-1}mbp^{-1}	$\Delta H_{v\,H}$ KJ m^{-1}	l$_o$
Poly r(G-C)	41	4.1	13.o	236	240
Poly r(G-m^{5}C)	52	4.6	14.1	253	230
Poly r(G-br^{5}C)	46	4.4	13.9	244	230

3.4.4 Energetics of Helix Disproportionation Reactions

It is readily accepted that three-strand polynucleotide structures
are less stable than related two-strand structures because of the
relatively higher charge density of the three-strand complex . It
was shown for poly rA poly rU (Neumann,1969) and poly dI poly dC
(Chamberlin,1965)e.g.that the double stranded conformation is the
thermodynamically stable conformation at low salt concentration
and low temperature.On the other hand it was shown that the 1:1
mixture of the two ribohomopolymers will, under certain salt
conditions (above 200 mM NaCl), rearrange to give a three-stranded
structure of (poly A)2(poly U) and free single-strand poly A when

the temperature is raised.The disproportionation reaction of poly
A poly U results in changes in the UV absorption spectrum in the
range of 280 nm which can be used to construct a phase diagram
which relates the stability of various helical forms in a 1:1
mixture of poly A and poly U to the sodium chloride concentration.
The temperature, at which disproportionation occurs, decreases
with increasing NaCl concentration. This behavior is also observed
for the other order/order transitions.
The inversion of the relative stabilities with increasing
temperature may be thought of as a reflection of the difference
between the enthalpies of the two reactions:

poly A + poly U === (poly A poly U)

2 poly A + 2 poly U === (poly A 2 poly U)+ poly A

The first of these two reactions has a more negative reaction
enthalpy and, consequently, a larger temperature coefficient of
the equilibrium constant associated with the formation of an
additional base pair in a partially formed helical structure.
Thus,the first reaction is disfavoured preferentially , when the
temperature is raised .Ross and Scruggs(1965) have shown by direct
calorimetric studies that the disproportionation reaction derives
its driving force from the increase in entropy (48.0 J K^{-1} mole
A^{-1})associated with the liberation of the partly disordered
single-strand of poly A. Entropic contributions in the first place
account for the stability of the "three-strand" complexes in 1:1
mixtures of polynucleotides with mono- or oligonucleotides , since
disproportionation results in the liberation of many small
molecules.(Klump,1975)
The two- to three-strand interconversion is also observed for the
deoxy-analogues of the poly A poly U system, poly dA poly
dT (Klump,1978) and poly dA poly dU.(cf. Table VII)

Table VII: Energetics of the Disproportionation Reactions

Polynucleotide	T. °C	ΔH$_{o,1}$ KJ/MBP	ΔS$_{o,1}$ J/MBP K	ΔH$_{v,H}$ KJ/M	<m> b
2(poly A poly U)-- (poly A 2 poly U) + poly A	45	5.3	16.7	723	136
2(poly dA poly dT)- (poly dA 2 poly dT) + poly dA	45	7.4	56	789	107
2(poly dA poly dU)- (poly dA 2 poly dU) + poly dA	45	7.1	53.5	820	115

The interconversion reaction exemplifies the danger of using the denaturation temperatures of polynucleotide complexes in deducing their relative stabilities at temperatures which are not near Tm.Under certain conditions , the value of Tm for(poly A poly U) is less than that for(poly A 2 poly U) , yet the equilibrium structure, present in a 1:1 mixture at room temperature, is the two-stranded complex because there is an inversion of relative stabilities with decreasing temperature. Table VII gives a compilation of the experimentally obtained data.

3.4.5 Energetics of Strand-Displacement Reactions

The correct measure for the stability of helical conformations is the free energy ΔG_s of a given state s, and, if all components are in their standard states , the standard free energy ΔG_s° can be used instead. Whether a given reaction will occur is exclusively determined by the change in the standard free enthalpy and neither by the change in enthalpy nor by the relative magnitude of Tm.
In the following we will illustrate the concept of a helix displacement reaction:

$$\text{poly A B + poly C === poly A + poly B C}$$

and give the experimentally obtained values for the individual steps.
We will not only demonstrate , however, that one strand within a helical complex can be replaced by a single strand, added to the system, but we will show that displacement reactions are possible between two helical complexes as well,according to the reaction scheme:

$$\text{poly A B + poly C D === poly A C + poly B D}$$

For the helix-single-strand exchange let us focus on the following reaction:

$$\text{poly rA poly dT + poly dA === Poly dA poply dT + poly rA}$$

Energetic states for this experimentally approachable system are the following(assuming $\Delta G_A^\circ = \Delta G_B^\circ = \Delta G_C^\circ$):

poly rA + poly dT + poly dA

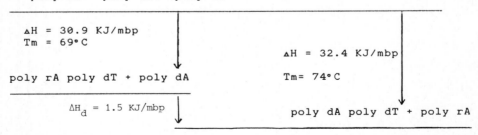

$\Delta H = 30.9$ KJ/mbp
Tm = 69° C

$\Delta H = 32.4$ KJ/mbp

poly rA poly dT + poly dA   Tm= 74° C

$\Delta H_d = 1.5$ KJ/mbp   poly dA poly dT + poly rA

We know the Tm values for the formation of the helix poly A B and poly B C from their component single strands if we assume that the transition temperature for the helix random coil transition is identical with the formation temperature. We can evaluate the free enthalpy change accompanying the displacement ΔG_d° because Tm= $\Delta H^\circ / \Delta S^\circ$, and both Tm and ΔH° are experimentally accessible quantities. The only assumption which has to be made is that the

transition enthalpy ΔH is temperature-independent. This assumption seems to be justified because of the almost undetectable small value of ΔCp. $\Delta S°$ is also given from the experimental values using Gibbs-Helmholtz equation.
So we obtain the standard free enthalpy of formation of poly rA dT according to the following equation:

$$\Delta G°_{dAdT} = \Delta H°_{rAdT} - T \Delta S°_{rAdT} \cdot \Delta H°_{rAdT}(1-T/Tm_{rAdT})$$

An exactly corresponding equation holds for $\Delta G°_{dTdA}$. Clearly it follows that

$$\Delta G°_d = \Delta G°_{dAdT} - \Delta G°_{rAdT}.$$

The quantitative evaluation of the displacement free enthalpy change at 25°C under standard conditions yields $\Delta G°_d = 0.6$ kJ per bp.
Because $\Delta H°$ is negative for all known duplex formations $\Delta G_d°$ is also negative if $Tm_{dAdT} > Tm_{rAdT}$, which in fact is correct. The double helix with the higher Tm will be favoured in a helix displacement reaction. The predictions, based on the formalism given above, are in good agreement with the observed results. Today it would be possible to evaluate all the thermodynamic data for the systems, introduced by Chamberlin and Patterson (1965).

Table VIII: Observed melting temperatures for duplexes involved in strand-displacement reactions

Reaction	Tm (reactants) °C	Tm (products) °C
poly dIdC + poly rI == poly dI + poly rIdC	46.1	52.3
poly rIdC + poly rC == poly dC + poly rIrC	52.3	60.2
poly dIrC + poly rI == poly dI + poly rIrC	35.4	60.2
poly dIrC + poly dC == poly rC + poly dIdC	35.4	46.1
poly rAdU + poly rU == poly dU + poly rArU	44.0	55.5
poly dArU + poly rA == poly dA + poly rArU	46.0	55.5
poly dArU + poly dU == poly rU + poly dAdU	46.0°	50.0
poly rAdU + poly dA == poly rA + poly dAdU	44.0	50.0

These results also demonstrate the general observation that RNA/RNA duplexes are more stable than DNA/DNA duplexes , while the relative stability of RNA/DNA hybrids is less predictable.
The same approach can also be made for a crosswise exchange of strands between two metastable helical complexes.
To demonstrate this case we will consider the following system:

poly dArU + poly rAdU === poly dAdU + poly rArU

In a 100 mM NaCl solution the hybrid helices have a transition temperature of 46°C / 44°C respectively and the poly rArU melts at 55.5 °C , 5.5 degrees higher then the DNA analogue poly dAdU.(Tm

50.0°C). The corresponding transition enthalpies are 31.6 KJ/mbp
for dAdU,   30.6 KJ/mbp for dArU , 28.6 KJ/mbp for rAdU , and
finally 32.2 KJ/ mbp for rArU.

The displacement reaction can be illustrated by the following thermodynamic cycle:

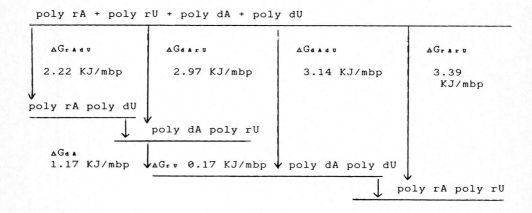

poly rA + poly rU + poly dA + poly dU

This scheme demonstrates that for the adenine/uridine system all
necessary thermodynamic quantities are known, which enable us to
correlate the $\Delta G_d°$ with the Tm parameters. The enthalpy change
around the thermodynamic cycle is obviously zero as the theory
requires.If $\Delta H_d°$ is not zero, an alternative approach is possible.
From a measured transition temperature Tm for the displace-
ment reaction, we can find the ratio $\Delta H_{BC}/$ ΔH_{AB} .Supposed now we
know a single enthalpy change for any of the three reactions in a
A,B,C reaction scheme the standard thermodynamic relations permit
us to estimate all the enthalpies and consequently also free
enthalpies based simply on Tm determinations.Thus even limited
thermodynamic data can be of much use.

3.5   Microcalorimetry as an Analytical Tool to Characterize DNAs

The idea to test calorimetry as an analytical tool emerged from
the observation, that the experimental transition curve of calf
thymus DNA (Privalov,1975) scanned in an adiabatic differential
microcalorimeter, showed a complicated but remarkably reproducible
pattern of individually distinguishable peaks.

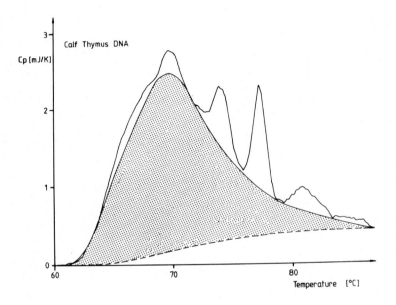

Fig.13: Apparent heat capacity change as function of the
temperature throughout the helix coil transition of calf thymus
DNA.

This pattern was obtained regardless of the source of the
sample(i.e. from freshly prepared DNA or from a commercial
supplier). It turned out to be identical with the pattern of the
same DNA, obtained by Blake et al.(Blake,1978) with the help of
differential UV melting. It was also very similar to the elution
profile from apparent buoyant density gradient centrifugation
(Kurnit,1973), which sorts the sheared DNA fragments according to
their GC content. The pronounced superstructure of the
calorimetric transition curve is far from accidental. It reflects
the relative population of the sequence conposition of DNA
stretches of the length of the cooperative units along the calf
genome.These sequences reveal a large variability of thermal
stability, and the superposition of the melting curves of the
individual sequences results in the envelope of the complete
melting curve. The presence of highly repetitive short and
homogeneous sequences such as the so-called satellite- DNAs, which
turn out to be related to the individual peaks on top of a broad
transition in our calf thymus DNA, is clearly demonstrated in this

scan (Klump,1987). To produce such seperate peaks the domain
(Blake,1978) (satellite) has to fulfil two conditions:
(i) the number of identical repeats of this particular domain has
to be considerably high, i.e. the concentration of this sequence
has to be of the order of 2-3% of the total DNA, and (ii) the length of
the domain must be of the order of the cooperative length of the
bulk DNA, namely 50 to 100 base pairs.We take the temperature at
the peak maximum as a measure for the base composition of the
domain underlying the peak, following Marmur and Doty's finding of
a linear dependence of the transition temperature of the gross GC
content of a sequence(1972). Thus from the number and from the
relative position of the extra peaks the number and composition of
the satellite DNAs of a given animal species can be determined,
even in a heterogeneous mixture of the unfractionated bulk DNA.
The large number of the different unique sequences , i.e. single
copies of genes, is represented in the large unsymmetric
denaturation peak, which stretches over the full range of the
transition interval.The enthalpy per base pair , the transition
temperature , the axial phosphate distance along the helix axis ,
and the free enthalpy difference for a given set of environmental
conditions can be utilized to identify a given DNA according to
its GC content. All four physical properties mentioned above are
linearly dependent on the base sequence and the next nearest
neighbour interactions.
There is a curious finding in the calorimetric results, which
first looked very puzzeling.The cooperativity or cooperative
length seems to be dependent on the source of a particular DNA.The
cooperative length is basically determined from the ratio of the
van't Hoff enthalpy of the transition, which can be computed from
any physical signal, which is strictly related to one of the
conformational states, and the calorimetrically model-free determined transition enthalpy per base pair.The eukaryotic DNAs show
the shortest cooperative length of all DNAs (15 to 20 base pairs),
theprokaryoticshave a distinctly greater cooperative length(25 to
35 base pairs) and the virus-DNAs as well as the linearized
plasmids show the largest cooperativ length.(40-60 base pairs).

The answer to this paradox is the different frequency of single
sequences Eukaryotes contain a large variety of closely related
sequences of low frequency . The superposition of the melting
curves of all these scarcely populated sequences spreads over the
broad transition range. The slope of the "transition curve" at T =
Tm(degree of transition vs. temperature) is rather flat, which
leads to a comparable low van't Hoff enthalpy. Virus DNAs on the
other hand consist of a relatively small number of gene sequences,
which are therefore relatively frequent in a sample of a given DNA
concentration.
There is an other interesting finding related to the magnitude
and the variation of the van't Hoff enthalpy and , thus , to the
cooperativity, namely the influence of counter ion concentration
dependence on the transition temperature as well as on the
width of the transition interval.
The van't Hoff enthalpy can be determined from the transition temperature and the width of the transition interval according to the
following relation

$$\Delta H_{vH} = 2(n + 1)R\ Tm^2\ d\theta/\ dT$$

where n is the number of strands of the ordered secondary
structure, R is the gas constant and dT is the width of the
transition interval.(Savoie,1978) Thus out of two sequences with
identical Tm , which means of identical sequence composition , the

sequence with a smaller transition interval will show the higher van't Hoff enthalpy. The transition interval is related to the character and the concentration of counter ions in the solution. Hence only van't Hoff enthalpies at identical environmental conditions should be compared. If we plot the van't Hoff enthalpies for a large number of different DNAs also at different cation concentrations of the same counter ion it turns out that almost all DNAs show a linear dependence of the van't Hoff enthalpy from the logarithm of the sodium concentration. The extrapolation of these individual plots to a very low sodium concentration gives a curious result. All the lines intersect in one point. (about $7,2 \times 10^{-5}$ m Na$^+$) This point of identical van't Hoff enthalpy has not been explained adequately at present.

It should be mentioned that the cooperative length under these conditions is ten base pairs , one helical turn , which is the shortest cooperative length to be reasonable in a helix with a ten base pair repeat. The polymer concentration in these experiments , expressed in mono-nucleotide concentration was in the same low range as the cation concentration, i.e. there are no free cations but all are bound to the phosphate residues in the DNA.

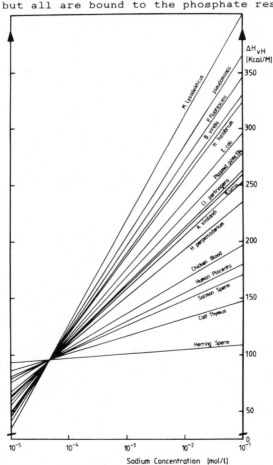

Fig.14: Calculated van't Hoff enthalpy as function of the cation concentration for various DNAs.

3.6   Molecular Forces that Stabilize DNA Sequences

3.6.1 Discrimination between Sequence and Solvent Effects

Up to now there is no clear understanding of the fundamental four-
ces that determine the thermal stability of the ordered secondary
structure of nucleic acids.The proposed interactions which are
presumed to contribute to the stability of the B-DNA structure are
the following (Cantor and Schimmel,1980):
(i)electrostatic interactions along the helix axis between
consecutive heterocyclic bases either of the same strand or of the
opposite strands (stacking interactions) (Rich,1982),
(ii) hydrogen bonding interactions within the plane of the
individual base pairs (Watson and Crick,1953),
(iii) and hydrophobic interactions , supposedly originating from
the energy required for the cavity formation in the water
structure in the surroundings of the large polymer helix.
(Sinanoglu,1964).
We will try to seperate the individual contributions from the
experimental data and in the best case quantify the partial
energies. We will also try to discriminate between the
contribution of sequence effects and of the solvent effects to the
stability of the DNA structure.
Since the helix formation is favoured by enthalpy changes, we will
base our arguments on thermodynamic data obtained by calorimetry
in the first place , and on absorption spectroscopy.As can be
safely assumed , outlined above, that it is sufficient to restrict
the considerations to next nearest neighbour interactions. The
polymer strands are assumed to be of
(i)infinite length,
(ii)to adopt the B-DNA conformation, and
(iii)to adopt the statistically coiled conformation at elevated
temperature with no intermediate conformations populated.
Infinite length in this context means that the average degree of
polymerization is at least one order of magnitude bigger than the
average cooperative length.This condition is readily fulfilled for
native DNAs but has to be checked for the polynucleotides. If a
property turns out to depend on the dinucleotide frequency, i.e.
of the individual neighbour bases within a sequence, we will term
this property a first-neighbour property, if on the other hand a
property depends on the bulk composition it is called a zero-order
property. The numerical values of the thermodynamic state
functions (ΔH, ΔS, ΔG)per base pair seem to belong into this
latter category.
To approach this task of evaluating the next nearest neighbour
influences it is theoretically sufficient to restrict the
investigation to eight linearly independent double-stranded
sequences, as was shown by Gray and Tinoco for RNAs. We neglect
only one of these sequences (poly d(A-C-G) poly dC-G-T))for obvious reasons;this sequence was not available to us. But as it will
turn out this is not a serious drawback to reach our goal.
The first point to discuss is the unforseable linear dependence
of the transition temperature Tm from the net base composition of
a DNA sequence. This fact was first discovered by Marmur and Doty,
(1972)and consequently used to predict the composition of any
newly isolated DNA from the melting temperature under standard
conditions (SSC buffer, i.e. 0.15 M Na⁺). Since the transition
temperature of some DNAs with high GC content were not obtainable
under the standard experimental conditions in the past it was
convenient to lower the buffer concentration by one order of

magnitude, thus lower the Tm by about 17 degrees, and extrapolate
to the Tm in SSC buffer. Schildkraut and Lifson(1965) give a
semiempirical equation which relates the three variables in the
system, Tm, %GC, and log Na⁺.

$$Tm = 16.6 \log Na^+ + 0.41 <GC> + 81.5$$

This equation implies for its validity that the slope of Tm vs.log
Na⁺ is independent of the net GC content, which i not correct, and
that the slope of Tm vs. the net GC content <GC> is independent
of the sodium concentration, which again is not valid. The
comparison of different experiments shows that the slope of Tm vs.
log Na⁺ decreases with increasing counter ion concentration. For a
fixed counter ion concentration there is a fixed Tm difference

$$\Delta Tm = Tm_{GC} - Tm_{AT}.$$

There are three different published data for this difference at
three different counter ion concentrations, but it is possible to
plot ΔTm vs. log Na⁺; the slope of this plot yields -9.7°C per
order of magnitude increase in sodium concentration, in other
words the slope changes about 20% when the cation concentration
changes for one order of magnitude.

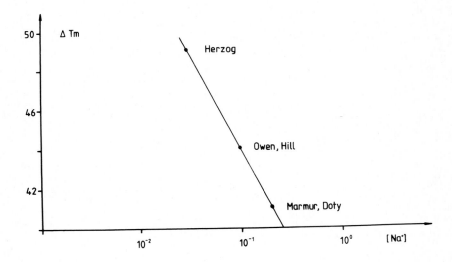

Fig.15: Tm difference (Tm$_{GC}$ - Tm$_{AT}$) determined from three different
counter ion concentrations.

A close inspection of the Tm vs. log Na⁺ for different DNA
sequences reveals a systematic decrease of this slope with
increasing GC content , which results in an intersection of all
these plots at Tm of 122°C.Herzog,1983)

A comparison of the same slopes from different polynucleotides on the other hand shows no such common transition temperature. These slopes run more or less parallel.

A note of caution is necessarily to be expressed here, for the extrapolation of Tm values to much higher or much lower cation concentrations is not supported by experimental results. At higher cation concentrations e.g. the anion influence cannot be neglected any more. It will lead to a decrease of Tm with still increasing sodium concentration . Dickerson (1981) , and more reacently Saenger (1986) have shown from X-ray analysis of DNA structures that the change in water activity dominates the preference for the actual DNA conformation, either for A- B-, or Z-DNA.

The Tm differences, easily detectable and impressively large in some cases , however, should not be overestimated , for the Tm is the most sensitive indicator for even very small conformational differences in the secondary structure.Tm differences of the order of 2°C are easy to detect and to reproduce under the same experimental conditions. The corresponding free enthalpy difference is of the order of 0.04 Kcal/mbp (0.17 KJ/mbp).

Let us now view the other thermodynamic state functions in some more detail.As mentioned earlier the superposition of the corresponding property planes for polynucleotides and for DNA sequences shows the almost complete correspondence of the two data sets. This is even more impressive if one tries to fit data from other laboratories into these schemes; all these data can readily be included. Scanning the literature for ΔH values of the transition of pure GC-containing sequences shows the complete absence of the necessary data. The data given here are the first to be published for these particular sequences, and they are only accessable because of the apparatus improvements mentioned in the introduction.

3.6.2 Deconvolution of Inter- and Intramolecular Forces (Stacking vs. H-Bonding).

It is now common place that base pairing between GC base pairs is favoured over base pairing of AT base pairs by a ΔH of about 1.7 Kcal/mbp (7.1 KJ/mbp) . There is a large number of values in the literature which refers to this difference. To mention only two of them: Klump and Ackermann (1971) in a very early paper give 1.8 Kcal/mbp, based on calorimetric measurements, and Ornstein and Fresco(1983) calculate 1.75 Kcal/mbp based on their empirical potential method. The extrapolated value from the complete data set , given here, results in 1.75 Kcal/mbp for 1 mM Na$^+$ and 1,50 for 100 mM Na$^+$.The second value is due to the decreasing slope of the transition enthalpy with increasing counter ion concentration, and Klump and Herzog give 1.51 Kcal/mbp , extrapolated from a series of ten different DNAs.(1984) Thus it seems that the order of magnitude of this difference is well documented.

Another observation to be discussed in this context is the variation of the transition enthalpy with the salt concentration.Raising the counter ion concentration from 1 mM to 200 mM causes the transition enthalpy of poly d(A-T) to rise to about 1.1 Kcal/mbp.On the other hand it is shown that the transition enthalpy decreases again when the sodium concentration is further increased to 1 M.The same holds for poly d(G-C) and there is no doubt that it will also hold for polynucleotides, which were not investigated in this respect. The slope of ΔH vs. log Na$^+$ is only marginally smaller for the latter polymer.(0.95 Kcal/mbp per order of magnitude change of the sodium concentration). Since both Tm and ΔH increase with increasing sodium concentration it is easy to calculate the temperature dependence of ΔH.It turns out that the transition enthalpy increases by 0.6 Kcal/mbp when the Tm increases for 10°C.

Let us now consider the increase of the transition enthalpy with increasing GC content in more detail. Viewing the geometrical overlay of the canonical base pairs it seems to be reasonable to assume that the vertical stacking of the four different base pairs (each base pair in two different orientations) is identical . The pitch of the helical secondary structure is, within the limits of experimental error sequence-independent for helical polymers in solution(Arnott,1970). At least the results of NOE-NMR-investigations show only one single sharp phosphate resonance(Marky,1981). Thus it is, as a first aproximation, reasonable to assume identical stacking enthalpy for the different base pairs. The increase of transition enthalpy is strictly correlated with the increase of %GC of a sequence. Since GC base pairs differ from AT base pairs by one extra H-bond the difference in transition enthalpy of 1.7 Kcal/mbp is related to the contribution of this H-bond to the stabilizing enthalpy of a GC base pair. Privalov has determined the enthalpy of a single H-bond in a biopolymer system to 1.5 Kcal/m H-bond, in good agreement with the value , extrapolated here.To subdivide the contribution of the H-bonds and the contribution of the stacking interactions is straightforward to subtract twice the enthalpy for a H-bond from the total amount of the stabilizing enthalpy of an AT base pair and three times the value of a H-bond from the total stabilizing enthalpy of a GC base pair.The difference amounts to 4.2 Kcal/mbp$_{AT}$ and 4.3 Kcal/mbp$_{GC}$ for the stacking enthalpy per base pair.

By an independent experimental approach we have determined the stacking enthalpy for a purine base (A) to yield 2.5 Kcal/m adenine, which was confirmed by Breslauer et al.(1977)for a similar system. Thus we can be confident about the order of magnitude of the stacking enthalpy.

To sum up the arguments of the deconvolution of the subdivision of the two basic contributions to the stabilizing enthalpy of a base pair within the helix structure it is justified to state that the two sources (stacking and H-bonding) contribute almost equally to the stability of a base pair. The interaction of the two polynucleotide strands cannot exclusively rely on the possible H-bonds between complementary base pairs. But the undisputed predominance of the H-bonds for the discrimination between the canonical base pairs remains , and it can be better

understood after this subdivision, for a pairing of improper bases leads to an unfavourable contribution , i.e. a decrease of stability of a given base pair.

To solve this problem of subdivision of different contributions to the integral stabilizing enthalpy on a quantitative basis we can try a different approach.If it is not for the different number of H-bonds of GC- and AT base pairs what else may serve to account for the enthalpy differences ?

We can calculate the stacking free enthalpy E from the equation

$$E = RT \ln \sigma$$

when we know the magnitude of the cooperativ parameter sigma. One way to evaluate sigma is from the average helix length at T = Tm by the following relation:

$$\langle n \rangle = \sigma^{-1/2}$$

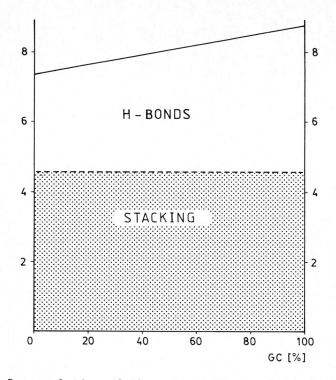

Fig.16: Deconvolution of the contributions of stacking and of H-
bonding to the stability of a double helical DNA sequence as function of the bulk sequence composition.

We have shown earlier that we have an independent approach to $<n>$
from the ratio of the van't Hoff enthalpy and the calorimetrically
determined enthalpy per base pair.
A   reasonable   value   for sigma ,   calculated   according   to   this
formalism is 16 x 10^4. The resulting stacking free enthalpy yields
-4.4 Kcal /mbp (- 18.4 KJ/mbp).   This value compares favourably to
a value ,published by Gruenwedel(1974) (- 5Kcal/mbp) for poly d(A-
T)   calculated   by a different formalism.   Go(1967)   introduced   a
statistical   mechanical   formalism ,   which relates the   degree   of
transition   to the transition enthalpy ΔH, to the   marginal   loop
size parameters alpha and beta   ,   and to the cooperativity sigma.

$$(d\theta/dT)_{T_m} = 0.177 \quad \Delta H/RT^2 (\alpha + \beta/ 2 \sigma)^{-1/2}$$

The   numerical   calculation for a polynucleotide helix   yields   the
change in stacking free enthalpy by stacking a H-bonded base   pair
on top of an existing helical stack, namely -20.9 KJ/mbp , a value
not far from the -18.4 KJ/mbp given above.
The   total   free enthalpy change per base pair   derived   from   the
experimental aproach as outlined above for 298 K amounts to -3.76
KJ   .   This value is in good agreement with other   free   enthalpy
values   based   on   a   similar   experimental   approach   ,   given   by
Privalov (1979) for T2 phage DNA (-5 KJ) and with the value given
by   Zimm(1964) (-4.2 KJ) ,derived from a   statistical   mechanical
approach.

We can demonstrate our evaluation of the different contributions in a comprehensive fashion.

Deconvolution of the Different Contributions to Helix Stability:

	$\Delta G°$ KJ/mbp	ΔH KJ/mbp	ΔS J/mbp
Integral value of all contributions	- 3.76	-32.2	-98.65
Stacking	-18.39	-17.56	+ 2.80
H-bonds ad hydro-phobic interactions	+14.63	-14.63	-101.45

The intergral value of the transition enthalpy is an experimental result. Since the helix to random coil transition is regarded as a reversible reaction also for DNAs (only kinetically slowed down) the thermodynamic formalisms and definitions are valid without restriction, i.e. ΔG equals zero at T = Tm. We can now calculate the free enthalpy change due to the transition under the assumption that the temperature dependence of ΔH is negligible as compared to the temperature dependence of ΔG. The free enthalpy of transition at room temperature is 3.76 KJ/mbp for calf thymus DNA. It is summed up from various contributions, including the stacking free enthalpy (-18.39 KJ) and the free enthalpy of base pairing via H-bonds and other contributions (+ 14.63).
The calculated entropy change due to stacking amounts to a positive value (+ 2.8o J/mbp), a reasonable positive change since we have to account for an increase of order in the system.
The third attempt to explain the increase of the enthalpy change as function of the net GC content follows a procedure introduced by Privalov and Ptitsyn(1967). This approach is focussed on the influence of the cation concentration on the transition enthalpy. The possibility that the increase of ΔH with temperature and the increase of Tm with temperature are unrelated cannot be dismissed. The increase of ΔH with the net GC content would be only apparent.
The property diagrams which show the ΔS-function derived from the polynucleotide data is completely superimposable over the corresponding diagram which represents this function , derived from the DNA measurements. As outlined before the course of the ΔS function reflects entirely backbone related properties due to gain in rotational freedom along the single bonds of the backbone and the N-glycosidic bond to the base, which are degenerated in the ordered helical structure. The ΔS value accounts also for the change in solvation upon going from the helical to the single stranded state and the release of cations during the denaturation process(Record,1968). Intuitively one would expect a much bigger contribution to the entropy change from the latter two changes, but the calculated ΔS value, derived from the Boltzmann approach and the experimentally observed value are so close that it is reasonable to assume that the model ,adopted to explain the entropy change, is correct.
There is an independent approach to decide whether the model assumption for the explanation of the entropy change is correct. Ross et al(1972) have measured the thermodynamic parameters for the association of a single strand to an existing double helix . The experimentally obtained entropy value is 13 cal per mol monomer , exactly half the value that the theory predicts for the

double strand. There are the corresponding entropy changes measured for the denaturation of triple stranded structures. The transition entropy per base for the triple strands is exactly 12.6, a value very close to the theoretical value, and finally the poly G system which forms a four-stranded helical structures .The denaturation can be measured by scanning microcalorimetry. The resulting transition entropy per subunit yields 12.5 Kcal. It is far from accidental that the transition entropy of the four-stranded helix relates to the transition entropy of the three-stranded ,the two-stranded and the single-stranded structures as 4:3:2:1. There is no better clue for the relation between the entropy change and the strandedness of the structure, or in other words , the entropy change is indeed completely independent from the base sequence.

The most rewarding result from all the calorimetric investigations of the polynucleotides and the different native DNAs is the complete ΔG function and its relation to the two most important parameters of the gene sequences, namely the net GC content of the sequence studied , and the counter ion contribution.On the basis of the ΔG values for a given sequence it will now be possible to calculate and compare the transition free enthalpy change for the order disorder reaction of the most interesting DNA sequences such as promoter sequences e.g.(von Hippel,1974).

To elucidate this advantage let us view the ΔG surface more closely:

At the low <GC> (<10%) and low sodium concentration (1 mM) the double helix is thermodynamically not the most stable structure any more,i.e. poly d(A-T) cannot be in the helix conformation.This prediction ,which follows from our state diagram is readily supported by the experimental results of Gruenwedel(1974).

Further lowering of the counter ion concentration to 0.1 mM Na⁺ leads to the destabilization of the sequences with a higher net GC content.Actually the normal eucariotic DNA sequences are at the limit of thermal stability at room temperature at such low sodium concentrations. Further lowering the counter ion concentration to a value of 0.01 mM requires a net GC content of 85% to retain a stable helical structure at room temperature, a value out of the range of all known native DNA sequences.

By shifting the reference temperature to body -temperature we can also influence the stability range of the different sequences.Raising the reference temperature to 37°C for a solution of 1 mM NaCl shifts the stability-limit with respect to the %GC of the sequences to 20%. Rising the reference temperature further to 60°C requires either 100 mM Na⁺ for poly d(A-T) or , at 1 mM Na⁺, 85% GC for a sequence to stay in the B-DNA conformation.

The presence of physiological salt concentration in the external environment of a eukaryotic DNA on the other hand allows elevation of environmental temperature to almost 100°C before the DNA denatures.

In the past there were several attempts to calculate the variation of the helix energy as function of the next nearest neighbour frequencies.De Voe and Tinoco took the sum of dipole/dipole ,dipole/ induced dipole, and London-energies to predict the stabilities of random and non-random sequences .All the calculated values for both alternatives fall on a straight line. Since for the sake of simplicity in the theoretical calculations solvent effects were completely dismissed, it is not surprising that De Voe's numerical values do not coincide with our experimental values.The differences between the two data sets shows that the calculated values are about a factor of ten higher than the model-

free data, obtained by calorimetry. The slopes, however, are
surprisingly similar in both plots.In the future the experimental
values should be utilized to scale the theoretical calculations
to obtain better numerical values.

3.7   Concluding Remarks

The progress of adiabatic biocalorimetry in the last ten years has
enabled us to gain insight into a new field of nucleic acid-
transitions.Besides the canonical B-DNA to coil transitions of
linear DNA sequences we have evaluated the conformational changes
of a series of new secondary structures and gained access to a
complete thermodynamic data set for all linear sequences as a
function of two fundamental system parameters, namely the net GC
content of the given sequence and the counter ion concentration.
This data set is tested to allow comparison of known thermodynamic
data of sequences taken from the literature and will allow
us to predict properties of new sequences , exclusively based on
the knowledge of their primary structure.
In this survey of new approaches to evaluate the thermal stability
of nucleic acids we can discuss now on a quantitative basis :
(i) the stability range of left-handed structures,
(ii) the transition of right-handed to left-handed conformations
for DNAs, and for the first time, for RNAs,
(iii) the full range of order/order transitions such as
disproportionations, strand-displacements, and helix inversions,
(iv) the denaturation of supercoils,
(v) and give the complete thermodynamic state functions for the
synthetic polynucleotides and the native DNAs.
The further expansion of this thermodynamic library has to include
the hairpins, the different forms of loops, the presence of odd
bases and non-B conformational states.Hopefully this expanded
themodynamic library will provide us with an experimentally based
data set to predict the complete secondary structure of any DNA
molecule based purely on its primary sequence. Such a predictive
ability will be an important tool to understand sequence specific
structural domains , which are assumed to serve as important binding sites for gene regulating proteins.

Acknowledgements

The Deutsche Forschungs-Gemeinschaft and the Fonds der Chemischen
Industrie have generously supported the investigations on which a
great part of this review is based.

Aboul-ela,F.,Koh,D.,Martin,F.,and Tinoco,I.Jr.(1985)Nucl.Acid Res.
13, 4811-4824
Ackermann,Th.,and Rüterjans,H.(1964)Ber.Bunsenges.Phys.Chem.68,85
Arber,W.,and Morse,M.(1965)Genetics,51,137-48
Arnott,S.(1970) in "Progress in Biophys.Mol.Biol." Buttler,J., and
Nadel,D. ed Pergamon Press,Oxford
Azorin,F.,Nordheim,A.,and Rich,A. (1983) EMBO J.2,649-656

Behe,M.,and Felsenfeld,G,(1981)PNAS 78,1619-23
Behe,M.,Szu,S.,Charney,E.,and Felsenfeld,G.(1985) Biopol.24,289
Blake,C.(1978)Nature 273,267-68
Blake,R., and Lefoley,S.(1978) Biochim.Biophys.Acta 518,233-246
Bloomfield,V.,Crothers,D., and Tinoco,I.Jr(1974)in "Physical Chem.
of Nucleic Acids",Harper and Row,New York

142

Breslauer,K., and Sturtevant,J.(1977)Biophys, Chem. 7,205-209
Breslauer,K.(1981)Biochemistry,21,437-44
Breslauer,K.(1986) in"Chemical Therm. Data for the Biol.Sci."Hinz,
H. ed.
Breslauer,K.,Frank,R.,Blocker,H.,and Marky,L.(1986)PNAS,83,3746-5
Bunville,L.,Geiduschek,E.,Rawitscher,M.,and Sturtevant,J.(1965)
Biopolymers 3,213-40

Cantor,C.,and Schimmel,P.(1980)in "Biopys.Chemistry,Part I"Free-
man,SanFrancisco
Cech,Th.(1981) Biochemistry 20,1431-37
Chaires,J.and Sturtevant,J.(1986)PNAS 83,5479-5483
Chamberlin,M,and Patterson,D.(1965)J.Mol.Biol.12,410-28
Crothers,D.,and Zimm,B.(1964) J.Mol.Biol.9.1-9

DeVoe,H., and Tinoco,I.Jr.(1962) J.Mol.Biol.4,500-517
Dickerson,R., and Drew,H.(1981) J.Mol.Biol. 151,535-556
Dickerson,R.,Drew,H.,Conner,B.,Wing,R.,Fratini,A.and
Kupka,M.(1982),Science 216,475-485

Frank-Kamenetzkii,M.,Luskin,A.,and Anshelevich,V.,(1985)J.Biomol.
Structure and Dynamics 3,35-42

Gruenwedel,D.(1974)Biochim.Biophys.Acta,340,14-26
Gray,D., and Tinoco,I.Jr.(1970) Biopolymers 9,226-29
Go,M.(1967) J.Phys.Soc.Japan 23,597-608

Haniford,D.,and Pulleyblank,D.(1983) Nature,302,632-34
Herzog,K.(1983) Thesis, Univ. Freiburg ,W.-Germany
Hinz,H.,Filimonov,V.,and Privalov,P.,(1977)Eur.J.Biochem.72,79-86
Hinz,H.,in "Biochem.Thermodynamics,(1979)Jones,M,ed.Elsevier,Publ.
von Hippel,P. Revzin,P.,Gross,C., and Wang,A.(1974) PNAS 71, 4808-
4821
Hirose,T.,Crea,R.,and Itakura ,K.(1978)Tetrahedron Let.28,2449-52

Irikura,K.,Tider,B.,Brooks,B.,and Karplus,M.,(1985)Science 229,571
Ivanov.V.,and Mingat,T.(1981) Nucl.Acid Res.9,4783-4798
Ivanov,V.,Krylov,D.,and Minyat,E.(1985)J.Biomol.Struc.Dyn. 3,43-55

Jovin,T.,McIntosh,and Arndt-Jovin(1983)J.Biomol.Struct.and Dynam.
1,2157

Kallenbach,N.(1968)J.Mol.Biol.37,445-456
Klump,H.(1975) Biochim.Biophys.Acta 383,117-122
Klump,H.,and Burkart,W.(1977)Biochim.Biophys,Acat 475,601-04
Klump,H.(1978)Ber.Bunsenges.Phys.Chem.82,805-809
Klump,H.,Beauvais,J.,and Devauchelle,J.(1983)Arch.Virology,75,269
Klump,H.,and Herzog,K.(1984)Ber.Bunsenges.Phys.Chem.88,20-24
Klump,H.,(1985)Thermochim.Acta 85,457-63
Klump,H.,and Löffler,R.(1985)Hoppe SeylersBiol.Chem.366,345-53
Klump,H.(1986)FEBS Letters 196,175-79
Klump,H.(1986) Ber.Bunsenges.Phys.Chem.90,444-47
Klump,H.,and Jovin,Th.(1987) Biochemistry in press
Klump,H.and Niermann,T.(1987)Biophys.Chem.submitted
Klump,H.(1987) in "Protein Structure and Stability and Related
Subjects" Twardowski,J. ed. Reidel Publ.
Klump,H.(1987)Ber.Bunsenges.Phys.Chem.91,206-11
Kurnit,B.,Shafit,B., and Maio,J.(1973)J.Mol.Biol. 81,273-284

Lathe,P.,Majunder,K.,and Bramachari,S.(1983)Curr.Sci. 52,907-910

Malfay,B.,Rousseau,N.,and Leng,M.(1982)Biochemistry 21,5463-67
Manning,G.,(1972)Biopolymers 11,937-951
Marky,L., Patel,D., and Breslauer,K.(1981)Biochemistry 20,1427-31
Marky,L.,and Breslauer,K.(1982)Biopolymers,21,2188-94
Marmur,J., and Doty,P.(1962) J.Mol. Biol. 5,109-119
Nelson,J.,Martin,F.,and Tinoco,I.Jr.(1981)Biopolymers 20,2509-31
Neumann,E.,and Ackermann Th.(1969)J.Phys.Chem. 73,2170-78
Nordheim,A.Lafer,E.,Peck,L.,Wang,J.,Stollar,B.,and Rich,A.(1982)

Owen,R., Hill,L., and LapangeS.(1969) Biopolymers 7, 503-516

Patel,D.,Kozlowski,S.,Marky,L.,Broka,C.,Rice,J.,Itakura,K.,and
Breslauer,K,(1981)Biochemistry 21,437-44
Pohl,F.,and Jovin,Th.(1972)J.Mol.Biol. 67,375-96
Pohl,F.(1974)Eur.J.Biochem. 42,495-504
Pohl,F. (1976) Nature 260,365-66
Privalov,P.,Kafiani,K.,and Monaselidze,D.(1965)Biophizika   10,393-
98
Privalov,P.,Ptitsyn,O.,and Birshtein,T.(1967)Biopolymers 10,513-22
Privalov,P.(1974)FEBS Letters Supl. 40,140-149
Privalov,P.(1975) J.Chem Thermodynamic 7,41-47
Privalov,P.,Plotnikov,V., anf Filimonov,V.(1975)J.Chem.Therodyn.
7,41-47
Privalov,P.(1978)in "Chemical Thermodyn."Rouquerol,J.,and Sabbah,R.
ed. Pergamon Press.Oxford

Rawitscher,M.,Ross,Ph., and Sturtevant,J.(1963) J.Am.Chem.Soc.85
1915-1923
Record,Th.(1967) Univ.California Berkeley
Record,Th. and Lohman,T.(1978)Biopys.Chem.9, 71-78

Relay,M.,Maling,B.,and Chamberlain,M.(1966)J.Mol.Biol.20,359-389
Rich,A.,(1982) in"Structure,Dynamics, Interaction, and Evolution"
Helene,C. ed.Reidel Publ.Dortrecht
Rich,A. Nordheim,A.,and Wang,A. (1984)Ann.Rev.Biochem. 53,791-846
Ross,Ph., and Scruggs,R.(1965)Biopolymers 3,491-498
Roy,K.,and Miles,T.(1983)Biochem.Biophys.Res.Comm. 115,100-105
Rüterjans,H.(1965) Thesis,Univ.Münster,W.-Germany

Ornstein,R., and Fresco,J.(1983) Biopolymers 22, 1979-200
Saenger,W.,Hunter,W.,and Kennard,O.(1986) Nature,324,385-88
Savoie,R.,Klump,H., and Peticolas,W.(1978)Biopolymers 17,1335-1345
Schildkraut,C., and Lifson,S.(1965) Biopolymers 3,195-203
Seidel,A.(1985) Thesis,Univ.Regensburg,W.-Germany.
Shiao,D.,and Sturtevant,J.(1973) Biopolymers 12,1829-36
Sinanoglu,O., and Abdulnur,S.(1964)Photochem.Photobiol.3,333-339
Singleton,C.,Kilpatrick,A.,and Wells,R.(1984)J.Biol.Chem.259,1963
Soumpasis,D.,and Jovin,Th.(1986)in "Nucleic Acids and Molecular
Biology,Eckstein F. and Lilley,D. ed. Springer,Heidelberg, FRG
Sturtevant,J.,and Geiduschek,E.(1969)J.Am.Chem.Soc.80,2911
Szu,S.,and   Charney,E.(1985)in "Mathematics and Computers   in
Biochemical Application,"Eisenfeld,D.,and Delis,C.ed.Elsevier,Publ.

Tinoco,I.Jr.,Uhlenbeck,O., and LevineM.(1971) Nature 230,362-367
Tinoco,I.Jr.,Bohrer,P.,Dengler,B.,Levine,M.,Uhlenbeck,O.,Crothers,
D.,and Gralla,J.(1873)Nature New Biol.246,40.41

Uhlenbeck,O.,Bohrer,P.,Dengler,B., and Tinoco,I.Jr.(1973)J.Mol.Biol.
73, 483-496

Vogel,A.,(1975) in "Textbook of Quantitative Inorganic Chemistry"
Longmans,London

144

Watson,D., and Crick,F.(1953) Nature 171, 737-39
Winnacker,R.(1985) in "Genes and Klones",Springer,Heidelberg,FRG.

CHAPTER 4

ANALYSIS OF ALLOSTERIC MODELS FOR MACROMOLECULAR REACTIONS

STANLEY J. GILL, CHARLES H. ROBERT & JEFFRIES WYMAN

4.1   INTRODUCTION

In the early 1900s Christian Bohr, K. A. Hasselbalch and August Krough (1904) reported the first accurate measurements of oxygen binding in dog blood. The oxygen binding curve did not follow the expected curve of simple binding to a single site, namely a rectangular hyperbola, but traced an S-shaped curve that rose slowly at low pressures, became steeper at intermediate pressures, and levelled off at saturation pressures. Such a sigmoidal curve has since been recognized as a hallmark of positive cooperativity, where the presence of bound ligand assists the binding of subsequently added ligand.[1]

What is the molecular origin of cooperativity? In the case of hemoglobin a surprising observation was made in 1938 by Felix Haurowitz, who found that deoxygenated Hb crystals shattered when exposed to oxygen. This implied that a marked structural change occurs upon oxygenation. It took over twenty years of work by Perutz and his collaborators (1960), employing x-ray structural measurements, to establish the nature of this structural change. Other protein crystals have since been found to break upon equilibration with ligand. These phenomena have given us a physical image of a conformational "switch" involved in the function of these macromolecules.

Sometime before the details of the x-ray structure studies were available, Pauling (1935) proposed a molecular model in which the ligand affinity of a given site is modified by ligand binding to adjacent sites by simple geometrical interaction rules. Wyman (1948) proposed that ligand-linked structural changes could explain the cooperative binding properties in Hb found by Bohr and his colleagues. The connection was amplified a few years later by Wyman and Allen (1951). The cooperativity of the oxygen binding could be explained by the shift in the equilibria of different structural forms upon oxygenation.

The role of conformational perturbations in enzyme control was an immediate extension of these concepts. The term "allosteric" was introduced by Monod and Jacob (1961) to describe the effect of binding an activator (or inhibitor) at one region of an enzyme upon the catalytic activity or substrate affinity at another region. The term was given more definite structural meaning with the proposal (Monod,

---

[1] John Edsall has traced the historical origins of the concept of allosterism (Edsall, 1980) and has gathered many details of the history of blood and hemoglobin research (Edsall, 1972).

et al, 1963) that the binding of ligand induced conformational changes. These are described as "allosteric transitions". The general effects of ligands on a conformational equilibrium constant was developed shortly thereafter (Wyman, 1964).

For the most part these ideas appeared quite esoteric. That changed with the introduction of a simplified two-state allosteric model which was presented by Monod, Wyman and Changeux (1965). Two classes of allosteric effects were recognized: "homotropic" effects as interactions between identical ligands, and "heterotropic" effects as interactions between different ligands. An example of heterotropic interaction is the effect of oxygen binding on hemoglobin's binding affinity for a different ligand, such as carbon dioxide. The presence of CO_2, also studied by Bohr, decreases the binding of oxygen to hemoglobin at a given oxygen partial pressure, and so is described as a negative heterotropic effect. These effects had been studied from a purely thermodynamic viewpoint for many years, but a model was now at hand to give a description based on plausible molecular events. The advantage of the simple Monod, Wyman, and Changeux (MWC) allosteric model over that proposed by Pauling earlier (1935) was the availability of structural change information. A different model, also based on conformational effects, was developed by Koshland, Nemethy, and Filmer (1966) and termed the induced fit model or KNF model. This model uses a combination of structural knowledge and hypothesized interaction rules (like those of Pauling) to characterize the binding properties of multi-site molecules. Both the MWC and KNF models are but special cases of a parent allosteric model (Wyman, 1972).

In the last fifteen years there has been strong effort to link the detailed structural knowledge gained by x-ray crystallography and spectroscopic studies with a thermodynamic description of hemoglobin. Perutz (1970, 1976, 1979) proposed a mechanism of cooperativity based on structural information and this has led to the incorporation of such features into statistical thermodynamic models developed by Szabo and Karplus (1972) and by Johnson and Ackers (1982). Extensive binding studies have been conducted, in particular by Imai (1982) and Ackers (e.g., Mills et al, 1976; Chu, et al 1984). Especially detailed functional studies by Smith and Ackers (1985) have uncovered a challenging twist to hemoglobin's function with the elucidation of "selection rules" in the reactivity of different sites at different degrees of saturation. Not surprisingly, the simple allosteric models are not able to explain such observations in detail. However extension of the simple picture to include hierarchies of allosteric processes (Wyman, 1984; Gill, et al, 1986; Decker, et al, 1986; DiCera, et al 1986) accounts for many of the complexities observed.

While the study of hemoglobin and other respiratory proteins remains a major focus and source of structural and functional information, the success of even the simple allosteric models for biological regulation processes is widespread in macromolecular systems. In this chapter we wish to examine the general properties of the allosteric concept. A basic framework is described which facilitates the translation step between a specific model and observed properties of a macromolecular system, such as extent of binding, heat of reaction, and heat capacity changes with temperature. Specific examples are then used to demonstrate each of these ideas.

4.2   GENERAL FORMULATION OF BINDING (Gill, et al, 1985)

The characterization of chemical ligand binding reactions and of temperature effects play key roles in the evaluation of the thermodynamic characteristics of macromolecular regulation. Consider a generalized macromolecular reaction

$$M_o \xrightarrow{\Lambda_1} M_1 \tag{1}$$

where Λ_1 is an equilibrium constant. In this general formulation the transition (1) can comprise changes in enthalpy and volume induced by ligand binding or by temperature or pressure shifts, as well as actual change in ligand saturation of the macromolecule (chemical binding). The purpose of writing (1) in this way is to allow us to describe physical and chemical binding in parallel terms. For the reaction of macromolecule in the reference state (M_o) to any state k we can write the $\Lambda(T,p,\mu_x)$:[2]

$$\Lambda_k = (\Lambda_k)_o \; e^{-(\Delta \bar{H}_k /R)(\frac{1}{T} - \frac{1}{T_o})} \; e^{-(\Delta \bar{V}_k /RT)(p-p_o)} \; e^{(\Delta \bar{X}_k /RT)(\mu_x - \mu_{x_o})} \tag{2}$$

in which the bar notation denotes the quantity on a per macromolecule basis. Here $\bar{X}_k$ is the number of ligand molecules bound to the macromolecule in state k. At equilibrium the concentration of species M_k is given by $[M_k] = \Lambda_k [M_o]$.[3]

A variety of relations describing the linkage between chemical ligands X, Y, etc. (each additional chemical ligand enters eq. (2) by an exponential factor like that for X) and the physical ligands heat and volume can be obtained from the **binding partition function** (Wyman, 1965; Hill, 1960) for the macromolecular system. This function represents the sum of the species concentrations relative to a reference species (or set of species), and will be denoted by Q:

$$Q = \sum_{k=0}^{n} \frac{[M_k]}{[M_{ref}]} = 1 + \frac{[M_1]}{[M_{ref}]} + \frac{[M_2]}{[M_{ref}]} + \cdots + \frac{[M_n]}{[M_{ref}]} \tag{3}$$

The choice of reference component is arbitrary. For descriptions of ligand binding,

---

[2] In the case where the heat capacity change $\Delta \bar{C}_{hi}$ is not zero an additional term,

$$(T_o / T)^{-\Delta \bar{C}_{hi}/R} \; e^{(\Delta \bar{C}_{hi} T_o /R)(\frac{1}{T} - \frac{1}{T_o})}$$

needs to be included in eq. (2).

[3] In writing this equation in terms of the ratio of concentrations we are strictly absorbing the activity coefficients of M_k and M_o into the constant Λ_k .

the unligated macromolecule is the most useful reference form. With an allosteric macromolecule this state may consist of more than one species, resulting in a sum for the denominator of equation (3). If we take the reference species as the form M_o the employment of the mass law relationships enables one to rewrite Q in terms of the equilibrium constants.

$$Q = \sum_k \Lambda_k \tag{4}$$

The thermodynamic properties of each reaction may be evaluated by partial differentiation of $\ln \Lambda_k$ with respect to the appropriate independent variables of the system:

$$\frac{\partial \ln \Lambda_k}{\partial \mu_x} = \frac{\Delta \bar{X}_k}{RT} \tag{5}$$

$$\frac{\partial \ln \Lambda_k}{\partial (1/T)} = \frac{-\Delta \bar{H}_k}{R} \tag{6}$$

$$\frac{\partial \ln \Lambda_k}{\partial p} = \frac{-\Delta \bar{V}_k}{RT} \tag{7}$$

In order to obtain the total amount of ligand, enthalpy, or volume change which occurs upon taking the macromolecular system from its reference state to any general state of the system, one differentiates $\ln Q$ with respect to the appropriate independent variable. For the case of chemical ligation this gives:

$$\frac{\partial \ln Q}{\partial \mu_x} = \frac{1}{Q} \sum_k \frac{\partial \ln \Lambda_k}{\partial \mu_x} \Lambda_k = \frac{1}{Q} \sum_k \Delta \bar{X}_k \Lambda_k = \bar{X} - \bar{X}_o \tag{8}$$

The term $\bar{X}_o$ represents the amount of X ligand bound to the reference state, usually chosen to equal 0. The fraction of macromolecules of the type represented by M_k (= MX_k) is given by Λ_k/Q. This expression allows one to see that the last term of equation (8) is obtained by taking the weighted sum of the $\Delta \bar{X}_k$ values over all species.

Similarly the amounts of enthalpy and volume changes involved in going from the reference state to a general state of the system, $\bar{H} - \bar{H}_o$ and $\bar{V} - \bar{V}_o$, are obtained by differentiation of $\ln Q$ with respect to 1/T and p. The inclusion of $\bar{H}_o$ and $\bar{V}_o$ is necessary because the enthalpy and volume for the M_o reference state are not zero. In summary we write:

$$\bar{X} - \bar{X}_o = RT \frac{\partial \ln Q}{\partial \mu_x} \tag{9}$$

$$\bar{H} - \bar{H}_o \quad = -R \ \frac{\partial \ln Q}{\partial(1/T)} \tag{10}$$

$$\bar{V} - \bar{V}_o \quad = - RT \ \frac{\partial \ln Q}{\partial p} \tag{11}$$

The parallel form of these expressions serves to emphasize how chemical and physical binding phenomena (namely heat and volume changes associated with macromolecular reactions) can be considered in the same general context. We will now place a variety of analyses into this general framework, starting with the phenomenological approach and leading into the allosteric picture.

4.3 PHENOMENOLOGICAL DESCRIPTION OF CHEMICAL LIGAND BINDING PROCESSES
The Adair scheme and the binding polynomial

The binding of ligand by a macromolecule is described in a purely phenomenological way by the Adair scheme (Adair, 1925). In this approach a separate binding constant is applied to each specifically ligated species, regardless of whether or not the individual-site binding affinities are equal when no ligand is present. Such a phenomenological description of a ligand binding process is an important standard against which models can be compared (Connelly, et al, 1986). Usually a model is not useful if the number of parameters to be determined from experiment is more than the number of Adair constants required to describe the system. We will here reformulate the generalized binding partition function in terms of the phenomenological mass law equilibria.

For a macromolecule with t binding sites for one type of ligand X a linear array of species can be represented

$$M \ \rightarrow \ MX \ \rightarrow \ MX_2 \ \rightarrow \ \dots \ \rightarrow \ MX_t$$

For the purpose of analyzing chemical ligation experiments, the generalized reaction shown in eq. (1) is usually written so that the overall reaction for adding iX ligands to the unligated macromolecule M_o to produce the species MX_i is

$$M \ + iX \ \overset{\beta_i}{\rightarrow} MX_i \qquad , \ i = 1 \text{ to } t \tag{12}$$

The generalized equilibrium constant Λ_i from eq. (2) is then seen to have the conventional form $\beta_i x^i$ if the ligand activity x is explicitly brought out of the ligand chemical potential $\mu_x = \mu_x^o + RT \ln x$ (and noting that $\Delta\bar{X}_i = i$ if the reference species is chosen to be the unligated macromolecule), leaving

$$\Lambda_i \ = \ \beta_i \ x^i \ = \ \frac{[MX_i]}{[M]} \tag{13}$$

Thus β_i is a function of T and p only. Again the activity coefficients of the macromolecular species are absorbed into the equilibrium constant.

In terms of the β's, Q in the general eq. (4) can be expressed as a polynomial P of degree t in the ligand activity x:[4]

$$P = \sum_{i=0}^{t} \beta_i x^i = 1 + \beta_1 x + \beta_2 x^2 + \dots + \beta_t x^t \qquad (14)$$

This expression is called the **binding polynomial**. An equivalent form of eq. (14) can be written for any thermodynamic model of a ligand binding process, allowing the values of the phenomenological mass law binding constants (the β_i) to be of aid in determining the parameters of a model formulation.

The average saturation of the macromolecule with ligand X, the expression for the ligand binding curve, can be written by evaluating the derivative relation eq. (9):

$$\bar{X} = \frac{\partial \ln Q}{\partial \mu_x} = \frac{\partial \ln P}{\partial \ln x} = \frac{\beta_1 x + 2 \beta_2 x^2 + \dots + t \beta_t x^t}{1 + \beta_1 x + \beta_2 x^2 + \dots + \beta_t x^t} \qquad (15)$$

Since the unligated macromolecule is chosen as the reference state, the polynomial is unity when no ligand is present; thus the term $\bar{X}_o$ is zero.

A convenient characterization of the binding curve is its median ligand activity (see Gill, 1979). This is defined as that activity x_m at which the area below the binding curve (plotted as $\bar{X}$ vs. the logarithm of x) from zero to the median activity is equal to the area above the binding curve (bounded by $\bar{X}=t$) from there to infinite ligand activity. In this case the median activity is found to be given by

$$\beta_t (x_m)^t = 1 \qquad (16)$$

It can be seen by comparison to the binding polynomial (eq. 14) that the relative concentrations of the unligated and fully ligated forms are equal at the median ligand

---

[4] The same information can be described in terms of stepwise reactions defined by the following processes:
$$MX_{i-1} + X \rightarrow MX_i, \quad i = 1 \text{ to } t \qquad (\alpha)$$
The stepwise reaction equilibrium constant, K_i, is given as
$$K_i = \frac{[MX_i]}{[MX_{i-1}] x} \qquad (\beta)$$
Either of these formulations, in terms of overall or stepwise reactions, describes the ligation reactions of the system. The overall (β_i) or stepwise (K_i) equilibrium constants are known as the Adair constants. The relation between the overall and stepwise constants is
$$\beta_i = K_1 K_2 K_3 \dots K_i \qquad (\gamma)$$
or
$$K_i = \frac{\beta_i}{\beta_{i-1}} \qquad (\delta)$$

activity. The standard free energy of saturating the macromolecule with ligand X is related to the median activity through eq. (16):

$$\Delta G^o_{sat} = -RT \ln \beta_t = t\,RT \ln x_m \qquad (17)$$

This free energy is seen by eq. (17) to be represented by the rectangular area from the logarithm of the standard state ligand activity (zero) to the logarithm of the median activity. This area can equivalently be determined from the area above the binding curve from zero to infinity minus the area below the binding curve from negative infinity to zero, allowing the median, and thus β_t and ΔG^o_{sat}, to be determined graphically.

A useful form of the binding polynomial results when the x activity is defined relative to the median x activity. This is equivalent to choosing the standard state › activity to be the median value. The change of variable from x to x′ is given by x′ = x/x_m so that the binding polynomial P_m is given as

$$P_m = 1 + B_1 x' + B_2 x'^2 + \ldots + B_{t-2} x'^{t-2} + B_{t-1} x'^{t-1} + x'^t \qquad (18)$$

where the last coefficient is unity by equation (16), and $B_i = \beta_i / x_m^i$. The form of this equation allows one to evaluate whether the binding polynomial is symmetrical, which will be the case when $B_i = B_{t-i}$ for all coefficient pairs i,t-i. A symmetrical binding polynomial gives rise to a symmetrical binding curve (Wyman, 1948). Evaluation of the symmetry can be useful in determining the applicability of certain models.

As mentioned earlier, with the binding partition function we can examine enthalpy changes due to ligation analogous to the way we examine ligation state changes. The average enthalpy of the macromolecular system is expressed by eq. (10) (with Q = P):

$$\bar{H} - \bar{H}_o = -R\,\frac{\partial \ln P}{\partial(1/T)} = \frac{\beta_1 x\,\Delta H_1 + \beta_2 x^2 \Delta H_2 + \ldots + \beta_t x^t \Delta H_t}{P} \qquad (19)$$

The temperature dependence of this equation is hidden in the equilibrium constants, so that at constant T the enthalpy of the system is formally a function of the ligand activity. The binding constants at a given temperature can be obtained from ligand saturation experiments (binding curves) employing eq. (15). Equation (19) then enables one to examine, at the appropriate temperature, titration calorimetry data or batch calorimetry data at varying ligand activities in order to resolve the enthalpy changes associated with the ligation steps. An example of the use of the above procedure in applying an allosteric model to hemocyanin oxygen binding is outlined in the Applications section.

4.4 INDEPENDENT-BINDING MODEL

With the previous analysis we applied the concepts of section 4.2 in a model-independent fashion to the binding reactions of a macromolecule. In this section we will apply those concepts to one model of macromolecular function, the independent binding model. The simplest a priori model for the functional organization of macromolecule is the assumption of functional independence of its composite domains. With respect to chemical ligand binding, this translates to independent ligand binding sites. In the case of the physical ligand enthalpy the interpretation is that the various parts of the macromolecule melt independently. For completeness we will outline here the independent binding model in the generalized context.

First we will consider a two-identical-sites macromolecule. Again the sites may function as chemical ligand binding sites or as independent melting domains ("enthalpy binding sites"). For simplicity we will consider only a single transition for each site, corresponding to "unligated and ligated" or "native and denatured", etc. The transition for a site is written as in the general case (section 4.2):

$$A \quad \overset{\lambda}{\rightarrow} \quad B$$

The equilibrium constant λ for a site or domain is a function in general of T, p, and μ_x and has a form analogous to that of Λ given in eq. (2). Since the domains are independent, the binding partition function Q for this system can be written by explicitly indicating each of the species, i.e., the macromolecule in each combination of the two domains in either form relative to the reference component.

$$Q = [M_{AA}] / [M_{ref}] + [M_{AB}] / [M_{ref}] + [M_{BB}] / [M_{ref}] \tag{20}$$

Choosing as the reference component the macromolecule in form AA (composed of a pair of A domains), the equilibrium concentrations of these species can be represented

$$[M_{AA}] / [M_{AA}] = 1 \, , \qquad [M_{AB}] / [M_{AA}] = 2\lambda \, , \qquad [M_{BB}] / [M_{AA}] = \lambda^2 \tag{21}$$

where the factor of two indicates that there are two ways, AB and BA, for the two-domain macromolecule to exist in a configuration with one domain in state A and one in state B. Substituting these quantities into eq. (20) we see that Q is a quadratic in λ allowing the factorization

$$Q_{2 \text{ identical}} = (1 + \lambda)^2 \tag{22}$$

Extending to t identical independent sites, the form of eq. (22) is

$$Q_{t \text{ identical}} = (1 + \lambda)^t \tag{23}$$

with $\lambda(T,p,\mu_x)$ again the equilibrium constant for the generalized reaction of a single site.

When the sites are nonidentical Q is a product of the terms in parentheses, one term for each site:

$$Q_{t \text{ nonidentical}} = \prod_{j = 1}^{t} (1 + \lambda_j)^t \qquad (24)$$

where now the λ_j serve for the reaction of site j.

If a second transition is assumed to occur for each site (domain), an additional term (e.g. ω_j) may be added into the parenthetical expression. In the case of the physical ligand enthalpy this would mean that the domain can exist in an additional enthalpy state, perhaps an intermediate in the melting process. In the case of a chemical ligand this means that an additional ligand binds to the same site, exclusive of the first. This competitive binding has been termed "identical linkage" (Wyman, 1968). It is helpful to define the quantity in parentheses as the individual binding partition function for each site (domain), denoted q_j.

For one assumed
domain transition: $\qquad q_j = 1 + \lambda_j$ $\qquad\qquad\qquad\qquad (25)$

For more than one: $\qquad q_j = 1 + \lambda_j + \omega_j + \ldots$ $\qquad\qquad (26)$

The independent site model forms a part of the MWC allosteric model. It has also been used extensively in the analysis of thermal transition data, obtained by scanning calorimetry (Privalov, 1979) as will be examined in the Applications section.

4.5 THE GENERAL ALLOSTERIC MODEL

We now wish to apply the general concepts of section 4.2 to the allosteric models. Again we will examine both chemical ligand binding and the binding of the physical ligand enthalpy. In the general allosteric model (Gill, et al, 1985) the macromolecule M has q + 1 conformers which will be designated by an h subscript having value of o,1,2...q. The macromolecule in each conformation h can bind a maximum of t_h ligands. Thus the zeroth form has t_o binding sites, etc.. $M_h X_i$ denotes the hth conformer with i ligands X bound. The general array of all species is depicted in Figure 4.1. The horizontal arrows denote conformational equilibria and the vertical arrows denote chemical ligand binding reaction. Each column in the array describes the

$$
\begin{array}{cccc}
M_{oo} & \rightarrow\; M_{1o} \;\rightarrow \ldots \rightarrow\; M_{ho} \;\rightarrow \ldots \rightarrow & M_{qo} \\
\downarrow & \downarrow \qquad\qquad\qquad \downarrow & \downarrow \\
M_o X & M_1 X \qquad\qquad\quad M_h X & M_q X \\
\downarrow & \downarrow \qquad\qquad\qquad \downarrow & \downarrow \\
M_o X_2 & M_1 X_2 \qquad\qquad M_h X_2 & M_q X_2 \\
\vdots & \vdots \qquad\qquad\qquad \vdots & \vdots \\
\downarrow & \downarrow \qquad\qquad\qquad \downarrow & \downarrow \\
M_o X_{t_o} & M_1 X_{t_1} \qquad\quad M_h X_{t_h} & M_q X_{t_q}
\end{array}
$$

Fig. 4.1. The general allosteric model.

successive reactions for addition of X ligand to a given allosteric form h, and it should be noted that the columns need not contain the same number of species, i.e., there may be a different number of binding sites t_h for each form h.

The reactions within the array are described by appropriate mass action laws, and the entire system of species is considered to be in equilibrium with ligand X at a given activity. The summation of the general binding partition function (eq. 4) can be performed to more explicitly reflect the square nature of the species array:

$$Q = \sum_k \Lambda_k = \underset{\text{columns}}{\sum_{h=o}^{q}} \underset{\text{rows}}{\sum_{i=1}^{t_h}} \Lambda_{hi} \qquad (27)$$

The form of the generalized reaction equilibrium constant Λ_{hi} ($= \Lambda_k$) for each element of this array can be factored into two parts-- a conformational part Λ' which has no chemical ligand dependence and a chemical ligation part Λ'' which applies only to the ligation process of that conformation. We will make our reference species the species represented by the upper left-hand corner of Figure 4.1, i.e. species M_{oo}, the unligated macromolecule in conformation h=0. The simplest choice for Λ' is the unligated conformational transition so that we can write for any reaction $M_{oo} \rightarrow M_{hi}$

$$\Lambda_{hi} = \Lambda'_{ho}(T,p) \cdot \Lambda''_{hi}(T,p,\mu_x) \qquad (28)$$

By convention we denote the equilibrium constant for the hth unligated form (Λ'_{ho}) by the symbol L_{ho}. The simple chemical-ligand-binding part (Λ'') of the equilibrium constant for the species hi will be written β to reflect its similarity to the reaction constants in the Adair formalism, i.e.,

$$\Lambda''_{hi} = \beta_{hi} x^i = \frac{[M_h X_i]}{[M_{ho}]} \qquad (29)$$

The binding partition function Q can now be written

$$Q = \sum_{h=o}^{q} L_{ho} \sum_{i=1}^{t_h} \beta_{hi} x^i \qquad (30)$$

The rightmost sum of eq. (30) has the same form as the phenomenological Adair binding polynomial, and we shall call it the sub-binding polynomial P_h. In its most general form the allosteric model is thus seen to be conformable into the Adair description and is thus necessarily able to describe mass-law binding to any t_{max} site macromolecule.

The partition function in the form of eq. (30) is still written with reference to the macromolecule in unligated conformation o (species M_{oo}). This reference species is the most convenient for comparing the general parallels between chemical and physical ligand binding, but further normalization is usually required. In problems of chemical ligation the partition function is usually taken relative to all the unligated species present in the system. This normalization factor corresponds to the sum of the top row species in the allosteric array of Figure 4.1. In the case of the physical ligands enthalpy and volume a more useful normalization is performed with respect to the array of species in a single conformation, usually the native conformation. Such a normalization typically includes a whole set of differently ligated species in that conformation, and corresponds to one of the columns of Figure 4.1.

For chemical ligand-binding applications the reference state is the unligated macromolecule regardless of conformation. For this choice of reference state the partition function must be normalized to the sum of species present in the absence of ligand. The result is a polynomial P of degree t_{max}

$$P = \frac{\sum_h L_{ho} P_h}{\sum_h (L_{ho})} \tag{31}$$

with the property that when the ligand X activity is zero the value of P is unity.

The binding polynomial (eq. 31) is the weighted sum of all sub binding polynomials and may be written in these terms as

$$P = \sum_h \alpha_h^o P_h \tag{32}$$

where the fraction of macromolecules in conformation h when no ligand is present is denoted α_h^o. α_h^o is thus given as

$$\alpha_h^o = \frac{L_{ho}}{\sum_h (L_{ho})} \tag{33}$$

The denominator of this expression is the relative sum of all the unligated species.

Again the expression for the binding curve, $\bar{X}$, is obtained from the logarithmic derivative (eq. 9) of this model's binding polynomial:

$$\bar{X} = \frac{d \ln P}{d \ln x} = \frac{1}{P} \sum_h \alpha_h^o P_h \frac{d \ln P_h}{d \ln x} \tag{34}$$

The fraction of macromolecules in a given allosteric form at an arbitrary ligand activity is an important characterizing parameter of an allosteric system. The fraction α_h of all macromolecules in form h at any given ligand activity is written

$$\alpha_h = \alpha_h^o \frac{P_h}{P} = \frac{\alpha_h^o P_h}{\sum_h \alpha_h^o P_h} \tag{35}$$

The binding curve expression (eq. 34) can be interpreted in terms of these fractions by substituting eq. (35):

$$\bar{X} = \sum \alpha_h \bar{X}_h \qquad\qquad\qquad (36)$$

The overall binding curve is thus seen to be the fractionally weighted sum of the individual allosteric form binding curves. However the fractions α_h are functions of the ligand activity (eq. 35) and thus change during the course of ligation. It is this feature which distinguishes the allosteric model from a mixture of molecules "frozen" in a given form (i.e., each form having a fixed fraction throughout ligation) and enables the allosteric macromolecule to exhibit cooperativity. This can be seen from a consideration of the free energy of ligand saturation, ΔG^o_{sat} for an allosteric system and will be examined in more detail in our treatment of the MWC model.

4.6 THE MWC AND KNF ALLOSTERIC MODELS FOR CHEMICAL BINDING PROCESSES

The general allosteric model is often not useful in a practical sense since it has more parameters than the phenomenological Adair formulation for a macromolecule binding chemical ligands. We will outline here two more-specific models that draw on the concept of conformational (allosteric) equilibria in quite different ways. Both models, that of Monod, Wyman, and Changeux (1965, the MWC model) and that of Koshland, Nemethy, and Filmer (1966, the KNF model), can be considered special cases of an allosteric square "parent" model (Wyman, 1972). The parent model is similar in form to the general allosteric model (Fig. 4.1) just presented, and we shall describe the differences here.

In the parent model the macromolecule is considered to be composed of t sites, each site existing in an either of two states: one state characterized by a high ligand affinity and the other characterized by a low ligand affinity. A particular conformation h is then identified with a particular combination of the subunits in these two states.[5] For example a t=2 macromolecule has three distinct conformations: both sites high affinity, both sites low affinity, and a doubly degenerate combination with one site high affinity, the other low affinity. The macromolecule in any single conformation h is then taken to bind ligand as if the sites were independent. The sub binding polynomials P_h are thus completely specified by the two affinities which we shall denote κ_A for the high affinity and κ_B for the low affinity. The unligated conformational equilibrium constants are free to be chosen in a more specific manner. The parent model thus has at most q + 1 variable parameters. Figure 4.2 shows the model for the case t=4 (q=4).

The MWC model is a particularly simple form of the parent allosteric model. The only nonzero conformational equilibrium constant is assumed to be that governing the concerted conformational change between the two homogeneous affinity conformations:

---

[5] The total number q+1 of distinct conformations (combinations) is then given by the number of terms in the t degree binomial expansion, or t+1.

one with all sites in the high affinity (A) state and the other with all sites in the
low affinity (B) state. The parent allosteric array is thus reduced to the two outside
columns. The number of variable parameters in the MWC formulation is three: the high
and low affinity binding equilibrium constants and one conformational equilibrium
constant.

In the KNF approach the binding affinity for the subunits in the low affinity (B)
state is considered to be so much lower than that of the subunits in the high-affinity
state (A) that any species containing ligated sites in the B state have essentially a
zero concentration. This constraint eliminates half of the species from the allosteric
array, leaving a triangular set of species above (and including) the diagonal
consisting of unligated species and those ligated species with subunits ligated in the
A state. The allosteric equilibrium constants, referred to the unligated high affinity
form, are then chosen in increasing order such that only one species of each row
remains-- that with each unligated subunit in the B state only. The resulting array of
species for the KNF induced-fit model is the diagonal from the "parent" allosteric
square. The KNF model requires in general t+1 parameters: the high affinity binding
constant and the q (=t) allosteric equilibrium constants. Usually the allosteric
equilibrium constants are obtained from geometrical modelling of the interactions
between binding sites, reducing the number of parameters required to two or three.

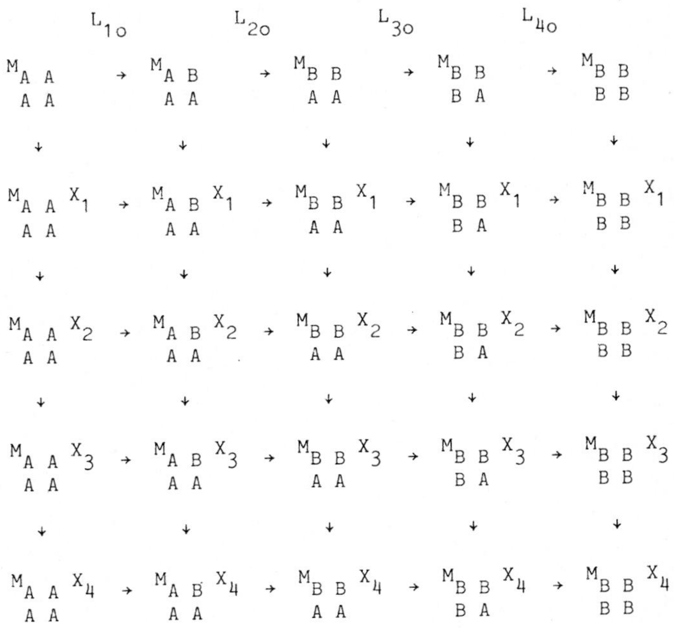

$$
\begin{array}{ccccccccc}
L_{10} & & L_{20} & & L_{30} & & L_{40} & & \\
M_{\substack{A\,A \\ A\,A}} & \to & M_{\substack{A\,B \\ A\,A}} & \to & M_{\substack{B\,B \\ A\,A}} & \to & M_{\substack{B\,B \\ B\,A}} & \to & M_{\substack{B\,B \\ B\,B}} \\
\downarrow & & \downarrow & & \downarrow & & \downarrow & & \downarrow \\
M_{\substack{A\,A \\ A\,A}}X_1 & \to & M_{\substack{A\,B \\ A\,A}}X_1 & \to & M_{\substack{B\,B \\ A\,A}}X_1 & \to & M_{\substack{B\,B \\ B\,A}}X_1 & \to & M_{\substack{B\,B \\ B\,B}}X_1 \\
\downarrow & & \downarrow & & \downarrow & & \downarrow & & \downarrow \\
M_{\substack{A\,A \\ A\,A}}X_2 & \to & M_{\substack{A\,B \\ A\,A}}X_2 & \to & M_{\substack{B\,B \\ A\,A}}X_2 & \to & M_{\substack{B\,B \\ B\,A}}X_2 & \to & M_{\substack{B\,B \\ B\,B}}X_2 \\
\downarrow & & \downarrow & & \downarrow & & \downarrow & & \downarrow \\
M_{\substack{A\,A \\ A\,A}}X_3 & \to & M_{\substack{A\,B \\ A\,A}}X_3 & \to & M_{\substack{B\,B \\ A\,A}}X_3 & \to & M_{\substack{B\,B \\ B\,A}}X_3 & \to & M_{\substack{B\,B \\ B\,B}}X_3 \\
\downarrow & & \downarrow & & \downarrow & & \downarrow & & \downarrow \\
M_{\substack{A\,A \\ A\,A}}X_4 & \to & M_{\substack{A\,B \\ A\,A}}X_4 & \to & M_{\substack{B\,B \\ A\,A}}X_4 & \to & M_{\substack{B\,B \\ B\,A}}X_4 & \to & M_{\substack{B\,B \\ B\,B}}X_4 \\
\end{array}
$$

Fig. 4.2.   MWC-KNF parent model (Wyman, 1972).

The MWC model can be seen to consist of more species from the parent model than the KNF induced-fit model, but the KNF formulation samples more allosteric forms. Additionally the KNF approach can provide for anticooperativity in ligand binding.

4.6.1 The MWC Model – Two Allosteric Forms (Monod, et al 1965)

Many properties of allosteric systems can be illustrated by the simple MWC model and we shall thus describe this case in greater detail than the treatment of the KNF model. The two homogeneous allosteric conformations of the macromolecule, one composed entirely of A-state (high affinity) sites and the other consisting of all B-state (low affinity) sites, are by convention denoted R and T, respectively, with R serving as the reference form. The species present in the presence of ligand X are represented by a two column array

$$
\begin{array}{ccc}
R_o & \rightarrow & T_o \\
\downarrow & & \downarrow \\
RX & \rightarrow & TX \\
\downarrow & & \downarrow \\
RX_2 & \rightarrow & TX_2 \\
\downarrow & & \downarrow \\
\vdots & & \vdots \\
\downarrow & & \downarrow \\
RX_t & \rightarrow & TX_t
\end{array}
$$

Similar to the general allosteric case, the equilibrium constant for the conformational transition of the unligated macromolecule

$$R_o \rightarrow T_o \qquad (37)$$

is denoted by L_o:

$$L_o = \frac{[T_o]}{[R_o]} \qquad (38)$$

so that the binding partition function can be written in a form similar to that of eq. (30).

The sub binding polynomial for each allosteric form, R and T, can be obtained from the binding partition function for the independent site model. The generalized equilibrium constant $\lambda(T, p, \mu_x)$ from section 4.4 for a chemical binding site can be written explicitly in terms of the ligand activity x:

$$\lambda = \kappa\, x \qquad (39)$$

where κ is the binding constant for a single site. The individual site binding constants for the R and T forms are designated by κ_R and κ_T, and the sub binding polynomials for each form are given from eq. (34) (where Q = P) and substituting equation (39),

$$P_R = (1 + \kappa_R x)^t \qquad (40)$$

$$P_T = (1 + \kappa_T x)^t \qquad (41)$$

The binding polynomial for this array can be written from eq. (30) with two terms in the sum with the R_o form serving as the reference species

$$P = P_R + L_o P_T \qquad (42)$$

For chemical ligation problems the normalization is carried out to all the unligated species as was done in the general case eq. (31):[6]

$$P = \frac{1}{1 + L_o} (1 + \kappa_r x)^t + \frac{L_o}{1 + L_o} (1 + \kappa_t x)^t \qquad (43)$$

The fractions of initially unligated R or T forms are given by

$$\alpha_R^o = \frac{1}{1 + L_o} \quad : \quad \alpha_T^o = \frac{L_o}{1 + L_o} \qquad (44)$$

Equation (35) allows us to write down the fraction in the R conformation at any ligand activity.

$$\alpha_R = \frac{P_R}{P_R + L_o P_T} = \frac{1}{1 + L_o \left(\dfrac{1 + \kappa_T x}{1 + \kappa_R x} \right)^t} \qquad (45)$$

and similarly for α_T.

$$\alpha_T = \frac{L_o P_T}{P_R + L_o P_T} = \frac{1}{1 + \dfrac{1}{L_o} \left(\dfrac{1 + \kappa_R x}{1 + \kappa_T x} \right)^t} \qquad (46)$$

These equations allow one to see how the fraction of macromolecule in a given conformation will depend on the ligand activity. The cooperativity of the system

depends on the shape of these functions and becomes more pronounced as the number t of sites increases.

---

[6] A frequently used form of the MWC binding polynomial is to refer the ligand activity to the R state affinity by letting $\alpha = \kappa_R x$, and defining the relative affinities by $c = \kappa_T / \kappa_R$. The binding polynomial is then

$$P = (1 + L_o)^{-1} \{ (1 + \alpha)^t + L_o (1 + c\alpha)^t \}$$

The symbol α used in this way will not be used in this chapter.

There is one situation where these fractions follow a curve symmetrical about 1/2 when plotted against the logarithm of the ligand activity x. This occurs when the asymptotes at zero ligand activity and at infinite ligand activity are equidistant from 1/2, that is, the value of α_R at infinite ligand activity, namely α_R^{∞}, is the same as the value of α_T at zero ligand activity, or α_T^{o}:

$$\frac{1}{1 + L_o c^t} = \frac{1}{1 + \dfrac{1}{L_o}} \tag{47}$$

where c is defined as the ratio κ_T/κ_R, resulting in the symmetry condition (for t>2)

$$(L_o)^2 c^t = 1 \tag{48}$$

Thus for macromolecular systems exhibiting symmetrical binding curves the MWC reduces to a two parameter model.

Expansion of the binding polynomial, P, and collection of terms with the same power of x yields the overall Adair binding constants:[7]

$$\beta_i = \binom{t}{i} \frac{\kappa_R^i + L_o \kappa_T^i}{1 + L_o} \quad \text{where} \quad \binom{t}{i} = \frac{t!}{(t-i)! \; i!} \tag{49}$$

enabling the determination of the Adair parameters for the system to be of aid in evaluating the MWC parameters.

The expression of the binding curve for the MWC model is obtained by the derivative eq. (9) of the binding polynomial (eq. 43)

$$\bar{X} = \frac{t \, \kappa_R x \, (1 + \kappa_R x)^{t-1} + t \, L_o \kappa_T x \, (1 + \kappa_T x)^{t-1}}{(1 + \kappa_R x)^t + L_o \, (1 + \kappa_T x)^t} \tag{50}$$

This expression may also be written as

$$\bar{X} = \alpha_R \, \bar{X}_R + \alpha_T \, \bar{X}_T \tag{51}$$

where α_R and α_T are the conformer fractions given in equations (45) and (46) which are similar to those written in eq. (35).

Various features of the MWC model are indicated by the specific properties of the binding curve. The median activity is computed from equation (49) for i = t and using the general eq. (27) $\beta_t x_m^t = 1$, one obtains

---

[7] The stepwise constants, $K_i = (\beta_i)/(\beta_{i-1})$, are then obtained as

$$K_i = \left(\frac{t-i+1}{i} \right) \frac{\kappa_R^i + L_o \, \kappa_T^i}{\kappa_R^{i-1} + L_o \, \kappa_T^{i-1}}$$

The intrinsic stepwise constants, κ_i, are the K_i's without the statistical factors.

$$\beta_t = \left(\frac{1}{x_m}\right)^t = \alpha_R^o \kappa_R^t + \alpha_T^o \kappa_T^t \tag{52}$$

and

$$\left(\frac{1}{x_m}\right)^t = \alpha_R^o \left(\frac{1}{x_{mR}}\right)^t + \alpha_T^o \left(\frac{1}{x_{mT}}\right)^t \tag{53}$$

where x_{mR} and x_{mT} are the median activities of the R and T form, i.e., $(\kappa_R)^{-t}$ and $(\kappa_T)^{-t}$.

The standard free energy of saturation for the MWC allosteric model can be obtained from eq. (17)

$$\Delta G_{sat}^o (MWC) = -RT \ln \left(\alpha_R^o \kappa_R^t + \alpha_T^o \kappa_T^t \right) \tag{54}$$

Contrasting this equation to that applying to the "frozen" case (each allosteric form locked in the initial fractions α_R^o and α_T^o

$$\Delta G_{sat}^o (frozen) = -t RT \left(\alpha_R^o \ln \kappa_R + \alpha_T^o \ln \kappa_T \right) \tag{55}$$

For any set of parameters it is found that the standard free energy for saturating an allosteric macromolecule is less than that of the frozen system. Since in both cases one starts from the same initial unliganded state this means that the binding curve for the allosteric case defines a smaller area to the vertical line at the standard state than that of the frozen system. Consequently the binding curve of the allosteric system must be steeper. An example is shown in the simulation shown in Figure 4.3. This result is not limited to the MWC model but applies to any system which can respond by equilibrium reaction between conformations or aggregation states. The cooperativity of the freely equilibrating system will always be greater than that of the fixed system.

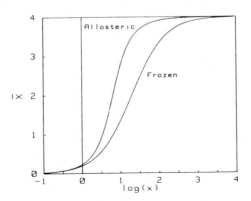

Fig. 4.3. Binding curves of the allosteric and "frozen" cases of a macromolecule with four binding sites and two conformations R and T: $\kappa_R = 0.5$; $\kappa_T = 0.05$, and $\alpha_R^o = 0.01$ and $\alpha_T^o = 0.99$ ($L_o = 100$)

$$0 \quad \text{———} \quad R_o \qquad\qquad T_o \quad \text{———} \quad -RT\left(\ln L_o\right)$$

$$-RT\left(\ln \kappa_R + \ln 4\right) \quad \text{———} \quad RX_1 \qquad\qquad TX_1 \quad \text{———} \quad -RT\left(\ln L_o + \ln \kappa_T + \ln 4\right)$$

$$TX_2 \quad \text{———} \quad -RT\left(\ln L_o + 2\ln \kappa_T + \ln 6\right)$$

$$-RT\left(2\ln \kappa_R + \ln 6\right) \quad \text{———} \quad RX_2$$

$$TX_3 \quad \text{———} \quad -RT\left(\ln L_o + 3\ln \kappa_T + \ln 4\right)$$

$$-RT\left(3\ln \kappa_R + \ln 4\right) \quad \text{———} \quad RX_3$$

$$TX_4 \quad \text{———} \quad -RT\left(\ln L_o + 4\ln \kappa_T\right)$$

$$-RT\left(4\ln \kappa_R\right) \quad \text{———} \quad RX_4$$

Fig. 4.4. Free energy levels of the two-state MWC model with t=4.

A free energy or chemical potential diagram of the standard state values for the RX_i and TX_i forms for a typical cooperative case (t = 4) is shown in Figure 4.4. The free energy difference from the R_o state is determined by the equilibrium constants for the reaction $R_o + iX \rightarrow RX_i$ given by $\binom{t}{i}\kappa_R^i$ for the R column or $Ro + iX \rightarrow TX_i$ given by $L_o\binom{t}{i}\kappa_T^i$ for the T column. The symmetry condition can be demonstrated in a graphical way by noting that equal spacings between the R_o to the T_o levels ($RT\ln(L_o)$) and between the R_4 and T_4 states ($RT\left(-\ln(L_o) - t\ln(c)\right)$) reflects the symmetry condition of eq. (48).

The enthalpy change for the process of binding ligand is obtained for the MWC model by equation (10):

$$\bar{H} - \bar{H}_o = -R \frac{\partial \ln P}{\partial(1/T)} = r(x)\,\Delta H_R + t(x)\,\Delta H_T + 1(x)\,\Delta H_{Lo} \qquad (56)$$

where

$$r(x) = \frac{t\,\kappa_R x\,(1 + \kappa_R x)^{t-1}}{(1 + \kappa_R x)^t + L\,(1 + \kappa_T x)^t}$$

$$t(x) = \frac{L_o\,t\,\kappa_T x\,(1 + \kappa_T x)^{t-1}}{(1 + \kappa_R x)^t + L\,(1 + \kappa_T x)^t}$$

$$1(x) = \frac{L_o(1 + \kappa_T x)^t}{(1 + \kappa_R x)^t + L\,(1 + \kappa_T x)^t} - \frac{L_o}{1 + L_o}$$

In practice the weighting factors r(x) and l(x) may be highly correlated.

4.6.2   The Pauling and KNF models

The Pauling (1935) model was extended by Allen et al (1950) and by Koshland, et al, (1966) to include a variety of rules that might govern the interaction properties between binding sites. These models are now commonly referred to as the KNF models, and were initially formulated to address the features of cooperative oxygen binding to hemoglobin. In these models one specifically describes the free energy of various X ligated states in terms of rules based on interaction contributions for particular geometrical forms. The approach is essentially one based on "bond" additivity for various structures. Hess and Szabo (.1979) give a discussion of this model in terms of stabilizing cross-links between unligated subunits. Such cross-links may correspond to salt bridges, as proposed by Perutz in his a stereo-chemical mechanism of subunit interaction in hemoglobin (Perutz, 1970).

The partition function for the KNF formulation is the sum of all species concentrations relative to a reference species. The pertinent species for the KNF formulation are given by the diagonal of the parent allosteric model shown schematically in Figure 4.2 for the case of t=4. The diagonal terms are given by

$$M_{\substack{B\ B\\B\ B}} \rightarrow M_{\substack{B\ B\\B\ AX}} \rightarrow M_{\substack{B\ B\\AXAX}} \rightarrow M_{\substack{AXB\\AXAX}} \rightarrow M_{\substack{AXAX\\AXAX}}$$

or

$$M_o \rightarrow M_1 \rightarrow M_2 \rightarrow M_3 \rightarrow M_4$$

Choosing the left (unligated) species as the reference component the overall reactions can be written

$$\Lambda_k = \frac{[M_k]}{[M_o]} \tag{57}$$

From the allosteric nature of the parent model we can break the overall equilibrium constants into a conformational equilibrium part and a ligand binding part, as was done in the general allosteric case:

$$\Lambda_k = L_k \kappa_A^k x^k \tag{58}$$

The general KNF binding partition function may then be written

$$Q = 1 + \Lambda_1 + \Lambda_2 + \Lambda_3 + \Lambda_4$$

or

$$P = 1 + L_1 \kappa x + L_2 \kappa^2 x^2 + L_3 \kappa^3 x^3 + L_4 \kappa^4 x^4 \tag{59}$$

164

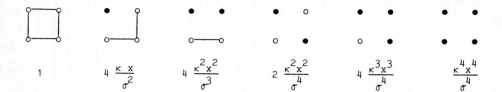

Fig. 4.5. Pauling "square" bond interaction model for hemoglobin. Species are depicted by number of ligands bound (filled circles) and by remaining bonds of strength $1/\sigma$. Statistical factors indicate the number of ways a given species can be formed.

where κ is equal to κ_A. The free energy of a given species is seen to be controlled by both the free energy of the allosteric conformation plus the binding free energy. Any cooperativity present system is then regulated by the values of the allosteric constants. In the event these have statistical values, the system behaves as T identical independent sites, with zero cooperativity.

A variety of schemes can be constructed to model the specific interactions among the binding sites which will determine the allosteric equilibrium constants. The main case considered by Pauling postulated that the binding sites are located at the corners of a square. One depicts the unligated form of this model as shown in Figure 4.5 by bonds drawn between the four subunits, represented as open circles. Two things happen with the addition of ligand: First, a new ligand bond is formed as depicted by the filled circle. The product of the intrinsic binding constant (κ) and the activity of the free ligand (x) raised to the power of the number of ligands bound gives the relative concentration of the ligated form compared to the unligated form, in the absence of site-site interaction. Second, the effect of site interactions is determined by the number of such interactions that are broken, each broken bond contributing a term to the free energy of a given ligated state. In the square model each site-site bond gives a factor $1/\sigma$ (the reciprocal reflects the fact that the bond is broken on ligation) to the equilibrium constant for the reaction between unligated and specifically ligated species. The number of ways an equivalent species can be represented is described by its statistical factor.

The binding polynomial for the square model is given as

$$P(\text{KNF square}) = 1 + 4\,\frac{(\kappa x)}{\sigma^2} + 4\,\frac{(\kappa x)^2}{\sigma^3} + 2\,\frac{(\kappa x)^2}{\sigma^4} + 4\,\frac{(\kappa x)^3}{\sigma^4} + \frac{(\kappa x)^4}{\sigma^4} \qquad (60)$$

We see that this model reduces to the case of four identical sites when there is no site-site interaction, i.e., $\sigma=1$. Furthermore, positive cooperativity occurs when $\sigma < 1$ and negative cooperativity occurs when $\sigma > 1$. Clearly only two parameters are needed to describe this model, κ and σ.

The median x_m for the KNF square model is given by the highest power term: $(\kappa/\sigma)^4 = \beta_4 = (1/x_m)^4$ or $x_m = \sigma/\kappa$. To investigate the symmetry properties of this model we may transform the activity to units of the median activity by letting $x' = x/x_m$ (eq. 29) and obtain

$$P_M(\text{KNF square}) = 1 + \frac{4}{\sigma}x' + (\frac{4}{\sigma^2} + \frac{2}{\sigma})x'^2 + \frac{4}{\sigma}x'^3 + x'^4 \qquad (61)$$

Since the second and fourth term have the same coefficient, this binding polynomial by necessity gives symmetrical binding curves. Oxygen binding curves to hemoglobin have a slight asymmetry and thus this model is not capable of describing the hemoglobin system.

A rectangular model with two long bonds of strength σ_1 and two short bonds of strength σ_2 leads by the same process, as seen in Figure 4.6, to the following binding polynomial.

$$P(\text{KNF rectangle}) = 1 + \frac{4\kappa x}{\sigma_1 \sigma_2} + \frac{2(\kappa x)^2}{\sigma_1 \sigma_2}(\frac{1}{\sigma_1} + \frac{1}{\sigma_2} + \frac{1}{\sigma_1 \sigma_2}) + \frac{4(\kappa x)^3}{\sigma_1^2 \sigma_2^2} + \frac{(\kappa x)^4}{\sigma_1^2 \sigma_2^2} \qquad (62)$$

This is a three parameter model. In this case $x_m = (\sigma_1 \sigma_2)^{1/2}/\kappa$ and the normalized polynomial has equal coefficients for the second and fourth terms, again indicating that only a symmetrical binding curve can be obtained for such a model.

The effect of different intrinsic binding constants for various sites is easily introduced. Suppose there are two α and two β subunits with different intrinsic affinities as might be the case for hemoglobin. In the case of the "rectangle" model let the diagonal corners be α and β with intrinsic binding constants of κ_α and κ_β, then the binding polynomial will be found to be

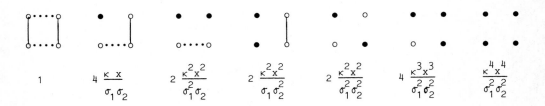

Fig. 4.6. Rectangular KNF bond model for hemoglobin partition function. Each species is represented by number of ligands bound and by number of interactions, with strengths $1/\sigma_1$ (——) and $1/\sigma_2$ (····). Statistical factors indicate the number of ways a given species can be formed.

$$P(\text{KNF rectangle, } \alpha\beta) = 1 + \frac{(\kappa\alpha + \kappa\beta)}{\sigma_1 \, \sigma_2}x + \frac{\kappa\alpha\kappa\beta}{\sigma_1\sigma_2}\left[\frac{2}{\sigma_1} + \frac{2}{\sigma_2} + (\frac{\kappa\alpha}{\kappa_\beta} + \frac{\kappa\beta}{\kappa_\alpha})\frac{1}{\sigma_1\sigma_2}\right]x^2$$

$$+ \frac{2(\kappa_\alpha^2\kappa_\beta + \kappa_\alpha\kappa_\beta^2)}{\sigma_1^2 \, \sigma_2^2}x^3 + \frac{\kappa_\alpha^2 \kappa_\beta^2}{\sigma_1^2\sigma_2^2}x^4$$

(63)

This four parameter model when written in median normalized form is also symmetrical.

Results for the median normalized form of the binding polynomial for several models (t=4) are collected in Table 4.1. If we introduce some feature of asymmetry such as the distorted square with one different interaction bond, or the square with one diagonal interaction bond, the coefficients of second and fourth terms of the binding polynomials are different, allowing modelling of macromolecules exhibiting nonsymmetric binding behavior. This is an example of a general principle pointed out previously by Allen et al, (1950) that symmetry of function (namely the shape of the binding curve) implies geometrical symmetry of structure and conversely.[8] This principle and various interpretations of the different models presuppose that the site-site interactions are represented by distances between the sites which are changed upon ligation. Thus a new structure (or in some cases structures) is produced for each specific degree of ligation with this model. These structures, uniquely . defined for each state of ligation, would correspond to the diagonal terms of the allosteric square matrix we have considered above.

The asymmetric binding properties of the hemoglobin binding curve have led Weber (1982) and Peller (1982) to use an extended version of the $\alpha_2 \, \beta_2$ rectangle (diagonal α-α interaction) shown given in Table 4.1. They obtain conditions for $\kappa_\alpha/\kappa_\beta$ and τ in which the asymmetry observed for hemoglobin can be rationalized by such a model. From an entirely different point of view the simple MWC model can also give asymmetric binding curves that match those observed for hemoglobin.

4.7   APPLICATIONS

4.7.1   Chemical ligand binding

In this section we will demonstrate the use of a generalized approach to ligand binding for determining thermodynamic parameters characteristic of a particular model. These are condensed results from a study (Parody-Morreale, et al, 1986) on the thermodynamics of oxygen binding to lobster (Homarus americanus) hemocyanin (Hc), a dodecameric protein made up of two identical hexamers. This study consisted of two types of experiments. The first type was a simple binding curve, in which we recorded changes in the average saturation of the macromolecule as we varied oxygen activity. Obtained from this data were values of the equilibrium constants governing the

---

[8] Note however that the linear models also lead in general to asymmetric binding polynomials.

Table 4.1

Median Normalized KNF Binding Polynomial Coefficients (β'_i) for Various Models of Hemoglobin (t=4)

Model	β_1'	β_2'	β_3'	Median x_m
Square	$4/\sigma$	$\dfrac{4}{\sigma}+\dfrac{2}{\sigma^2}$	$4/\sigma$	$\dfrac{\sigma}{\kappa}$
Tetrahedral	$\dfrac{4}{\sigma^{3/2}}$	$\dfrac{6}{\sigma^2}$	$\dfrac{4}{\sigma^{3/2}}$	$\dfrac{\sigma^{3/2}}{\kappa}$
Rectangular (Identical Sites)	$\dfrac{4}{\sqrt{\sigma_1\sigma_2}}$	$2\left(\dfrac{1}{\sigma_1}+\dfrac{1}{\sigma_2}+\dfrac{1}{\sigma_1\sigma_2}\right)$	$\dfrac{4}{\sqrt{\sigma_1\sigma_2}}$	$\dfrac{\sqrt{\sigma_1\sigma_2}}{\kappa}$
Rectangular (α,β sites)	$\dfrac{2\left(\sqrt{\dfrac{\kappa_\alpha}{\kappa_\beta}}+\sqrt{\dfrac{\kappa_\beta}{\kappa_\alpha}}\right)}{\sqrt{\sigma_1\sigma_2}}$	$2\left[\left(\dfrac{1}{\sigma_1}+\dfrac{1}{\sigma_2}\right)+\left(\dfrac{\kappa_\alpha}{\kappa_\beta}+\dfrac{\kappa_\beta}{\kappa_\alpha}\right)\dfrac{1}{\sigma_1\sigma_2}\right]$	$\dfrac{2\left(\sqrt{\dfrac{\kappa_\alpha}{\kappa_\beta}}+\sqrt{\dfrac{\kappa_\beta}{\kappa_\alpha}}\right)}{\sqrt{\sigma_1\sigma_2}}$	$\dfrac{\sqrt{\sigma_1\sigma_2}}{\sqrt{\kappa_\alpha\kappa_\beta}}$
Rectangular (α,β sites)	$\dfrac{2\left(\sqrt{\dfrac{\kappa_\alpha}{\kappa_\beta}}+\sqrt{\dfrac{\kappa_\beta}{\kappa_\alpha}}\right)}{\sqrt{\sigma_1\sigma_2}}$	$2\left(\dfrac{1}{\sigma_1}+\dfrac{1}{\sigma_2}\right)+\left(\dfrac{\kappa_\alpha}{\kappa_\beta}+\dfrac{\kappa_\beta}{\kappa_\alpha}\right)\dfrac{1}{\sigma_1\sigma_2}$	$\dfrac{2\left(\sqrt{\dfrac{\kappa_\alpha}{\kappa_\beta}}+\sqrt{\dfrac{\kappa_\beta}{\kappa_\alpha}}\right)}{\sqrt{\sigma_1\sigma_2}}$	$\dfrac{\sqrt{\sigma_1\sigma_2}}{\sqrt{\kappa_\alpha\kappa_\beta}}$
Distorted Square	$(\tau\sigma^3)^{1/4}\left(\dfrac{2}{\sigma^2}+\dfrac{2}{\sigma\tau}\right)$	$(\tau\sigma^3)^{1/2}\left(\dfrac{3}{\tau\sigma^2}+\dfrac{1}{\sigma^3}+\dfrac{2}{\tau\sigma^3}\right)$	$4(\tau\sigma^3)^{-1/4}$	$\dfrac{\sqrt{\tau\sigma^3}}{\kappa}$
Rectangle Diagonal α–α (τ)	$\dfrac{2\tau^{1/4}\left(\sqrt{\dfrac{\kappa_\alpha}{\kappa_\beta}}+\sqrt{\dfrac{\kappa_\beta}{\kappa_\alpha}}\right)}{\sqrt{\sigma_1\sigma_2}}$	$\tau^{1/2}\left[\dfrac{1}{\sigma_2\tau}+\dfrac{1}{\sigma_1\tau}+\dfrac{1}{\sigma_1\sigma_2}\left(\dfrac{\kappa_\alpha}{\kappa_\beta}+\dfrac{\kappa_\beta}{\kappa_\alpha}\right)+\dfrac{\kappa_\beta}{\kappa_\alpha}\right]$	$\dfrac{2}{\tau^{1/4}\sqrt{\sigma_1\sigma_2}}\left(\sqrt{\dfrac{\kappa_\alpha}{\sigma}}+\dfrac{\sqrt{\kappa_\beta}}{\sqrt{\kappa_\alpha}}\right)$	$\left(\dfrac{\sqrt{\sigma_1\sigma_2}\,\tau^{1/4}}{\sqrt{\kappa_\alpha\kappa_\beta}}\right)$
Square (Diagonal) (τ)	$\dfrac{2\tau^{1/4}}{\sigma}\left(\dfrac{1}{\tau}+1\right)$	$\dfrac{4\tau^{1/2}}{\sigma}+\dfrac{\tau^{1/2}}{\sigma^2}\left(\dfrac{1}{\tau}+1\right)$	$4/\left(\tau^{1/4}\sigma\right)$	$\dfrac{\sigma\tau^{1/4}}{\kappa}$
Linear	$2\left(1+\dfrac{1}{\sigma}\right)\sigma^{-1/4}$	$3\left(1+\dfrac{1}{\sigma}\right)\sigma^{-1/2}$	$4\sigma^{-3/4}$	$\dfrac{\sigma^{3/4}}{\kappa}$
Linear	$2\left(\dfrac{1}{\sigma_1}+\dfrac{1}{\sigma_1\sigma_2}\right)\left(\sigma_1^2\sigma_2\right)^{1/4}$	$\left(\dfrac{1}{\sigma_1}+\dfrac{2}{\sigma_2}+\dfrac{3}{\sigma_1\sigma_2}\right)\sigma_1\sigma_2^{1/2}$	$4\left(\sigma_1^2\sigma_2\right)^{-1/4}$	$\left(\dfrac{\sigma_1^2\sigma_2}{\kappa}\right)^{1/4}$

binding of oxygen. The second type of experiment was a calorimetric measurement of the average—enthalpy change of a hemocyanin solution as a function of oxygen activity. Analysis of this data gave ΔH's corresponding to the reactions involved in oxygen—binding.

The large number of binding sites in the lobster hemocyanin precludes an Adair analysis, but the relative simplicity of the binding curves suggests that one can use MWC modelling. From section 4.6, the binding polynomial for that model is written (eq. 43):

$$P = \frac{1}{1+L_o} \left[(1 + \kappa_R x)^t + L_o (1 + \kappa_T x)^t \right] \tag{64}$$

where x is the activity of the ligand X; κ_R and κ_T are the intrinsic binding constants for ligand X to the high (R) and low (T) affinity forms of the macromolecule, and L_o is the equilibrium constant between the R and T forms when no ligand is present. In our analysis, the dodecamer is treated as two independent hexamers, so that the allosteric unit size t=6. The average saturation of a macromolecule with ligand X is then given by eq. (50). In these measurements we dilute the oxygen atmosphere above the sample (thin layer) with nitrogen in logithmic steps and measure the change in optical density at equilibrium. The change of optical signal (proportional to the average saturation) as we change ligand (X) activity from x_{i-1} to x_i is given by

$$\Delta OD_i / \Delta OD_{TOT} = \overline{X}_i - \overline{X}_{i-1} \tag{65}$$

in which ΔOD_{TOT} is the optical—density change for the theoretical process of going from zero to infinite ligand activity, and must therefore be a parameter as well. Least—squares fitting to the data provides estimates of all of the parameters in equation (65).

A typical experimental result is shown in Figure 4.7. The data are obtained in the form of optical—density changes as the pure oxygen atmosphere is diluted in logarithmic steps. The standard error of a point from the MWC model are not significantly different from those expected for each data point, although a slight systematic deviation indicates the model may not be entirely adequate. The fitted MWC model prediction and the data are plotted in the more conventional $\overline{X}$ versus logarithm of oxygen pressure in Figure 4.8. Had the MWC model not fit the data reasonably a more complicated model would be called for. The binding polynomial for such a model could then be formulated and used with eq. (65) for the fitting procedure.

Analogous to the average saturation is the average enthalpy of a macromolecule at equilibrium with ligand X, and this quantity is expressed by equation (10). We have evaluated the derivative in eq. (10) with the assumption that the temperature dependence of each equilibrium constant in the MWC binding polynomial can be expressed in integrated van't Hoff form as $K=K^o \exp[-\Delta H/R(1/T-1/T^o)]$, as was done in the MWC

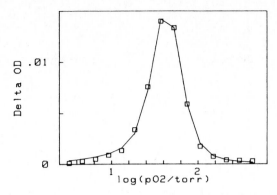

Fig. 4.7. MWC model fitted to changes in optical density vs. $\log(pO_2/\text{torr})$ for <u>Homarus</u> <u>americanus</u> hemocyanin (Parody-Morreale, et al, 1986). Parameters are:

κ_R = 0.239 ± 0.025 torr^{-1} (= 1.42×10^5 M^{-1})

κ_T = 0.000535 ± 0.000015 torr^{-1} (= 3190 M^{-1})

L_o = (1.57 ± 0.5) x 10^6.

section above (eq. 56). The reference-state enthalpy of the macromolecule is as usual chosen to be that of the unligated macromolecule. Our calorimetric data was obtained by titrating a solution of <u>H</u>. <u>americanus</u> Hc with oxygen-saturated buffer. The heat evolved during a titration step as we move from ligand activity x_{i-1} to x_i at Hc concentration c is thus

$$q_i = c \left[(\bar{H} - \bar{H}_o)_i - (\bar{H} - \bar{H}_o)_{i-1} \right] \qquad (66)$$

where $\overline{H} - \overline{H}_o$ is given by equation (56). Just as evaluation of equation (9) enabled us to determine equilibrium constants for a specific model, so does equation (10) allow the determination of enthalpy changes for those equilibria.

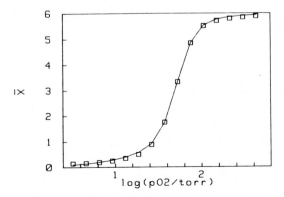

Fig. 4.8. Binding curve for <u>H</u>. <u>americanus</u>. Theoretical curve generated from MWC constants given in Fig. 4.7.

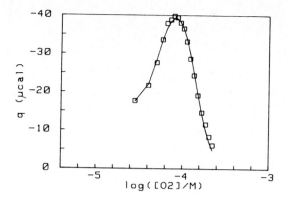

Fig. 4.9. Calorimetric data for oxygen titration of H. americanus hemocyanin.
Theoretical line obtained using MWC reaction constants of Figure 4.7 along with
fitted ΔH values as discussed in text.

The calorimetric data, heat evolved as ligand titrates the macromolecule, is shown
in Figure 4.9. The solid line in Figure 4.9 is the MWC model prediction based on
equation (66). The similarity between average saturation and average enthalpy is
evident from a plot of $\overline{H} - \overline{H}_o$ vs logarithm of oxygen concentration shown in Figure
4.10. The data plotted in this form can be considered as a "heat binding curve". The
parallel curves of Figures 4.8 and 4.10 indicate that the stepwise stoichiometric
reaction heats are practically the same. The MWC analysis of the calorimetric data
employing equation (56) shows that the terms r(x) and l(x) weighting ΔH_R and ΔH_{Lo} (see
eq. 56) are strongly correlated, preventing individual determination of those heats.
What can be obtained are heats for the κ_T binding reaction and for the overall binding
process, permitting a calculation of the allosteric

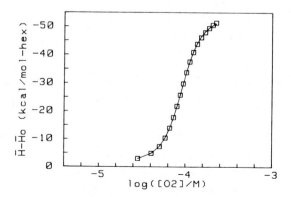

Fig. 4.10. Calorimetric data for oxygen titration of H. americanus hemocyanin
plotted in the form of a "heat binding curve. Theoretical line obtained using
MWC reaction constants of Fig. 4.7 along with fitted ΔH values from Fig. 4.9.

TABLE 4.2
Thermodynamic parameters governing oxygen binding to <u>Homarus americanus</u>
hemocyanin.   Standard state is 1 M, 25 C.

Reaction	ΔG° [a]	ΔH° [a]	$-T\Delta S^{\circ}$ [a]	$\Delta n^{\circ}_{H^+}$ [b]
$Hc_R(O_2)_{i-1} + O_2 \xrightarrow{K_R} Hc_R(O_2)_i$	-7.0	---	---	---
$Hc_T(O_2)_{i-1} + O_2 \xrightarrow{K_T} Hc_T(O_2)_i$	-4.7	-8	3	0.2
$Hc_R \xrightarrow{L_o} Hc_T$	-8.4	---	---	---
$Hc + 6 \, O_2 \xrightarrow{\beta_6} Hc(O_2)_6$	-33	-9	-24	-4.0
$Hc_R(O_2)_6 \xrightarrow{L_6} Hc_T(O_2)_6$	5.1	-39	44	5

a.   Units for ΔG°, ΔH° and $-T\Delta S^{\circ}$ given in kcal/mole reaction.
b.   Number of protons absorbed per mole reaction.

transition heat for the fully ligated macromolecule. These heats are shown in Table
4.2. By conducting calorimetric experiments in two different buffers we removed
specific buffer—heat effects and calculated the number of protons absorbed, Δn_{H^+}, for
each reaction. As a rough check on one of these numbers, an additional binding
experiment at pH 8.0 together with the pH 7.4 equilibrium constants permitted
evaluation of Δn_{H^+} for the overall reaction from $6 \cdot \partial \log p_m / \partial pH$. The value, -4.0,
fortuitously agrees exactly with the result determined calorimetrically. These values
are shown along with the other thermodynamic parameters in Table 4.2.

   This study has demonstrated how many or all of the thermodynamic parameters
governing the ligand binding reactions within the framework of an appropriate
molecular model can be determined directly.

4.7.2   Thermal transitions
   The analysis of experiments which consider thermal effects and chemical binding
separately has been developed for thermally induced transitions (Privalov, 1979,
Privalov and Filimonov 1978, Friere and Biltonen, 1978a) and the temperature
dependence of heats of reaction of specific ligand binding. A thermodynamic
development of linked binding effects was used to describe the effect of pH and NaCl
concentration on the melting of DNA double helix (Privalov and Ptitsyn, 1969) and to
validate the concept of macromolecular state functions for lysozyme thermal
transitions and guanidinium hydrochloride binding reactions (Pfeil and Privalov,
1976a,b,c). As shown in the previous sections, the most useful representation of the
system is given in terms of the partition function (Wyman 1964; Friere and Biltonen,

1978a; Schellman, 1975; Hill 1960), from which the specific effects of heat and volume changes accompanying macromolecular transitions and any associated ligand binding reactions can be described.

We wish to show here how recognizing the common thermodynamic features of thermal effects due to macromolecular reactions and ligand binding reactions enables the analysis of real macromolecules which are capable of undergoing a number of allosteric transitions to different states, each of which can specifically bind chemical ligands.

4.7.2A tRNA melting--the independent transition model

Independent ligand binding reactions and independent thermal transitions are the simplest ways to model the thermodynamics of a macromolecular system. Thermal melting curves of tRNA have been analyzed in great detail using this approach. The independent transition model, introduced in section 4.4, assumes that the macromolecule consists of a set of t independent segments or domains, each of which exists in only two conformations (such as native or denatured). The total number of allosteric states available to such a macromolecule then consists of all combinations of the states of the independent domains. Assuming each state is uniquely characterized by a set of thermodynamic parameters then the total number of allosteric states will be 2^n. The partition functions for many such cases have been obtained in section 4.4. For thermal transition involving nonidentical sites (melting domains) equation (24) applies with λ_j expressed in van't Hoff form using ΔH_j as the enthalpy change for the jth transition. The assumption of zero heat capacity change is approximately valid for nucleic acid denaturations.

For this case, it is usual to define the mid-point temperature T_j^m as the reference temperature chosen at the mid-point temperature of the ith transition so that the reference equilibrium constant for the transition of the jth domain becomes 1, the temperature dependence of the partition function is given by

$$Q^{Indep} = \prod_{j=1}^{t} (1 + e^{-\Delta H_j(\frac{1}{T} - \frac{1}{T_j^m})/R}) \qquad (67)$$

The total enthalpy per mole of macromolecule $\bar{H}$ is given from eq. (10) as:

$$\bar{H} = \bar{H}_o + \sum_{j=1}^{n} \frac{\Delta H_j e^{-\Delta H_j(\frac{1}{T} - \frac{1}{T_j^m})/R}}{1 + e^{-\Delta H_j(\frac{1}{T} - \frac{1}{T_j^m})/R}} \qquad (68)$$

where $\bar{H}_o$ is the enthalpy of the unmelted native state. As might be expected, this result shows that the enthalpy of the system is the sum of the segment enthalpy changes weighted by the fraction of domains with that particular segment melted.

The theoretical heat capacity is found by differentiation of equation (68)

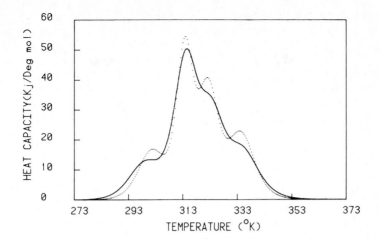

Fig. 4.11. Heat capacity versus temperature for tRNA^{Phe} in 20mM NaCl. Solid line is experimental data (Privalov and Filimonov, 1978). The fit of the independent domain model was coincident with the data. Dotted line is the result of a five allosteric state model. The inclusion of a sixth allosteric state results in a fit indistinguishable from the experimental data.

with respect to T:

$$\bar{C} = \bar{C}_o + \frac{1}{RT^2} \sum_{j=1}^{n} \frac{(\Delta H_j)^2 e^{-\Delta H_j(\frac{1}{T} - \frac{1}{T_j^m})/R}}{(1 + e^{-\Delta H_j(\frac{1}{T} - \frac{1}{T_j^m})/R})^2} \tag{69}$$

where $\bar{C}_o$ is the heat capacity of the reference state. The simplicity of this equation results from the fact that each segment transition contributes separately to the total heat capacity.

Equation (69) was used by Privalov and Filimonov (1978) to fit the thermal denaturation of tRNA. They found that a total of five independent domains were required to account for the complex melting curve which is shown by the solid line in Figure 4.11. The dotted line in this figure is the result of an allosteric analysis (Gill, et al, 1985) which will be outlined in the next section.

4.7.2B tRNA melting--the allosteric model

We shall next analyze the tRNA melting data as a set of allosteric transitions with large enthalpy changes relative to the native form. In general, the native state of the macromolecule consists of a number of species in the native conformation characterized by different amounts of bound ligand. We need a new reference state which includes all such species of the zeroth allosteric form. Thus the partition function Q given by equation (19) must be normalized by dividing by the sum of all zeroth form species, namely $L_{oo} P_o$, or just P_o since the reference component in eq. (19) is the macromolecule in the oo th conformation. An asterisk (*) will be used to denote quantities (e.g. Q^*) based on this total zeroth form reference state. For any given ligand (salt, proton, etc.) activity it turns out that the enthalpy changes associated with the allosteric transitions in tRNA are much larger than the enthalpy changes for ligations. This enables one to approximate the temperature dependence of each allosteric constant by a single van't Hoff equation with T-independent ΔH_{Lh}: The normalized partition function Q^* is then given by

$$Q^* = \frac{Q}{P_o} = \frac{\sum\limits_h L_{ho} P_h}{P_o} = \sum L_h^* \qquad (70)$$

The temperature dependence of the averaged allosteric constants is given by

$$L_h^* = L_h^{*o} \; e^{(-\overline{\Delta H}_h^* / R)(\frac{1}{T} - \frac{1}{T_o})} \qquad (71)$$

where now L_h^{*o} is the standard state $(1/T = 1/T_o)$ equilibrium constant at $1/T_o$ and $\overline{\Delta H}_h^*$ is the change in enthalpy for the reaction between the zeroth and h forms.

The total enthalpy $\overline{H}$ minus the enthalpy of the zeroth form is given from eq. (10):

$$\overline{H} - \overline{H}_o = \frac{\sum\limits_h \overline{\Delta H}_h^* \; L_h^{*o} \; e^{-\overline{\Delta H}_h^*(\frac{1}{T} - \frac{1}{T_o})/R}}{\sum\limits_h L_h^{*o} \; e^{-\overline{\Delta H}_h^*(\frac{1}{T} - \frac{1}{T_o})/R}} \qquad (72)$$

The apparent excess heat capacity of the system as found by differentiating the enthalpy with respect to T is:

$$\overline{C} - \overline{C}_o = \frac{1}{RT^2} \left[\frac{\sum (\overline{\Delta H}_h^*)^2 \; L_h^{*o} e^{-\overline{\Delta H}_h^*(\frac{1}{T} - \frac{1}{T_o})/R}}{\sum L_h^{*o} \; e^{-\overline{\Delta H}_h^*(\frac{1}{T} - \frac{1}{T_o})/R}} - \frac{(\sum \overline{\Delta H}_h^* \; L_h^{*o} \; e^{-\overline{\Delta H}_h^*(\frac{1}{T} - \frac{1}{T_o})/R})^2}{(\sum L_h^{*o} \; e^{-\overline{\Delta H}_h^*(\frac{1}{T} - \frac{1}{T})/R})^2} \right] \qquad (73)$$

Equation (73) reflects the heat capacity effects which are associated with alterations in the distribution of allosteric forms brought about by a change in temperature. Even in this simplified case, where the enthalpy changes between forms are assumed to be independent of temperature, two parameters are needed to describe each allosteric transition in the system: the relative enthalpy of the ith form $(\overline{\Delta H}_h^*)$ and the value of L_h^{*o} at $1/T_o$. It is also worth noting that both the equilibrium constant L_h^{*o} and the

TABLE 4.3
Enthalpy and Temperature Parameters for Thermal State Treatment of tRNA[Phe] Melting in 0.020M Nacl[a]

State i	4 State Model		5 State Model		6 State Model		7 State Model	
	$\overline{\Delta H}^*_h$	T^c_h	$\overline{\Delta H}^*_h$	T_h	$\overline{\Delta H}^*_h$	T_h	$\overline{\Delta H}^*_h$	T_h
1	457773	312.95	223232	301.72	190961	299.19	187998	299.09
2	886084	317.16	617757	309.27	453993	306.64	381719	307.10
3	1252000	321.89	968116	313.90	722787	310.00	498676	307.82
4			1252000	318.34	1010530	313.82	729380	310.80
5					1252000	317.75	1011390	313.80
6							1252000	317.71
Error[b]	.17		.046		.00012		.00010	

a. Parameters determined by least square analysis of Heat Capacity data from Privalov and Filimonov (Privalov and Filimonov 1978) on tRNA[Phe].

b. This is the standard errors of a point for the fit divided by the maximum value of the heat capacity of the data.

c. The units for $\overline{\Delta H}^*_h$ are joule/mole and for T_h are °K.

enthalpy change $\overline{\Delta H}^*_h$ depend on the chemical ligand activity.

In the general allosteric situation, the heat capacity (equation 73) cannot usually be factored into contributions for each specific allosteric state. The thermal properties of the various states are thus interdependent.

Using equation (73) Gill, et al (1985) determined that six allosteric states were able to represent the heat capacity data obtained for tRNA melting shown in Figure 4.11. The results of the fitting process are shown in Table 4.3 for the various models examined. Within experimental error the six state model is seen to fit the data.

The question arises as to how the six form allosteric model can represent the same data which was originally obtained from the independent transition model with 2^5 different allosteric forms. Figure 4.12 illustrates the comparison between the enthalpy states for the independent transition model for tRNA[Phe] and the six energy levels of the six-state allosteric model. There is no obvious relation between these two enthalpy level diagrams. However, since the two schemes both fit the experimental data, one expects to find an underlying connection. This is found by considering the fractions of the various allosteric species as a function of temperature. The fraction of macromolecules in the hth allosteric form, α_h, is given by:

$$\alpha_h = \frac{L^{*o}_h\, e^{-\overline{\Delta H}^*_h(\frac{1}{T} - \frac{1}{T_o})/R}}{\sum L^{*o}_h\, e^{-\overline{\Delta H}^*_h(\frac{1}{T} - \frac{1}{T_o})/R}} \tag{74}$$

176

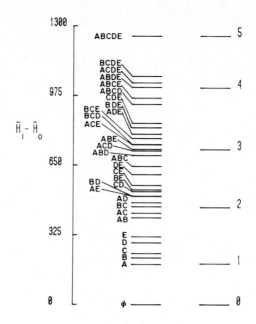

Fig. 4.12. Enthalpy level diagram for the independent transition model (left) and for the six allosteric state treatment (right) of tRNA^{Phe} melting in 20 mM NaCl (Privalov and Filimonov, 1978). The independent thermal transition enthalpies are denoted as follows: A transition, 188 KJ; B transition, 218 KJ; C transition, 239 KJ; D transition, 289 KJ; E transition, 318 KJ. The allosteric state levels are given by the six state model of Table 4.3.

The fractional occupancy of any level in the independent model is found by expanding the partition function given by equation (35) and noting that each term is proportional to the concentration of that particular species. The fraction is computed by normalizing to the sum of all species. For example a two-domain macromolecule has four different species each given by a term in the expansion of the partition function, or for this case

$$Q = (1 + \lambda_1)(1 + \lambda_2) = 1 + \lambda_1 + \lambda_2 + \lambda_1\lambda_2 \qquad (75)$$

The species fractions of each species in this example are then given by the appropriate term divided by the binding partition function Q. The underlying species fractions for both schemes are shown in Figure 4.13.[9]

---

[9] It is worth noting here that the maximum amount of any species occurs at the temperature corresponding to the point at which the enthalpy added to the system is equal to the enthalpy of that state. This result is analogous to the general rule in ligand binding that the maximum amount of ith species occurs at the extent of reaction corresponding to i moles of ligand bound.

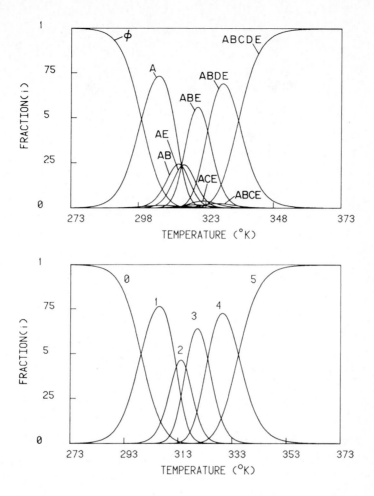

Fig. 4.13. The species fractions as a function of temperature computed for the independent transition model (top) and the six allosteric state formulation (bottom) of tRNA^{Phe} melting. The top figure labels correspond to the combinations of independent transitions given in Fig. 4.12 and the bottom figure labels correspond to the states given in Table 4.3.

4.7.2C   Two state protein denaturation

We have concentrated upon the key ideas of thermal binding with the assumption that all heat capacities of the various allosteric forms are the same. This leads to heat binding curves with horizontal asymptotes, similar to what we have found to occur in chemical ligand binding. However for many biopolymer transitions, for example proteins, significant heat capacity differences between native and denatured forms occur and it becomes necessary to consider the consequences of such changes (Wyman et al, 1979).

The equilibrium constants governing the allosteric reactions between the columns of Figure 4.1 are given by the L_h^* as expressed by equation (70). For the present systems the heat capacity changes for the reactions are not negligible, but may be assumed to the first approximation as constant. We let ΔH_h^{*o} denote the enthalpy of reaction at the reference temperature T_o so that

$$\Delta H_h^* = \Delta H_h^{*o} + \Delta C_h^* (T - T_o)$$ (76)

The temperature dependence of the equilibrium constant L_h^* is expressed as

$$L_h^* = L_h^{*o} \left(\frac{T_o}{T}\right)^{-\frac{\Delta C_h^*}{R}} e^{-\frac{\Delta H_h^{*o}}{R}\left(\frac{1}{T} - \frac{1}{T_o}\right)} e^{\frac{\Delta C_h^* T_o}{R} \cdot \left(\frac{1}{T} - \frac{1}{T_o}\right)}$$ (77)

The partition function Q^* given by the sum of all species concentrations in the system, normalized to the ground state species and is given by equation (70). The enthalpy relative to the ground state is then found from the derivative equation (10), or

$$\bar{H} - \bar{H}_o = \frac{\sum L_h^* \Delta H_h^*}{\sum L_h^*} = \sum \alpha_h \Delta H_h^*$$ (78)

where α_h is the fraction of h form species in the system. Differentiating with respect to T, the heat capacity $\bar{C} - \bar{C}_o$ is then

$$\bar{C} - \bar{C}_o = \frac{1}{RT^2} \{ \sum (\Delta H_h^*)^2 \alpha_h - (\sum \Delta H_h^* \alpha_h)^2 + \sum \Delta C_h^* \alpha_h RT^2 \}$$ (79)

This expression has the same form as (73) except for the inclusion of ΔC_h^*. A system of

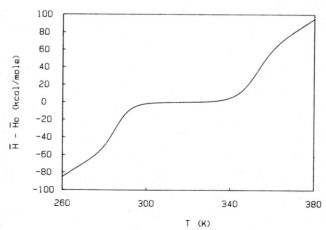

Fig. 4.14. Excess enthalpy binding curve for a two state $\Delta C_1 = 1500$ cal/mole-deg, $\Delta H_1^o = 50,000$ cal/mole, $L_1^o = 1$ and $T_o = 350°K$.

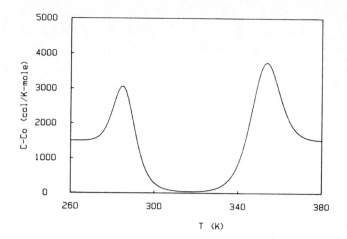

Fig. 4.15. Excess heat capacity for a two state transition with parameters given in Fig. 4.14.

two allosteric species is characterized by L_o^*, ΔH_o^* and ΔC_o^*, and a system of t allosteric forms is determined by $3(t-1)$ parameters. Some unusual heat binding curves can arise through the influence of ΔC_h. Experimentally, large changes in the heat capacity between the native form and the denatured form can be seen to occur.

In Figure 4.14 we show a hypothetical situation for molar enthalpy relative to the ground state for a two state transition. The flat region of this curve spans the range of temperatures in which the the native form dominates. At higher and lower temperatures the macromolecule denatures. The molar heat capacity of the macromolecule above that of the ground state for the same situation is shown in Figure 4.15. The

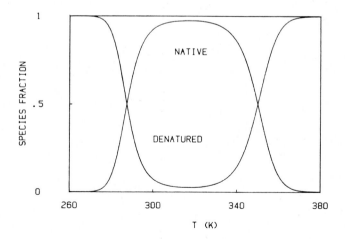

Fig. 4.16. Fractions of the two state transition forms as a function of T for parameters given in Fig. 4.14.

elevated heat capacities at the high and low temperatures describe the denatured state of the macromolecule. The heat capacity of the system is highest where the conversion from one form to the other is most temperature sensitive. The trough region of this plot describes the system where the native form is the dominant species. The fractions of the two forms as a function of temperature are shown in Figure 4.16.

Myoglobin is a system which has been studied by Privalov (1986) that shows behavior similar to this example. As these figures illustrate the ground state is stable at intermediate temperatures and as either temperature is raised or lowered the molecule is converted to the denatured form. This behavior is caused by the large heat capacity difference between forms so that the enthalpy change of the transition reverses sign at some temperature.

Large heat capacity differences between allosteric forms obviously complicates the analysis of scanning calorimetric data in all but the simplest cases.

ACKNOWLEDGEMENTS

This work was supported by National Science Foundation Grant PCM 772062 (J.W.) and National Institutes of Health Grant HL 22325 (S.J.G. and C.H.R.).

REFERENCES

Adair, G.S., J. Biol. Chem. 63 (1925), 529.
Allen, D.W., Guthe, K.F., and Wyman, J., J. Biol. Chem. 187 (1950), 393-410.
Bohr, C., Hasselbach, K.A. and Krogh, A., Skand. Arch. Physiol. 16 (1904), 402.
Brunori, M., Coletta, M., and DiCera, E., Biophys. Chem. (1986), in press.
Chu, A.H., Turner, B.W., and Ackers, G.K., Biochemistry 23 (1984), 604-617.
Connelly, P.R., Robert, C.H., Briggs, W.E., and Gill, S.J., Biophys. Chem. (1986), in press.
Decker, H., Robert, C.H., and Gill, S.J. in B. Linzen (Ed.), Invertebrate Oxygen Carriers Symposium Proceedings, Springer-Verlag, Berlin, 1986, in press.
Edsall, J.T., J. Hist. Biol. 5 (1972), 205-257.
Edsall, J.T., Fed.Proc. 39 (1980), 226-234.
Freire, E., Biltonen, R.L., Biopolymers 17 (1978), 463-479.
Gill, S.J. in M.N. Jones (Ed.), Biochemical Thermodynamics, Vol. 1, Elsevier Scientific Publishing Company, Amsterdam, 1979, 224-255.
Gill, S. J., Richey, B., Bishop, G., and Wyman, J., Biophys. Chem. 21 (1985), 1-14.
Gill, S. J., Robert, C. H., Coletta, M., Di Cera, E., and Brunori, M., Biophys. J. (1986), in press.
Haurowitz,F., Physiol. Chem. 254 (1938), 266-274.
Hess, V.L. and Szabo, A., J. Chem. Ed. 56 (1979), 289-292.
Hill, T. L., Introduction to Statistical Thermodynamics, Addison-Wesley, Reading 1960.
Imai, K., Allosteric effects in haemoglobin, Cambridge University Press, Cambridge, 1982, pp. 129-137.
Johnson, M.L., and Ackers, G.K., Biochemistry 21 (1982), 201-211.
Koshland, D.E., Jr., Nemethy, G., and Filimer, D. Biochemistry 5 (1966), 365-385.
Mills, F.C., Johnson, M.L., and Ackers, G.K., Biochemistry 15 (1976), 5350-5362.
Monod, J., Changeux, J. P., and Jacob, F., J. Mol. Biol. 6 (1963), 306-329.
Monod, J., Wyman, J., and Changeux, J.P., J. Mol. Biol. 12 (1965), 88-118.
Monod, J., and Jacob, F., Cold Spring Harb. Symp. Quant. Biol. 26 (1961), 389-401.
Parody-Morreale, A., Robert, C.H., Bishop, G., and Gill, S.J., in B. Linzen (Ed.), Invertebrate Oxygen Carriers Symposium Proceedings, Springer-Verlag, Berlin, 1986, in press.
Pauling, L., Proc. Natl. Acad. Sci. USA 21 (1935), 186-191.

Peller,L., Nature 300 (1982), 661-662.
Perutz, M.F., Rossman, M.G., Cullis, A.F., Muirhead, H., Will, G., and North, A.C.T., Nature 185 (1960), 416.
Perutz, M.F., Nature Lond. 228 (1970), 726-734.
Perutz, M.F., Br. Med. Bull. 32 (1976), 195-208.
Perutz, M.F., Ann. Rev. Biochem. 48 (1979), 327-386.
Pfeil, W. and Privalov, P.L., Biophys. Chem. 4 (1976), 23-32.
Pfeil, W. and Privalov, P.L., Biophys. Chem. 4 (1976), 33-40.
Pfeil, W. and Privalov, P.L., Biophys. Chem. 4 (1976), 41-50.
Privalov, P.L., Ptitsyn, O.B., Biopolymers 8 (1969), 559-571.
Privalov, P.L., Adv. Protein Chem. 33 (1979), 167-241.
Privalov, P.L. and Filimonov, V.V., J. Mol. Biol. 122 (1978), 447-464.
Privalov, P.L., J. Mol. Biol. (1986), in press.
Schellman, J.A., Biopolymers 14 (1975), 999-1018.
Smith, F.R. and Ackers, G.K., Proc. Natl. Acad. Sci. USA 82 (1985), 5347-5351.
Szabo, A. and Karplus, M., J. Mol.Biol. 72 (1972), 163-197.
Weber, G., Nature 300 (1982), 603-607.
Wyman, J., Adv. Prot. Chem. 4 (1948), 407-531.
Wyman, J., Adv. Prot. Chem., 19 (1964), 223-286.
Wyman, J., J. Mol. Biol. 11 (1965), 631-644.
Wyman, J., Quarterly Reviews of Biophysics 1 (1968), 35-80, (1968).
Wyman, J., Curr. Top. in Cell. Reg. 6 (1972), 207-223.
Wyman, J. and Allen, D.W., J. Polymer Sci. 7 (1950), 499-518.
Wyman, J., Gill, S.J., and Colosimo, A., Biophysical Chem. 10 (1979), 363-369.
Wyman, J., Quart. Rev. Biophys. 17 (1984), 453-488.

CHAPTER 5

THE THERMAL BEHAVIOUR OF LIPID AND SURFACTANT SYSTEMS

M.N. JONES

5.1. INTRODUCTION

The importance of lipids as one of the major components of cell membranes has led to detailed thermodynamic studies of their behaviour when dispersed in an aqueous environment. The amphipathic structure of membrane lipids which necessitates simultaneously satisfying the conflicting requirements of the hydrophobic acyl chains and the hydrophilic head groups to minimise the Gibbs energy of the system leads to structures based on bilayer arrangements of the lipid. The lipid bilayer has formed the basis of models of all membranes since the first tentative postulates of Danielli in 1934 [1] and is still the essential matrix into and onto which proteins and glycoproteins are incorporated in the current fluid mosaic model of the plasma membrane as proposed by Singer and Nicolson in 1972 [2].

The detailed structural chemistry of naturally occuring lipids has been considered elsewhere [3,4]. The main types of lipid are phospholipids, sphingolipids, glycolipids and sterols. Much of the thermodynamic work in recent years has been concerned with the thermal properties of phospholipids.

The chemical structure of phospholipids is as follows

$$^1CH_2 - O - R_1$$
$$|$$
$$^2CH \ - O - R_2$$
$$| \qquad\qquad O$$
$$\qquad\qquad\quad \|$$
$$^3CH_2 - O - P - OX$$
$$\qquad\qquad\quad |$$
$$\qquad\qquad\quad O_-$$

where R_1 and R_2 are acyl residues of fatty acids. They are thus diesters of sn-glycerol 3-phosphoric acid. Some of the most commonly occuring R groups are shown in Table 1 and the common X substituents are given in Table 2 where the charges shown pertain to physiological pH (7.4).

TABLE 1 Acyl chains in phospholipids

R group	Systematic name	Trivial name	Letter symbol
$CH_3(CH_2)_{10}CO-$	n-dodecanoyl	lauroyl	L
$CH_3(CH_2)_{12}CO-$	n-tetradecanoyl	myristoyl	M
$CH_3(CH_2)_{14}CO-$	n-hexadecanoyl	palmitoyl	P
$CH_3(CH_2)_{16}CO-$	n-octadecanoyl	stearoyl	S
$CH_3(CH_2)_5CH=CH(CH_2)_7CO-(cis\Delta^9)$		palmitoleoyl	
$CH_3(CH_2)_7CH=CH(CH_2)_7CO-(cis\ \Delta^9)$		oleoyl	
$CH_3(CH_2)_7CH=(CH_2)_7CO-(trans\ \Delta^9)$		elaidoyl	
$CH_3(CH_2)_4CH=CH\ CH_2\ CH=CH(CH_2)_7CO-(cis\ \Delta^9,\ \Delta^{12})$		linoleoyl	
$CH_3CH_2\ CH=CH\ CH_2\ CH=CH\ CH_2\ CH=CH(CH_2)_7CO-(cis\ \Delta^9,\Delta^{12},\Delta^{15})$		linolenoyl	
$CH_3(CH_2)_4(CH=CH\ CH_2)_4(CH_2)_2CO-(cis\ \Delta^5,\Delta^8,\Delta^{11},\Delta^{14})$		arachidonoyl	

TABLE 2 Phospholipid polar head groups

X groups	Name	Letter symbol
$-H$	Phosphatidic acid	PA
$-CH_2CH_2N^+(CH_3)_3$	Phosphatidylcholine (lecithin)	PC
$-CH_2CH_2NH_3^+$	Phosphatidylethanolamine (cephalin)	PE
$-CH_2-CH-CO_2^-$ $\quad\quad NH_3^+$	Phosphatidylserine	PS
(inositol ring structure with OH groups)	Phosphatidylinositol	PI

We will use the commonly adopted letter notation to denote phospholipid structures some examples of which are as follows[*]

1,2-dimyristoyl-sn-glycero-3-phosphocholine
(dimyristoylphosphatidylcholine)   DMPC

1,2-dipalmitoyl-sn-glycero-3-phosphoserine
(dipalmitoylphosphatidylserine)   DPPS

1-myristoyl-2-palmitoyl-sn-glycero-3 phosphoethanolamine C(14):C(16)PE or   MPPE

[*]Footnote   Recommendations for phospholipid nomenclature have been given by the IUPAC-IUB Commission on Biochemical Nomenclature [Eur.J.Biochem. 1977, 79, 11-21].   The suggested abbreviation for the phosphatidyl groups (ie 1,2-diacyl-sn-glycero (3) phospho-) is Ptd.

It should be noted however that the above commonly used nomenclature is inconsistent with the notation of Sundaralingam [5] used in conformational studies. In this notation the glycerol carbon atom to which the polar head group is attached is designated as C(l). While the Sundaralingam convention is more consistent for comparing the configurations of different lipids such as the structurally related glycerophospholipids and sphingolipids and has been recommended for conformational studies [6] it is rarely used in other areas and to adopt it here would lead to considerable confusion. The relationship between the two systems can be seen from the following example for the phosphatidylcholines. Thus

1,2-diacyl-sn-glycero-3-phosphorylcholine would be named either
2,3-diacyl-D-glycero-l-phosphorylcholine or 2R,3-diacylglycero-l-phosphorylcholine
(depending on whether the D/L or R/S system is used for the centres of asymmetry) according to the Sundaralingam numbering sytem.

In this chapter we will be primarily concerned with the thermodynamic behaviour of phospholipids when dispersed in aqueous media either alone or in combination with proteins or glycoproteins. The thermodynamics of the interaction between lipids and macromolecular components of cell membranes is of increasing interest. In relation to the problem of lipid-protein/ glyco-protein interaction we will also consider the thermodynamics of the interaction between soluble globular proteins and surfactant molecules.
The emphasis will be largely on developments during the years since the first edition of this volume was published.

5.2. INSTRUMENTATION

The study of the thermodynamics of lipid systems like many biochemical systems has been greatly facilitated by instrumental developments in calorimetry. In general the energy changes measured are very small either because the process itself does not result in a large amount of heat or because the amounts of material involved are small. Energies much less than 1 J arising from only micromoles of material need to be measured and this calls for the use of differential microcalorimeters [7].

The general principle involved in a differential microcalorimeter is that the heat evolved in the reaction vessel is rapidly conducted into a heat sink or if the heat change is endothermic heat flows into the reaction vessel. Embedded in the heat sink adjacent to the reaction vessel is a thermopile and the heat flow is monitored from a voltage-time curve. The area under the voltage-time plot (the thermogram) gives the total heat change in the process. In order to balance out thermal effects not directly associated with the process being studied eg. frictional heats on mixing, heats of dilution and heats of vaporisation or condensation resulting from small changes in vapour pressure, the twin principle is used. Heats of dilution etc. are cancelled in a reference vessel and only the differential voltage-time curve is recorded from reaction and reference vessels. The general principle of the heat conduction calorimeter was originally used in the Tian-Calvet calorimeter [8] and applied to the development of an instrument for biochemical studies by Benzinger and Kitzinger [10]. However one of the most widely used microcalorimeters manufactured by LKB-Produkter, Sweden was designed by Wadso [10] initially in a batch mode (ie for mixing aliquots of two solution) and also in a flow mode [11,12]. For relatively slow reactions batch calorimetry is usually the most accurate but is slow compared with the flow method where two solutions are continuosly mixed in the reaction vessel. The flow method however requires relatively large volumes of solutions. Both speed and economy can be achieved by the development of a titration system and methods for converting the commercial LKB 10700-1 batch microcalorimeter for thermometric titrations have been described [13,14]. These use motor-driven Hamilton syringes for delivery of titrants and heat quantities of the order of 10mJ on addition of $10\mu l$ of titrant can be determined to better than 1%. The limiting factor for the precision of the measurements is the differential heat evolution from compression-mixing when liquid is injected into a closed calorimetric vessel. Compression of the gas phase gives rise to a heat evolution of $0.2\pm0.1mJ$ on addition of $10\mu l$ to 5ml of an aqueous solution. Compression mixing can be avoided by elimination of vapour space as in the design proposed by McKinnon et al. [15] or by using an oil-seal which allows the volume of the vessel to increase at virtually constant pressure [16].

The other techniques of particular value in the study of lipid systems are differential thermal analysis (DTA) and differential scanning calorimetry (DSC). In DTA the temperatures of the sample and a thermally inert reference substance enclosed in the same thermal environment is changed according to a pre-set programme (ie temperature range and rate of heating). If the sample undergoes a transition with an accompanying exothermic or endothermic enthalpy change then its temperature will transiently exceed or be less than the temperature of the reference substance. It is this difference of temperature which is recorded as a positive or negative deviation as a function of the temperature of the thermal environment. Although this technique can give useful qualitative data it will not yield rigorous thermodynamic data because the difference in temperature that is recorded depends on the rate of heating, the thermal conductivity and the density of the sample, and the geometry of DTA cell employed. In contrast DSC is capable of giving thermodynamic data of high precision. In this technique sample and reference temperatures are continuously monitored and maintained equal by means of a feedback circuit which supplies electrical energy to them. If an endothermic transition occurs in the sample then energy is supplied to maintain its temperature equal to that of the reference. The amount of energy that is supplied is recorded and gives a direct measure of the energy change associated with the transiton. In order to raise the temperature of the calorimeter over a wide range large amounts of energy must be supplied while at the same time the extra energy supplied to the sample is often only very small. Despite this however, very high precision has been achieved [17].

5.3. THERMAL BEHAVIOUR OF ANHYDROUS LIPIDS

5.3.1 The solid state

A general characteristic of lipids in the solid state is the occurrence of more than one crystalline form. This phenomenon, which is known as polymorphism, commonly arises with long-chain compounds which crystallize with the molecules arranged in layers bounded by planes of polar head groups and terminal methyl groups. The hydrocarbon chains usually take up an extended all trans configuration so that the chains are either perpendicular to or tilted with respect to the planes of the head and terminal groups. Such arrangements occur to maximize the van der Waals interactions and minimize the Gibbs energy. Because slight changes in the mutual orientation of adjacent chains can arise with only minor changes in the interactions between them and the corresponding lattice energies, more than one stable crystalline state can exist. The crystalline modifications of lipids and their classification have been discussed by Chapman [18] and Williams and Chapman [19]. The hydrocarbon chain packing modes are classified by means of sub-cells which describe the symmetry relations between equivalent positions of one chain and its neighbours. Briefly four types of sub-cells can be identified. These are (1) when the planes containing the hydrocarbon chains are parallel (triclinic T//, orthorombic 0// and 0´ //, monoclinic M//) which are classified as β-crystals; (2) when the planes containing the chains are perpendicular to each other (O $\perp$ and O´ $\perp$), classified as β´-crystals; (3) when the chains are crossed and the interaction between them is small compared with the polar head group interactions; and (4) when the chains are packed to form a hexagonal lattice giving what are termed α-crystals. A combination of X-ray short spacings and infra-red absorption frequencies forms the basis of a practical system for identifying the three crystal forms α, β and β' [20]. The hexagonal polymorph is of higher energy than the two β forms and it is believed that the transition from the β forms to the α form is associated with the onset of cooperative movements of the hydrocarbon chains [21].

Due to the difficulties of growing well-ordered single crystals of a size (0.2 to 0.5mm) suitable for X-ray analysis accurate structural data on many complex lipids is unavailable. However detailed structural studies on single crystals of phospholipids have been made [6] and the preferrred conformation and molecular packing of some phosphatidylethanolamines and phosphatidyl-cholines are know. The structural notation used for these molecules as defined by Hauser et $al.$ [6] is shown in Fig 1 and the structural parameter relting to DMPC dihydrate are given in the legend. Fig 2 shows the molecular packing in the crystal and the positions of the water molecules in the head group region. The unit cell contains four molecules arranged tail-to-tail in pairs. Each pair consists of two crystallographically independent molecules which have approximate minor image symmetry with respect to the head group conformation. In the bilayer the two DMPC molecules are displaced by 2.5A$^\circ$ (on one zig-zag unit) with respect to the layer normal. This allows the phosphorylcholine groups to pack more tightly giving a molecular area of 38.9A^2 which is almost identical to that for DLPE single crystals in which the acyl chains have a zero angle of tilt.

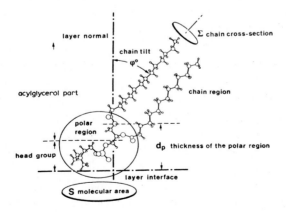

Fig. 1. Structural notation for phosphatidylcholines. For DMPC φ (tilt angle) = 12°, Σ (chain cross-section) = 19.0 A^2, dp (thickness of the polar region) = 10.4 A, S (molecular area) = 38.9A^2. Reproduced from reference [5]

188

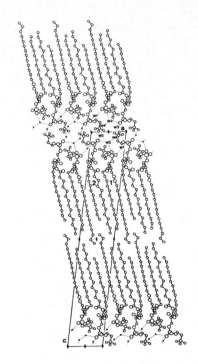

Fig. 2. Molecular packing of 2,3-dimyristoyl-D-glycero-1-phosphorylcholine dihydrate (DMPC) projected onto the a-c plane. The two molecules in the asymmetric unit are labelled 1 and 2. The position of the water molecules is indicated either by W1-W4 or by small open circles. The hydrogen bonds are represented by dotted lines. Reproduced from reference [5].

5.3.2   Thermotropic mesomorphism

On heating, crystalline lipids do not pass directly to the liquid state, intermediate states are formed which are called mesomorphic or liquid crystalline so that the transition from the crystal to the vapour can be represented by the following series of equilibria:

$$\text{crystal} \rightleftharpoons \text{mesomorphic state} \rightleftharpoons \text{liquid} \rightleftharpoons \text{vapour} \qquad (1)$$

The transition from the crystal to the mesomorphic state is endothermic and reversible and occurs at a well defined temperature known as the chain-melting temperature T_c.   As the name implies T_c is the temperature at which the hydrocarbon chains of the lipid molecules undergo rotation about the C-C bonds while the long-range order of the head groups remains.   As a consequence of the onset of twisting and flexing of chains above T_c they are no longer in the fully extended all _trans_ configuration. The evidence for this conclusion has been discussed by Williams and Chapman [19]. Many mesomorphic phases of lipids have been identified [22]. In the case of anhydrous lipids, liquid crystalline phases coresponding to the structures of the β, β´ and α crystalline states are found. Above the chain melting temperture the β form gives rise to the lamellar mesomorphic phase while the hexagonal lattice of the α form gives a hexagonal mesophase. The chain-melting temperature is possibly one of the most important parameters associated with the thermal behaviour of lipids and its relevance to biochemical phenomena will be discussed in more detail later in this chapter.

The value of the chain melting temperature depends on the length and degree of unsaturation of the hydrocarbon chains and also on the head group structure. In general the lower the chain length and the greater the degree of unsaturation the lower the value of T_c. Table 3 shows the thermodynamic data for a range of pure 1,2-diacyl-sn-glycero-3-phosphorylcolines determined by DSC. The enthalpy measurements have a reproducibility within 5%.

TABLE 3. Thermodynamic data for the crystalline → liquid-crystalline transition of 1,2-diacyl -sn-glycero-3 phosphorylcholine

Acyl chain and crystalline form	T_c °C	ΔH kJ mol^{-1}	ΔS J K^{-1} mol^{-1}	ΔS (configurational) J K^{-1} mol^{-1}
Behenoyl (C_{22},β)	120	72.4	184	99.6
Stearoyl (C_{18},β)	115	57.7	149	78.2
Stearoyl (C_{18},α)[a]	78	22.0	62.8	32.3
Palmitoyl (C_{16},α)[a]	65	19.0	56.6	28.5
Myristoyl (C_{14},α)[a]	51	15.3	47.3	24.7

[a]The α crystals are of the monohydrate (Data taken from Phillips _et al._ [23]).

The thermodynamic data is pertinent to the problem of the fluidity of the hydrocarbon chains in the crystalline and liquid-crystalline states. It is seen that both the enthalpy and entropy changes from the β-form to the liquid crystal are larger than from the α-form which is consistent with the transition from the β to the α-crystals being associated with some type of premelting. The values of T_c for the β-crystals are considerably larger than for the α-crystals. From this data Phillips _et al_ [23] calculated the enthalpies and entropies per methylene group. These incremental changes can be compared with the corresponding values for the melting of hydrocarbons and fatty acids as shown in Table 4.

TABLE 4

Enthalpies and entropies of transitions per methylene group

Compound	Transition	ΔH kJ mol^{-1}	ΔS J K^{-1} mol^{-1}
n-hydrocarbons	$\beta x^{le} \to$ liquid	4.1	11
Fatty acids	$\beta x^{le} \to$ liquid	4.1	11
PC	$\alpha x^{le} \to$ liquid-x^{le}	0.84	1.9
PC	$\beta x^{le} \to$ liquid-x^{le}	1.7	4.6

N.B. The calculations are based on the number of methylene in one chain of the phospholipid.

The constant values of the incremental changes per CH_2 group arise from the linearity of the plots of ΔH and ΔS for the transitions as a function of chain length. Since the temperature dependence of ΔH is approximately 300 times larger than that of ΔS, direct comparison of the ΔH values is not possible. Comparison of the entropies shows that for the $\beta \to$liquid-crystal transition the entropy change is about half of the value for melting. If it is assumed that only the configurational entropy is proportional to the number of carbon atoms in the chains then from the incremental changes per CH_2 group an estimate can be made of the configurational entropy change in the transitions. These values are shown in the fifth column of table 3, and amount to approximately 50% of the total change. This increase in entropy is attributable to the increase in the mobility of the hydrocarbon chains at T_c. Direct evidence for this conclusion in hydrated systems will be discussed below.

At higher temperatures the liquid-crystalline state melts to a liquid; for practical purposes this transition corresponds to the normal melting point. Phospholipid melting points can be broadly divided into two classes. Firstly, compounds with high melting points in the region of 200°C which include the 1,2-diacylphosphatidylcholines (230°C), ethanolamines (200°C), and sphingomyelines (210°). In these cases the melting points are practically independent of acyl chain length. In the second class are the 1,2-diacylphosphatidylglycerols and acids with chain length dependent melting points in the region of 70°C. The 1,2-diacylphosphatidylserines and inositols have melting points intermediate between these extremes although they have not been studied in such detail. The high melting materials have a liquid crystal structure which is predominantly held together by an ionic network so that the stronger the basicity of the head group the higher the melting point, which is consistent with the decreasing melting points in the series cholines >

ethanolamines > serines. In the case of compounds with chain length dependent melting points (glycerols and acids) the liquid crystal must be partially held together by hydrogen bonding between the head groups. Since hydrogen bonds are weaker than ionic bonds melting occurs at lower temperatures.

5.4. THERMAL BEHAVIOUR OF HYDRATED LIPIDS

5.4.1 Structure of lipid phases

The addition of water to anhydrous crystalline lipids results in the formation of a variety of lyotropic mesophases. The structure of these phases depends on the chemical structure of the lipid, the temperature and the amount of available water. With increasing water content a binary mixture of lipid and water can pass from a gel state in which the lipid is arranged in planes similar to those in the anhydrous crystal but in which water has penetrated between the head group planes, to a dilute solution of lipid aggregates normally classified as a micellar solution. Details of the range of distinct structural phases which lipids may adopted have been illustrated previously [24]. In this section we will deal with recent developments which relate specifically to phospholipids and some related membrane lipids. The most comprehensively studied lipids are the phosphatidylcholines on which both X-ray and calorimetric studies have been made for water contents upto approximately 60 wt %. As will be apparent it is probably unwise to attempt to make too wide generalizations about the thermotropic behaviour of lipids so we will consider some specific examples.

With increasing temperature hydrated DPPC goes through the following sequence of bilayer phases

$$
\begin{array}{ccccc}
& 27^{\circ}C & & 35^{\circ}C & & 41^{\circ}C \\
L_c & \underset{\text{slow}}{\rightleftharpoons} & L_{\beta}{'} & \rightleftharpoons & P_{\beta}{'} & \rightleftharpoons & L_{\alpha}
\end{array} \qquad (2)
$$

in which the molecular and chain-packing becomes increasingly less ordered [25,26]. The low temperature crystalline phase (L_c) observed only in recent years [27,28] in multilamellar liposomes exists at all hydration levels in the range 10.1 to 93.9 wt%. The formation of the L_c phase is kinetically very slow and is formed only after incubation for an extended period (~ days) at near 0°C. The transition $L_c \rightarrow L_{\beta}{'}$ has been termed the sub-transition [27] and there is increasing evidence that the formation of a stable L_c phase may well be a general phenomena. The gel phases ($L_{\beta}{'}$) of stearoylsphingomyelin [29] bilayers and cerebrosides [30,31] have been found to be convertible to a

192

more stable bilayer structure. Fig 3 shows the structures of the four bilayer phases as proposed by Ruocco and Shipley [26] together with the lattice dimensions derived from X-ray data. The transition from the L_c phase to the L_{β}' phase occurs with increasing head group hydration from 11 to 19 water molecules per molecule of DPPC and involves a tilting of the chains in the bilayers. The so-called "pretransition" from the L_{β}' phase to the "rippled" P_{β}' phase occurs without an increase in hydration. The hydrocarbon chains are no longer titlted but are still mainly in the all <u>trans</u> conformation. The main transition P_{β}' → L_{α} occurs with an increase in hydration from 19 to 25 water molecules per DPPC molecule and corresponds to the melting of the hydrocarbon chains by the introduction of increasing numbers of <u>gauche</u> bonds and the loss of short-range order. The main transition at the chain-melting temperature (T_c) is undoubtedly the most significant thermotropic event undergone by lipid systems. The bilayer periodicity in each of the L_c, L_{β}' and L_{α} phases changes only slowly with water content at a given temperature upto a limiting value, after which it remains constant (Fig 4).

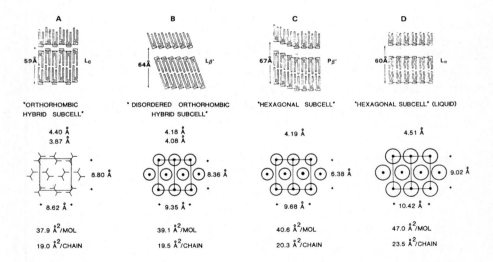

Fig. 3. Proposed molecular (top) and hydrocarbon chain (bottom) packing arrangements in L_c "crystal", L_{β}', gel, P_{β}', gel and L_{α} liquid crystal bilayer phases. The wide angle diffraction data, derived sub-cell parameters and areas/hydrocarbon chain corresponding to the different phases are listed. Reproduced from reference [26]

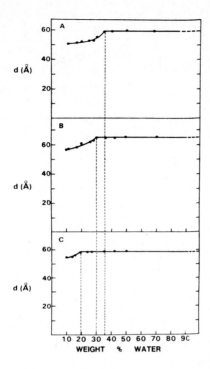

Fig. 4. Plot of bilayer periodicity, d(A), of DPPC as a function of hydration (wt % H_2O). (A) At 60°C in the L_α liquid crystal bilayer phase; (B) at 20°C in the $L_{\beta'}$, gel bilayer phase; (C) at 4°C in the L_c "crystal" bilayer phase. Vertical dashed lines represent the hydration limits of the L_α, $L_{\beta'}$, and L_c bilayer phases. Reproduced from reference [26].

Fig 5 shows the phase diagram for DMPC and phase behaviour for other phosphatidylicholine [32]. As for the anhydrous phosphatidylcholines the chain-melting temperatures increase with chain length. Below T_c the phase behaviour of the phosphatidylcholines are identical with respect to hydration (11 molecules of water per molecule). The phases $L_{\beta'}$ and $P_{\beta'}$ swell to a value of ~15 molecules of water per molecule. The swelling behaviour of the L_α phases is however chain length dependent.

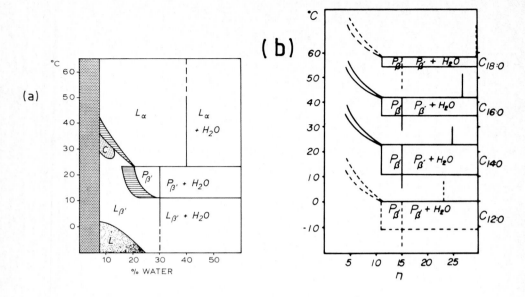

Fig. 5. (a) Temperature-composition phase diagram of hydrated DMPC, (b) Proposed generalized phase behaviour of synthetic phosphatidylcholines (PCs), n is the number of water molecules/PC molecule. The solid lines are phase boundaries based on experimental data; the dashed lines represent assumed phase boundaries. Reproduced from reference [32] with permission of Professor G.G. Shipley.

Studies on the positional isomer 1,3 dipalmitoyl-glycerol-2-phosphatidylcholine (hereafter refered to as β-DPPC) have revealed that this molecules exhibits distinctly different thermotropic behaviour and phase structure [33]. The phase behaviour can be described as follows

$$L_c \xrightarrow[\text{slow}]{27°C} L^i{}_\beta \xrightarrow{37°C} L_\alpha \qquad (3)$$
$$L_{\beta'}$$

On heating hydrated β-DPPC, the crystalline phase L_c passes to the $L^i{}_\beta$ phase and then directly to the liquid crystalline (lamellar) phase, L_α. There is no $P_{\beta'}$ rippled phase. The electron density prolifes of the L_c, $L^i{}_\beta$ and L_α phases are shown in Fig. 6. The periodicity of the phases are 58Å, 47Å and 51Å respectively. The relatively small periodicity of the $L^i{}_\beta$ phase has been interpreted in terms of the interdigitation of the acyl chains in adjacent monolayers to produce a "monolayer" structure. The transition $L_c \rightarrow L^i_\beta$ occurs with an increase in hydration from 14 to 22 molecules of water per molecule of β-DPPC and the water layers between the head group planes increases in thickness from 16A° to 17A°. On further heating the L^i_β gel phase the L_α phase forms with a hydration limit corresponding to 48 molecules of water per molecule of β-DPPC. The "monolayer" structure for β-DPPC in the L^i_β phase is similar to that found for hydrated dipalmitoylphosphatidylglycerol (DPPG) [34] and 2-deoxylysophosphatidylcholine [35]. On cooling the L_α phase a lamellar gel phase is formed with a larger periodicity (54A°) than the L^i_β phase. This $L_{\beta'}$ phase is thought to be a non-interdigitated bilayer phase.

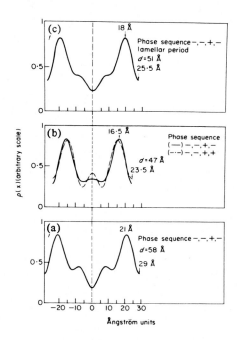

Fig. 6. (a) Electron density profile of the L_c phase of β-DPPC, 32 weight % water at -3°C ($^-$d/2 to $^+$d/2 represents one unit cell). (b) Electron density profile of the L^i_β phase of β-DPPC, 32 weight % water, at 32°C. (c) Electron density profile of the L_α phase of β-DPPC at 40 weight % water at 50°C. Reproduced from reference [33] with permission of Academic Press and the authors.

Interdigitation has also been observed for mixed-chain phosphatidylcholine bilayers when the acyl chain lengths are sufficiently different [36]. The mixed-chain lipid 1-stearoyl-2-caproyl-sn-glycero-3-phosphocholine (SCPC) in the gel phase is more tightly packed than symmetric saturated phosphatidylcholines and the hydrocarbon chains are not tilted. The X-ray structure is consistent with an area per SCPC molecule 3 times the area of the hydrocarbon chain (Fig.7a). Thus the C(18) chain spans the hydrocarbon region of the bilayer and the C(10) chains from apposing molecules are "butted-up", giving a tighter packing than expected. On passing to the liquid-crystalline phase the mismatch in chain lengths is accommodated by interdigitaion and in consequence the bilayer thickness increases by 12-16A° on chain-melting. In contrast to SCPC hydrated SMPC forms the usual tilted gel-state L_β' phase presumably because the difference is chain length between the stearoyl and myristoyl acyl chains is insufficiently large for the chains to pack in an energetically favourable configuration as in Fig. 7a.

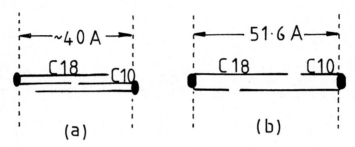

Fig. 7.   Hydrocarbon chain configurations of SCPC in the gel (a) and liquid-crystalline (lamellar) phase (b).   Adapted from reference [36].

The conformation of the head groups in phospholipid bilayers has been studied in some detail, [6,37-40]. For both the phosphatidylethanolamine and the phosphatidylcholines the glycerol group lies approximately parallel to the bilayer surface. For the ethanolamines the area per molecule in the gel state is approximately equal to the sum of the cross-sections of two close-packed hydrocarbon chains (35-42A^2) for the cholines the parallel arrangement of the phosphorylcholine groups requires an area of 47-54A^2. A tighter packing of the hydrocarbon chains can be acheived if the phosphorylcholine group is tilted with respect to the bilayer plane or alternate molecules are displaced giving a 'saw-tooth' pattern of head groups. Packing can also be tightened by tilting the hydrocarbon chains. In crystalline DMPC both inclination of the phosphorylcholine group and mutual displacement occurs [41].

5.4.2   Calorimetric studies on aqueous dispersion of lipids

5.4.21   Dipalmitoylphosphatidylcholines (α and β-DPPC)

1,2-Dipalmitoyl-sn-glycero-3-phosphocholine (α-DPPC) is the most extensively studied synthetic phospholipid. The thermotropic properties of aqueous dispersions of α-DPPC and the positional isomer β-DPPC are summarised in table 5. The measurements were made on hydrated dispersions of multilamellar liposomes by use of a high precision Privalov DSC [17,27] a Perkin-Elmer DSC 2 [25,28,33] or a Beckman DSC-2 [42]. Mabrey-Gaud [54] has discussed the specific application of the DSC to liposomal dispersions.

Table 5.   Thermotoropic properties of 1, 2-dipalmitoyl-sn-glycero-3-phosphocholine (α-DPPC) and 1, 3-dipalmitoyl-sn-glycero-2-phosphocholine (β-DPPC)

Transition	$\dfrac{T}{{}^{\circ}C}$	$\dfrac{\Delta H}{kJ\ mol^{-1}}$	(Ref)
	α-DPPC		
Subtransition ($L_c \rightarrow L_\beta{}'$)	18.4	13.5[a]	(27)
	19.8	15.5[b]	(28)
	18.0	12.6[c]	(25)
Pretransition ($L_\beta{}' \rightarrow P_\beta$)	35.1	4.56	(27)
	35.0	5.65	(28)
	34.3	7.11	(25)
Main transition ($P_\beta{}' \rightarrow L_\alpha$)	41.1	28.9	(27)
	41.3	33.1	(28)
	40.3	38.1	(25)
	β-DPPC		
Subtransition ($L_c \rightarrow L^i{}_\beta$)	27	38.1[d]	(33)
Main transition ($L^i{}_\beta \rightarrow L_\alpha$)	37	43.9	(33)
	37–38	39.3	(42)

a   Samples incubated for 3.5 days at 0°C

b   Samples incubated for 5 days at 4°C

c   Samples incubated for 13 days at -4°C

d   Samples incubated for approximately 1 year at -3°C

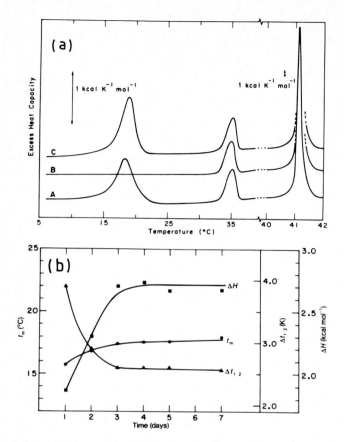

Fig. 8 (a) Smoothed calorimetric transition curves for a multilamellar suspension of high purity DPPC observed at a scan rate of 0.5 K min⁻¹. The noise level in the original recordings was about 0.04 kcal K⁻¹ mol⁻¹. Curve A, scan of suspension after being cooled at 0°C for 3.5 days. Curve B, reheating after cooling at 1°C for 3 hr. A barely detectable endothermic peak that was observed at 14°C does not show up in the figure. Curve C, reheating after cooling to 2°C for 3 days. Main transitions (only one shown in detail) are plotted on different scales. (b) Effect of cooling time at 0°C on the subtransition of a multilamellar suspension of commercial DPPC observed at a scan rate of 0.5 K min⁻¹. The various transition parameters are plotted against the cooling time in days. Reproduced from reference [27] with permission of Professor J.M. Sturtevant.

Fig. 8a shows the excess heat capacity of multilamellar suspensions of α-DPPC as a function of temperature. Three transitions are very clearly seen in samples that have been stored at a low temperature for an extended period (curve A), however after heating and only brief incubation at 1°C the subtransition is not detectable. The subtransition returns after several days incubation and the enthalpy, peak width ($\Delta t_{\frac{1}{2}}$) and transition temperature all reach limiting values (Fig. 8b). The enthalpy of the transition amounts to approximately 40% of the enthalpy change on chain-melting ($P_{\beta}' \rightarrow L_{\alpha}$). It is thus surprising that the ($L_c \rightarrow L_{\beta}'$) transition remained undetected (or unreported) until 1980. Whereas the pretransition was reported two decades ago. The main transition of α-DPPC occurs at approximately 41°C. There are considerable variations between the enthalpies of chain-melting found by different laboratories and even by the same group at different times [27]; these discrepancies most probably reflect differences in the thermal history of the samples and probably do not arise from the precision of the calorimetry. Measurements made with a high precision Privalov DSC would be expected to be the most reliable and free of the well known problems of heating rates associated with commercial DSC equipment.

Kodama *et al* [55] have studied the development of the main transition as controlled amounts of water are added to completely dehydrated (vacuum dried (10^{-4}Pa) above the anhydrous chain melting temperature) α-DPPC. The thermograms and resulting phase diagram are shown in Fig. 9. The chain melting transition for the anhydrous material is 97.8°C (NB this is very significantly higher than previous reports (see table 3)). The main transition (T_c) does not decrease smoothly with increasing water content but stepwise, the thermograms showing the composite thermal behaviour of five endothermic peaks. The final peak (T_c=42.6°C) corresponds to the limiting behaviour at a water content above 50g%. A further significant observation is that the pretransition appears for the first time at a water content of 17g% which is at the same water content where ice-melting is first observed suggesting that free water is a requirement for the appearance of the pretransition.

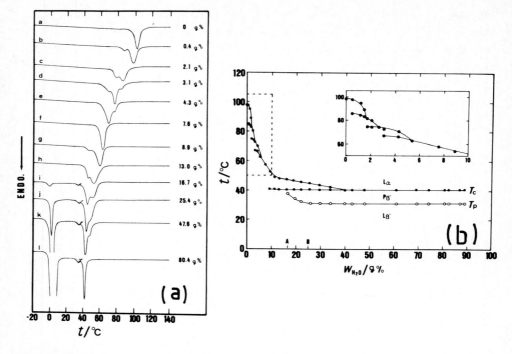

Fig. 9. (a) Typical DSC curves at the scanning rate of 1 K min⁻¹ for 12 samples with increasing amounts of water (g%) indicated at the right hand side of each curve. The endothermic peak with dotted area corresponds to the pretransition.   (b) Phase diagram of the DPPC-water system. ● , T_c curve; O , T_p curve.   At the water content (17%) indicated by mark A the phase separation to L_β´ and the excess water phase takes place.   At mark B (25 g%) the pretransition temperature reaches the limiting temperature of 32.3⁰C.   Reproduced from reference [55].

The thermotropic behaviour of β-DPPC as described above (4.1) is characterised by only two transitions according to the scheme of Serrallach et al [33].   The pretransition has an enthalpy almost equal to that of chain-melting.   There is a discrepancy between the observations of Serrallach et al [33] and the previous study of Seeling et al [42] on β-DPPC who reported a "pretransition" at 17-23⁰C with a transition enthalpy of 7.1 to 17.6 kJmol⁻¹. The latter authors stress that the temperature and enthalpy of the pretransition was strongly dependent on sample pre-history and in the light of the work of Serrallach et al [33] it is conceivable that they were perhaps observing the subtransition of β-DPPC.

5.4.22   Phospholipids

The   transition   temperatures,   enthalpies   and   entropies   of   the   chain transition   (L_β´   → L_α)   for   a   range   of   phosphatidylcholines   and   other phospholipids are given in Tables 6 and 7. It is appropriate to consider how chemical structure affects the thermotropic properties such as acyl chain length, head groups structure and the associated affects of hydration, pH and ion binding.

Table 6. Transition temperatures, enthalpies and entropies of the gel to liquid-crystalline phase transition ($L_\beta \to L_\alpha$) of phosphatidylcholines.

Liquid	$\begin{array}{c}T\\ °C\end{array}$	$\begin{array}{c}\Delta H\\ kJ\ mol^{-1}\end{array}$	$\begin{array}{c}\Delta S\\ J\ mol^{-1}K^{-1}\end{array}$	(Ref)
C(12):C(12)	-1.8	7.11	26.2	(43)
C(14):C(14)	23.7	26.2	88.3	(44)
C(14):C(14)	24.0	27.2	91.5	(50)
C(16):C(16)	41.1	28.9	131	(27)
C(16):C(16)	41.5	36.4	115	(50)
C(18):C(18)	54.2	33.0	101	(47)
C(18):C(18)	55.1	43.1	131	(46)
C(18):C(18)	54.3	45.6	139	(50)
C(20):C(20)	64.1	51.5	152	(50)
C(22):C(22)	75	61.9	178	(45)
C(10):C(18)	10.1	33.1	117	(46)
C(14):C(18)	38.6	29.0	93.0	(47)
C(14):C(18)	42	34.3	109	(48)
C(16):C(18)	49	34.9	108	(47)
C(18):C(10)	19.7	42.3	144	(46)
C(18):C(12)	18.5	32.2	110	(46)
C(18):C(14)	29.9	23.4	77.2	(46)
C(18):C(14)	29.4	21.8	72.1	(47)
C(18):C(14)	34	25.1	81.7	(48)
C(18):C(16)	43.9	30.4	95.6	(47)
C(18):C(16)	44.1	30.1	94.9	(46)
C(18):C(18 cis Δ^9, Δ^{12})	-16.2	13.8	53.7	(56)
C(18):C(18 cis Δ^9, Δ^{12},Δ^{15})	-13.0	27.6	106	(56)
C(18):C(18 cis Δ^5 $\Delta^8,\Delta^{11},\Delta^{14}$)	-12.6	22.2	85.2	[56]

Table 7 Transition temperatures, enthalpies and entropies of the gel to liquid-crystalline phase transition ($L_\beta \rightarrow L_\alpha$) of selected phospholipids.

Lipid system	T $^\circ C$	ΔH kJ mol^{-1}	ΔS J mol^{-1} K^{-1}	(Ref)
Phospatidylethanolamines (PE)				
C(12):C(12) PE	30.5	14.6	48.1	(49)
C(12):C(12) PE	30.5	17.9	58.9	(50)
C(14):C(14) PE	49.1	23.8	73.9	(49)
C(14):C(14) PE	49.9	27.6	85.4	(50)
C(16):C(16) PE	63.1	36.8	109	(49)
C(16):C(16) PE	63.9	36.0	107	(50)
C(18):C(18) PE	70.4	43.9	128	(50)
C(20):C(20) PE	81.1	51.0	144	(50)
Phosphatidic acids (PA)				
C(14):C(14) PA(pH6)	52.2	23.8	73.1	(50)
C(14):C(14) PA(pH12)	22.4	17.2	58.2	(50)
C(14):C(14) PA(pH7)	48-50	24.3	75.4	(51)
C(14):C(14) PA(pH7, 0.1 M NaCl)	45-48	29.7	92.9	(51)
C(14):C(14) PA(pH6 0.1 M NaCl)	51	20.1	62.0	(52)
C(16):C(16) PA(pH6)	65.0	33.1	97.9	(50)
C(16):C(16) PA(pH12)	43.1	23.8	75.3	(50)
Phosphatidylserines				
C(14):C(14) PS(pH1)	48	25.2	78.5	(53)
C(14):C(14) PS(pH7)	36	29	93.8	(53)
C(14):C(14) PS(pH13)	15	25	86.8	(53)
C(16):C(16) PS(pH1)	65	35.3	104	(56)
C(16):C(16) PS(PH7)	54	37.4	114	(56)
C(16):C(16) PS(PH13)	32	33.6	110	(56)

(a) Acyl chain length

The transition temperatures of the $L_\beta' \rightarrow L_c$ phase change increase with increasing acyl chain length. Fig. 10 shows data for the phosphatidylcholines and the phosphatidylethanolamines. The trends are not linear and the slopes of the lines decrease with increasing chain length. Extrapolation of the curves suggests that for acyl chain lengths of 24 or greater there would be no difference between the chain-melting temperatures of the cholines and ethanolamines. Assuming chain-melting is a first-order thermodynamic transition the entropy change can be calculated from the measured enthalpy and the transition temperature ($\Delta S = \Delta H/T_c$). Both the enthalpies and the entropies of chain-melting increase with temperature. Fig. 11 shows ΔH and ΔS taken from Tables 6 and 7 for the cholines and ethanolamines plotted as a function of acyl chain length. There is unfortunately a relatively large scatter when data from different sources are compared, in contrast to measurements of the transition temperature which show more consistency. The transition enthalpies and entropies for the cholines and ethanolamines are not significantly different for a given chain length and can be represented by the linear equations.

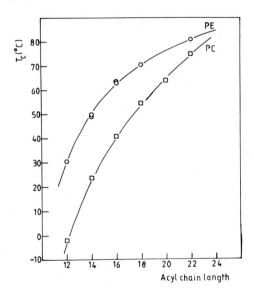

Fig. 10. Dependence of chain melting temperature (T_c) for hydrated phosphatidylcholines (☐, PC) and phosphatidylethanolamines (○, PE) on acyl chain length.

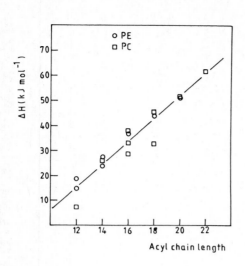

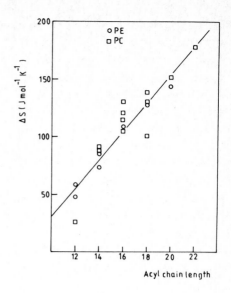

Acyl chain length   Acyl chain length

Fig. 11.   Dependence of the enthalpy (ΔH) and entropy (ΔS) of chain melting on acyl chain length for hydrated phosphatidylcholines (□ , PC) and phosphatidylethanolamines (○ , PE).

$$\Delta H(kJmol^{-1}) = 4.57(\pm0.32)N - 39.5(\pm5.2) \qquad (4)$$

$$\Delta S(Jmol^{-1} \ K^{-1}) = 12.4(\pm1.1)N - 92.2(\pm18.1) \qquad (5)$$

where N is the number of carbon atoms per acyl chain.   The methylene increments in ΔH and ΔS are thus 2.29 kJ mol^{-1} per CH$_2$ group and 6.20 J mol^{-1} K^{-1} per CH$_2$ group respectively.   These values may be compared with chain melting of solid hydrocarbon (β crystal → melt) for which ΔH and ΔS are 4.1 kJmol^{-1} per CH$_2$ group and 11 J mol^{-1} K^{-1} per CH$_2$ group respectively (table 4).

A method of correlating the thermodynamic parameters with the structure of phosphatidylcholines based on the division of their structure into three regions (R1, R2 and R3) has been proposed by Mason and Huang [57].   R1 consists of the carbonyl and α carbon atom of each acyl chain, R2 the acyl chain segments and R3 the terminal region of the sn2 chain which is 1.5 carbon bond lengths shorter than the sn1 chain.   They argue that the non equivalance of the two acyl chains in the gel state and the fact that the terminal methyl groups are out of register by ~ 2.7A^0 disrupts the all-<u>trans</u> packing configuration. The magnitude of the thermodynamic properties is largely determined by the carbon units within R2 to maximize van der Waals contact. The restriction imposed by regions R1 and R3 on region R2 indirectly affect the phase transition. It is suggested that the disruption of packing is expressed by a perturbation parameter (P) given in terms of the relative C-C bond lengths of the three regions i.e. $P=(R_1+R_3)R_2$ x 100. Good linear correlations are obtained for T_c, ΔH, ΔS and ΔV for chain-melting when expressed as a function of P. The most convincing aspect of this approach is the correlation between T_c and P, since the scatter in measured values of ΔH, and ΔS was not taken into account in their treatment. An interesting corollary of this approach is that for a value of P corresponding to acyl chain lengths less than 11 $\Delta H=\Delta S=\Delta V=0$ and no bilayer phase transition should be observed. This prediction is consistent with studies on short chain phophatidylcholines (C_6 - C_8) which form ionic micelles in aqueous solution [58,59].

Studies on mixed chain phosphatidylcholines [46-48] (Table 6) show that for the series in which the *sn*-2 chain is the same (stearoyl) and the *sn*-1 chain length is increased the transition temperature increases but the enthalpy of chain-melting remains constant. When the shorter acyl chain is in the *sn*-2 position the temperature of the transition also increases with chain length but there is no obvious consistent variation in ΔH. Mason *et al* [46] have interpreted the trends in ΔS for these mixed chain phosphatidylcholines in terms of the effect of the conformational inequivalence of the two acyl chains. Fig. 12 (a) shows the parameters used to define the molecular conformations. ΔC is the absolute difference in chain length of the two acyl chains given by $n_1-n_2+1.5$ where n_1 and n_2 are the number of carbon atoms in the *sn*-1 and *sn*-2 chains respectively. The figure of 1.5 allows for the displacent between the two chains in terms of C-C bond lengths (i.e. ~ 3.7/2.54A^0) which arise because the *sn*-2 chain must be bent at the C(2) atom in order to run parallel to the *sn*-1 chain. CL is the length of the longer of the two chains i.e. either n_1-1 or $n_2-2.5$ which ever is largest. The parameter $\Delta C/CL$ represents the chain inequivalence relative to the overall chain length. Fig. 12(b) shows $\Delta S/\Delta Sexp$ as a function of $(\Delta C/CL)x10^2$. $\Delta Sexp$ is the expected transition entropy calculated from the methylene increment ($\Delta Sexp=(n_1+n_2)x7.9$ Jmol^{-1}K^{-1}). The oscillation in $\Delta S/\Delta Sexp$ reflects the disruptive effect of the terminal methyl groups on bilayer packing in the gel state coupled with progressively greater interdigitation as the chain-length difference is increased.

Unsaturation in mixed chain phosphatidylcholines leads to substantial decreases in both chain-melting temperature and the enthalpy of the transition [56] (Table 6). Introduction of upto three double bonds in the acyl chain in the 2-position of C_{18} cholines brings the chain-melting temperature down by almost $70\,^{\circ}C$. It is suggested [56] that the overall effect of multiple double bonds on T_c is a compromise between the increase in area per molecule which sterically restricts molecular packing and the loss of rotational freedom of the chains imposed by the double bonds. Increasing unsaturation of a single acyl chain does not necessarily lead to a systematic decrease in T_c or as is generally assumed an increase in fluidity at a given temperature above T_c.

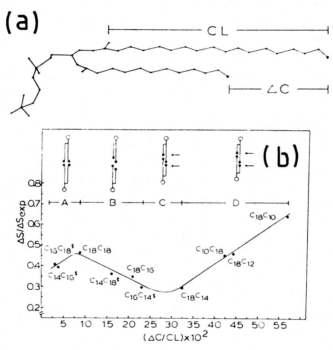

Fig. 12 (a) Molecular diagram of 1-stearoyl-2-lauroyl-*sn*-glycero-3-phosphorylcholine demonstrating the parameters used in the calculation of the chain parameter ($\Delta C/CL$). $\Delta C = |n_1 - n_2 + 1.5|$ where n_1 and n_2 are the number of carbons in the *sn*-1 and *sn*-2 acyl chains, respectively. CL is the length (in carbon bonds) of the longer of the two fatty acyl chains; thus, the larger of the two values $n_1 - 1$ or $n_2 - 2.5$ is employed.
(b) Plot of the magnitude of the thermal phase transition as a function of the chain parameter for the saturated mixed-chain phosphatidyl cholines. The trends displayed by the plot have been divided into four regions (A-D), and the acyl chain packing suggested to exist within these regions are diagramatically represented above each region. The size of the acyl chain terminal methyl groups has been exaggerated in order to emphasize their perturbing effect upon the acyl chain packing. Reprinted with permission from Biochemistry 20, 6086-6092, 1981. Copyright American Chemical Society.

(b) Head group structure, pH and ion binding

At physiological pH for a given acyl chain length the structure of the head group affects the chain-melting temperature but has little affect on the enthalpy of the transition. Thus for dipalmitoylphospholipids chain-melting increases in the series PC<PS<PE≈PA. Since the transition enthalpies remain almost independent of head group it follows that the major affects of changing the head group are entropic. High-sensitivity DSC has been used to study head group effects on the apparent molar heat capacities in gel and liquid-crystalline phases of PC^S, PE^S and PA^S [50]. The methylene increment in the heat capacities (ΔC_p) for PC^S in the gel state at 20 °C is 54 J K^{-1} mol^{-1} and for PE^S in 44 J K^{-1} mol^{-1}. The corresponding figures for the liquid-crystalline state at 80 °C are 69 and 31 J K^{-1} mol^{-1} respectively. For the gel phase the methylene increment is considerably larger than for n-alkanes (~19 J K^{-1} mol^{-1}) for both PC^S and PE^S. In the liquid-crystalline phase only PE^S have the expected behaviour (the methylene increment for n-alkanes at 80 °C is 33 J K^{-1}mol^{-1}). The differences in behaviour between the two types of head group is believed to relate to differences in hydration, PE^S are less hydrated than PC^S. It is possible that for the PC^S, and to perhaps a lesser extent for the PE^S, that there may be significant amounts of water in the bilayer. Estimates of the heat capacities based on the summation of the expected heat capacities of groups in the structure give C_p values which are considerably lower than experimentally observed values. For example, for DPPC at 50°C Blume [50] estimates that C_p= 1351 J K^{-1} mol^{-1} in the liquid-crystalline state compared with the experimental value of 1757 J K^{-1} mol^{-1}. This difference being due to a structural contribution to C_p through hydrophobic hydration of 5 to 6 methylene groups.

The state of head group inoisation is important for the phophatidic acids (PA^S) and phosphatidylserines (PS^S). For the PA^S changing the pH from 6 to 12 changes the head group from being singly charged to doubly charged the transition temperature drops more than 20 °C and the transition enthalpy is significantly reduced. For singly charged DMPA the transition temperature goes through a maximum at pH 3.5 where the bilayer is particularly stable and the transition enthalpy goes through a maximum [51]. At lower pH when the lipid is fully protanated it forms a suspension of microcrystals. Ionic strength also effects gel state stability and the cooperativity of chain-melting. At pH 7 the transition enthalpy increases on addition of 0.1M sodium chloride and the size of the cooperative unit decreases from ~100 to ~30.

The phosphatidylserines show complex thermotropic behaviour dependent on thermal history, pH and extent of hydration [53]. For DMPS at pH 1 hydration is low and three different transition temperatures have been observed depending on the method of preparation and history of the samples. A dispersion in 0.1M hydrochloric acid on first heating from 0 to 100°C gives an endothermic transition at 52°C, an exothermic transition at 60°C and a further endothermic transition at 84°C. Cooling to 0°C followed by reheating from 0°C to 100°C shows only one endotherm at 52°C. This transition which corresponds to chain-melting drops to the lower temperature of 48°C after prolonged storage. The complex behaviour observed on initial heating reflects effects associated with hydration. The endotherm at 84°C is associated with chain-melting in unhydrated bilayers. PS[s] are zwitterionic at pH 1, singly negatively charged at pH 7 and doubly negatively chasrged at pH 13. On passing through this sequence, the transition temperature is reduced from 48°C to 15°C for DMPS and from 65°C to 32°C for DPPS.

Cation binding to bilayers have been shown to markedly affect thermal properties in a number of systems including phosphatidic acids [51,52], phosphatidylglyercols [60,61,62] and phosphatidylserines [53,63].. Both Ca^{2+} and Cd^{2+} ions complex with DPPC and DMPA [52]. Addition of Ca^{2+} to DMPA reduces the enthalpy of chain-melting at 51°C but not the transition temperature implying that the DMPA-Ca^{2+} complex forms a separate phase in equilibrium with uncomplexed DMPA. The transition enthalpy goes to zero when the DMPA - Ca^{2+} ratio is 1:1. At pH 6 DMPA has a single negative charge and it might be expected that the divalent cation would neutralize two phosphatidic acid head groups, however, it is believed that on binding Ca^{2+} a proton is released consistent with the formation of a complex 1:1 stoichiometry [64]. In contrast the interaction of Ca^{2+}, Mg^{2+} and Mn^{2+} ions with phosphatidylglycerols results in the formation of a stable dehydrated phase at a bound cation to lipid mole ratio equal to or greater than 0.5:1 indicating that all the lipid is neutralized by the cations [60]. There are however specific effects of individual cations on the thermotropic behaviour as illustrated in Fig. 13(a) for DPPG as well as kinetic effects associated with heating rate. The sequence of events which give rise to the thermal effects is shown in Fig. 13(b). At a low temperature a stable dehydrated state (B) is formed which probably involves "trans-type" bridging with the divalent cation neutralizing the charge on two head groups across the bilayer as has been found for Ca^{2+}-PS complexes [65]. State B undergoes a transition to a hydrated liquid-crystalline phase (C) on heating. The differences in the transition temperature for the

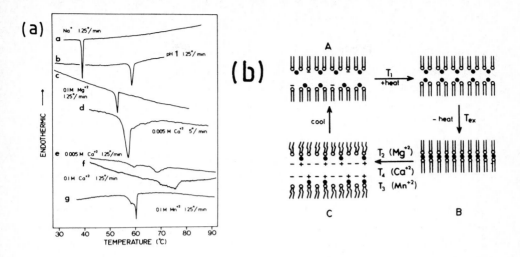

Fig. 13 (a) DSC cooling thermograms of DPPG in the presence of (a) 0.1M Na+, pH 7.4, (b) 0.1M Na+, pH 1, (c) 0.1M Mg^{2+}, (d & e) 0.005M Ca^{2+}, (f) 0.1M Ca^{2+}, and (g) 0.1M Mn^{2+}. Cooling rates were 1.25°C/min for all except (d) which was 5°C/min. Different amounts of lipid were present in the pans so the relative peak areas cannot be compared except for (d) and (e) where the sample were scanned at different rates.

(b) Diagrammatic representation indicating the structure of the hydrated metastable state A, dehydrated stable state B, and hydrated liquid-crystalline state C and the mechanism of conversion between them. The transition from state B to state C is reversible at slow cooling rates for Ca^{2+} and Mn^{2+}. The unlettered state in the upper right-hand corner may not actually exist but indicates that the cation must rebind and neutralize the bilayer surfaces before state B can occur. The transition from state A to state B may occur during the exothermic transiton T_{ex}, at temperatures below T_1 at slow heating rates in the case of Ca^{2+} and Mn^{2+}. (●) Divalent cation. Reprinted with permission from Biochemistry 22, 5425-5435, 1983. Copyright (1983) American Chemical Society.

different cations probably reflects the state of dehydration in state B. For example, in the presence of Mg^{2+} the transition temperature is lower than in the presence of Ca^{2+} ions because the Mg^{2+} complex is more hydrated. Rapid cooling of the liquid-crystalline phase results in a metastable hydrdated phase (A) in which "cis-type" bridging by the cations between adjacent lipid molecules in the phase of the bilayer occurs. In state A the bilayer separation is relatively large due to incomplete charge neutralization and electrostatic repulsion. The hydrated metastable state (A) to the stable dehydrated state (B) requires further ion binding before charge neutralization is achieved and this is possibly responsible for the exothermic peaks seen in the thermograms.

The effects of the trivalent lanthanum ion (La^{3+}) on the thermal properties of DPPC have been investigated [66,67]. Initially, it was reported [66] that the main transition was unaffected by the presence of La^{3+} ions below a concentration of 1M (lipid/ion ratio 1:1). Using high precision DSC Chowdhry et al. [67] find that at a lipid to ion ratio of 3:1 the enthalpy and specific heat of the pretransition are reduced by 10 and 25% respectively. The enthalpy of the main transition is unaffected by La^{3+} ions but the transition temperature is increased by 0.93°C. X-ray data on DPPC indicates that there is a change in the configuration of the acyl chains from the $L_{\beta}{'}$ to the L_{β} state on addition of La^{3+} ions ie. the chain tilt angle (~30°) is lost. However, Chowdhry et al. [67] argue that if the change in tilt angle from the $L_{\beta}{'}$ to the P_{β} phase were to make a major contribution to the pretransition then larger changes in the thermal properties would occur. It is thus concluded that the existence of a tilt angle different from 0° is neither a prerequisite for nor a major cause of the pretransition in DPPC.

5.4.23 Phospholipid mixtures

The thermotropic behaviour of mixtures of different phospholipids is of direct relevance to the behaviour of biological membranes. Most studies have been concerned with binary mixtures of lipids in water; while strictly such a system is a ternary mixture, the water is generally in great excess and has an approximately constant activity. Studies of such systems by differential scanning calorimetry can yield a partial or complete phase diagram for the system. From a thermogram for a particular lipid composition, showing a gel to liquid-crystalline phase transition, the onset temperature corresponding to the formation of the first trace of liquid-crystalline phase defines a point on the solidus curve and the completion temperature corresponding to the melting of the last trace of gel phase defines a point on the liquidus curve.

The phase behaviour of lipid mixtures has been reviewed by Lee [68,69] and some of the types of phase diagrams encountered are illustrated in Fig. 14. Fig. 14(a) is characteristic of an ideal lipid mixture in which the two lipids are miscible over the entire composition range in both the fluid and solid phases. Such behaviour is observed or approximated to if the two lipids have identical head groups and similar acyl chain lengths. If the acyl chain lengths are different, partial miscibility in the solid phase occurs (Fig 14 (b)) and part of the solidus line approaches the horizontal or shows a distinct kink. Complete immiscibility in the solid phase coupled with ideal mixing in the fluid phase results in a phase diagram with a eutectic point (Fig. 14 (c)), while partial miscibility in the fluid phase results in a diagram as shown in Fig. 14 (d).

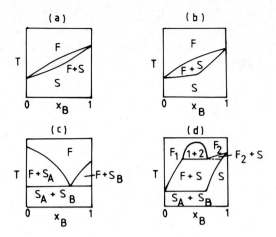

Fig. 14. Schematic phase diagrams for phospholipid mixtures F and S denote fluid and solid phases. The different types of phase diagram (a-d) are discussed in the text.

Table 8 summarises the phase behaviour of a selection of binary lipid mixtures. For the phosphatidylcholines an acyl chain length difference of four carbon atoms is sufficient to lead to partial miscibility in the solid phase. If one of the lipids has an unsaturated acyl chain then immiscibility occurs when the chain length difference is only two carbon atoms. If unsaturation is

Table 8. Phase behaviour of some binary mixtures of phospholipids.

System	Phase Diagram Type	(Ref)
DPPC + DMPC $(C_{16} + C_{14})$	ideal (a)	(70)
DPPC + DSPC $(C_{16} + C_{18})$	ideal (a)	(70)
DMPC + DSPC $(C_{14} + C_{18})$	solid phase immiscibility (b)	(70, 43)
DLPC + DSPC $(C_{12} + C_{18})$	solid phase immiscibility (b)	(43)
DPPC + POPC $(C_{16}, C_{16} \text{ cis}\Delta^9 C_{18})$	solid phase immiscibility (b)	(71)
DPPC + SOPC $(C_{16}, C_{18} \text{ cis}\Delta^9 C_{18})$	solid phase immiscibility (b)	(71)
DPPC + DPPE	solid phase immiscibility (b)	(70)
DPPC + DPPS	solid phase immiscibility (b)	(72)
DPPE + DEPC $(C_{18}, \text{trans}\Delta^9 C_{18})$	solid & fluid phase immiscibility (d)	(73)
DPPC + PI	ideal (a)	(74)
$DC_{15}PG$ + DMPC	non-ideal (a/b)	(75)

coupled with a difference in head group then both fluid and solid phase immiscibility can occur as in the case of mixtures of DPPE and dielaidoylphosphatidylcholine (DEPC) which have the phase diagram shown in Fig. 14 (d) where component B is DPPE.

More detailed studies on the phase behaviour of mixtures of DMPC+DPPS [72] and DPPC+DPPE [76] show that there is an additional complication in the phase diagram arising from the P_β phase. Fig. 15 shows the phase diagram for DPPC+DPPE. The initial work [70] on this mixture showed a solidus line from $x_{DPPE} = 0$ at ~ 41°C with a kink at x_{DPPE} ~ 0.3 indicative of partial miscibility. The more detailed phase diagram determined from both calorimetry and a variety of other physical techniques shows the region of existence of the pretransition below x_{DPPE} of 0.6 (NB while there is good agreement between the results obtained for the liquidus curve using different physical methods the onset of melting is less well defined).

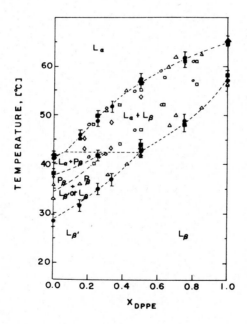

Fig. 15. Experimental temperature vs. composition diagram for aqueous dispersions of DPPC/DPPE mixtures: (●) data from ^{13}C NMR; (■) data from ^{2}H NMR. Included are data obtained with other physical techniques: (O) Tempo spin-label; (Δ) calorimetry; (□) chlorophyll a fluorescence; (◊) Raman spectroscopy. Reprinted with permission from Biochemistry 24, 6243-6253. Copyright (1982) American Chemical Society.

Phase separation in binary mixtures of phospholipids in which one component is an acidic phospholipid can be induced by divalent ions, typically Ca^{2+}, such lateral phase separation occurs in mixtures of DPPC+DPPS [77], egg PC + bovine brain PS [78] but to a much lesser extent for egg PC + yeast PI [79]. Barium and strontium ions induce similar phase separation in PC+PS mixtures but Mg^{2+} ions are ineffective. The difference in behaviour between Ca^{2+} and Mg^{2+} ions probably relates to the different degrees of hydration of the PS-cation complexes as discussed above (Section 4.22 (b)).

Lipid-cholesterol interaction has been the subject of considerable research over many years [80] since it was first shown by Ladbrooke et al. [81] that cholesterol addition to PC bilayers resulted in the loss of the gel to liquid crystalline phase transition. More recent work has been concerned with the precise nature of the endotherms observed for binary mixtures of cholesterol with a variety of single acyl PC[s], mixed acyl PC[s] [82] and with PE[s] [83]. Detailed examination of the endotherms observed for several mixtures of PC[s] and cholesterol suggests that at low cholesterol concentrations (<25mol %) the endotherms can be deconvoluted into a broad and narrow peak. One interpretation of these peaks is that the narrow (high temperature) peak is associated with transitions occurring in almost pure lipid domains and the broad (low temperature) peak is associated with a cholesterol-rich phase. A similar conclusion was reached in studies on stearoylsphingomyelin-cholesterol dispersions, here a cholesterol-rich phase co-exists with a cholesterol-poor phase in which the spingomyelin molecules may exist in two different gel states [84]. In contrast cholesterol appears not to interact strongly with the structurally related ganglioside GM [85]. For sphingomyelin, cholesterol reduces the enthalpy of the narrow peak at zero at a concentration of ~ 20 mole % [84] whereas the enthalpy of melting of GM_1 ganglioside in the presence of cholesterol is never completely eliminated even at a concentration of 50 mole % cholesterol [85].

There have been numerous theoretical models of the phase behaviour of pure lipids [86] and lipid mixtures [87-90]. Merajver et al. [86] have explained the gel to liquid crystalline (lamellar) phase transition of pure sphingomyelins, N-lignocerylsphingosinephosphorylcholine ($n-C_{24}$) and N-palmitoyl- sphingosine-phosphorylcholine ($n-C_{16}$) on the basis of a statistical mechanical analysis of the conformations of the acyl chains constrained on a three-dimensional lattice by generating a random walk on a two dimensional square lattice. The transition temperature and enthalpies predicted are in good agreement with calorimetric data. Molecular field theories [87,90] have been used to predict the phase behaviour of binary mixtures of phosphatidylcholines and Scott et al. [88] have used the scaled particle theory.

5.5.   PROTEIN-LIPID INTERACTIONS

The structure and function of biological membranes depends on the properties of phospholipid bilayers, intrinsic and extrinsic proteins and glycoproteins and the subtle interactions between these macromolecules and phospholipids.   The vast field of macromolecule-lipid interactions has been investigated by a wide range of physical techniques [91,92] in which thermodynamic measurements form an important part.   Calorimetric studies on natural membranes have revealed thermal transitions in both the lipid and macromolecule components [93-95]; however because of the complexity of   such systems and the difficulty in obtaining high precision calorimetric data many workers have chosen to study model membrane systems in which particular purified macromolecules have been incorporated into the lipid bilayers of vesicles or liposomes.   Results obtained on such "reconstituted" systems are more readily interpreted and the information gained should be transferable to natural membranes.   The effects of membrane proteins on the thermotropic properties of membrane lipids have been extensively investigated [96-98].

The earliest classification of the effects of proteins on the enthalpy (ΔH) and transition temperature (T_c) of the gel to liquid-crystalline phase transition was given by Papahadjopoulos *et al.* [96] who recognised three groups of interactions as follows:-

Group 1 interactions characterised by an increase in ΔH accompanied by either an increase in T_c or no change.   Such effects are observed when the protein or macromolecule is bound electrostatically to the lipid bilayer and are exemplified by ribonuclease and polylysine.

Group 2 interactions characterised by a drastic decrease in both ΔH and T_c. Such effects arise as a consequence of surface binding followed by partial penetration and/or deformation of the bilayer, examples of which are cytochrome c and basic myelin protein.   The interactions are initially electrostatic and depend strongly on the negative charge of the bilayer.

Group 3 interactions characterised by a linear decrease in ΔH as a function of macromolecule to lipid ratio with no significant effect on T_c.   Such effects arise as a result of penetration of the marcomolecule into the bilayer as a consequence of strong hydrophobic interactions.

The interpretation of thermotropic behaviour typical of group 3 interactions has been considered by numerous investigators and amongst the macromolecules studied are included synthetic polypeptides [191,102], glucagon [103], human apolipoprotein A-II [104,105], cytochrome b_5 [106], bacteriorhodopsin [110-112], human glycophorin [113,114], glycoprotein G [115] and peripheral matrix protein M [116] from the vesicular stomatitis virus and cytochrome c oxidase [118-120]. The origin of the decrease in ΔH at approximately constant T_c has been described in terms of a number of theoretical models for protein-lipid interaction [121-124]. At the simplest level the decrease in ΔH which is also usually accompanied by loss of the pretransition is represented by the equation

$$\Delta H = \Delta H^o \left(1 - N \frac{[P]}{[L]} \right) \qquad (6)$$

where ΔH^o is the enthalpy change for the pure lipid gel to liquid-crystalline transition, and N is the number of lipid molecules "withdrawn" from participation in chain melting as a result of incorporation of the macromolecules. The N lipid molecules withdrawn per macromolecule have been regarded as "boundary" lipids which are motionally restricted by interaction with the macromolecule. However, in many systems there is a rapid exchange ($\sim 10^6$-10^7 s^{-1} or greater) between bulk bilayer lipid and lipid adjacent to the macromolecule surface [91,125]. Discrepancies between results relating to motional restrictions arise because of differences in the time scales of the physical methods. On the e.s.r. time scale exchange rates can appear slow leading to the idea of motional restriction wheras using n.m.r. exchange appears fast. Below T_c the macromolecule can segregate from the bulk lipid to form a separate phase which incorporates lipid which does not participate in chain-melting but is in rapid equilibrium with the bulk lipid.

Table 9 gives parameters derived from DSC and microcalorimetric measurements for a range of lipid systems. Peptides and relatively small polypeptides withdraw lipid from participation in chain-melting. However, relatively small changes in the structure of peptides can give rise to significant changes in thermotropic properties. For example, pentagastrin (N-t-Boc-β-Ala-Trp-Met-Asp-Phe-NH$_2$) alters the phase transition behaviour of DMPC in an atypical way [99,100]. The DSC scans can be resolved into two well defined components one above and one below the transition temperature of the pure lipid. The molecular basis for the two components is not fully understood; the high temperature transition could be attributed to the melting of lipids belonging to peptide-rich domains and the lower temperature transition to the remaining bulk lipid. The small downward shift and broadening of the lower temperature transition compared with that of pure DMPC suggests that the bulk

TABLE 9. Parameters derived from DSC and microcalorimetry for the interaction between peptides, macromolecules and phospholipids

Peptides/Macromolecules	Lipid	Conditions	N	Enthalpy Change	Ref
Pentagastrin-related pentapeptides					
(1) N-t-Boc-β-Ala-Trp-Met-Gly-Phe-NH$_2$	DMPC	Pipes pH 7.4	2.4		[99,100]
(2) N-t-Boc-β-Ala-Trp-Met-Phe-Phe-NH$_2$	DMPC	Pipes pH 7.4	3.2		[99,100]
Amphiphilic peptides					
(1) Lys$_2$-Gly-Leu$_{16}$-Lys$_2$-Ala-Amide	DPPC-d$_{62}$	Phosphate pH 7.0	15		[101]
(2) Lys-Gly-Leu$_{24}$-Lys$_2$-Ala-Amide	DPPC-d$_{62}$	Phosphate pH 7.0	13		[101]
Glucagon (29 amino acid residues) M=3485	DMPC	Pipes pH7.4, 23°C	51	335kJ (mol glucagon)$^{-1}$	[103]
"	DMPC	Pipes pH7.4, 25°C	51	-628 kJ (mol apo A-II)$^{-1}$	[103]
High density apolipoprotein A-II	DMPC	Tris,pH7.4, 23.45°C	75	+377 kJ (mol apo A-II)$^{-1}$	[105]
"	DMPC	Tris,pH7.4, 24.5°C	75	-1087 kJ (mol apo A-II)$^{-1}$	[105]
"	DMPC	Tris,pH7.4, 30.0°C	75	-259 kJ (mol apo A-II)$^{-1}$	[105]
Cytochrome b$_5$ (M=16700)	DMPC	Tris,pH 8.1	35±4		[106]
α-Lactalbumin (M=17400)	DMPC	0.1M NaCl, pH 7.4, 20°C		-160 kJ (mol lactalbumin)$^{-1}$	[107-109]
	DMPC	0.1M NaCl, pH 7.4, 30°C		-20 kJ (mol lactalbumin)$^{-1}$	[107-109]
	DMPC	0.1M NaCl, pH 4.0, 20°C		-1190 kJ (mol lactalbumin)$^{-1}$	[107-109]
	DMPC	0.1M NaCl, pH 4.0, 30°C		-430 kJ (mol lactalbumin)$^{-1}$	[107-109]
Bacteriorhodopsin (M=25000)	DMPC	Acetate, pH 5	41		[110-112]
Glycophorin (M=31000)	DMPC	Tris, pH 7.4	42±22		[113-114]
	DPPC	Tris, pH 7.4	197±28		[113-114]
	DSPC	Tris, pH 7.4	240±64		[113-114]
Vesicular Stomatitis virus (G) glycoprotein (M=69,000)	DPFC	0.05M KCl	270±150		[115]
Vesicular Stomatitis virus (M) protein (M=29,000)	DPPC/DPPG(1:1)	Tricine, pH 7.5	0		[116]
Apo HDL (M=23000)	DMPC	25°C		-920.5 kJ (mol apoprotein)$^{-1}$	[117]
Apo A-1 (M=283000)	DMPC	25°C		-899.6 kJ (mol apoprotein)$^{-1}$	[117]
Apo A-2 (M=17400)	DMPC	25°C		-962.3 kJ (mol apoprotein)$^{-1}$	[117]
Apo C1 (M=8000)	DMPC	25°C		-2720 kJ (mol apoprotein)$^{-1}$	[117]
Cytochrome C oxidase	DMPC	Tricine	99±5		[118-120]

lipid fraction is affected by the peptide-rich domains. Interestingly, the replacement of the Asp residue in position 4 by either Gly or Phe (see Table 9) gives rise to behaviour characteristic of intrinsic membrane proteins; the transition temperature of the DMPC is almost unchanged and a single DSC peak is found with an area which decreases with increasing peptide level, which at low peptide to lipid molar ratios can be interpreted according to equation (6). At high peptide to lipid ratios deviation from equation (6) occurs due to self association of the peptides.

The values of N for many of the intrinsic membrane proteins correspond to several concentric layers of lipid around the proteins. For example, for the glycophorins, the major sialoglycoproteins of the human erythrocyte membrane, N values of 42 to 240 are found depending on the acyl chain length of the lipid. This observation can be interpreted in terms of the bilayer thickness relative to the length of the 22 hydrophobic amino acid transbilayer segment of glycophorin. The length of the shorter DMPC acyl chains is close to half the length of the transbilayer segment assuming it to be α-helical. The acyl chains can thus pack around the α-helix with less disruption than is the case for DPPC and DSPC chains. For the latter in order to bring the bilayer thickness down to the α-helical segment length, trans-gauche "kinks" must be introduced with leads to greater packing disruption and hence larger values of N.

To interpret the effects of bacteriorhodopsin on the thermotropic properties of DMPC Heyn et al. [110,111] required a more sophisticated model than that based on equation (6). Bacteriorhodopsin (BR) crystallizes in the bilayers of vesicles well below T_c. A transiton is observed centred at 17.5°C which corresponds to the breakdown of the BR lattice to form BR aggregates and monomers. In the course of lattice disaggregation the BR molecules acquire a layer of "solution" lipid which is assumed to be in a state intermediate between that in the gel and liquid-crystalline phases. The enthalpy change (ΔH_{1cal}) for the partial melting of m gel phase lipids per BR molecule required to solvate the BR aggregates and monomer at 17.5°C is less than the enthalpy of the main transition at T_c (ΔH_{2cal}). The total enthalpy (ΔH_{cal}) of the overal broad transition is given by equation (7)

$$\Delta H_{cal} = m \frac{[BR]}{[L]} \Delta H_{1cal} + \left(1-m \frac{[BR]}{[L]}\right) \Delta H_{2cal} \tag{7}$$

where $\Delta H_{1cal}/\Delta H_{2cal}$ = 0.7 and m=60. Interpretation of the data in terms of the simpler equation gives a value of N of 41.

Morrow *et al.* [102] in a study of the thermodynamics of amphiphilic peptides containing blocks of leucine residues argue that a two-phase equilibrium may be too complex to be summarized in a single parameter equation and that even relatively small peptides do not remove lipid from the main transition but rather influence the transiton thermodynamics for the bilayer as a whole. Furthermore, they suggest that the boundary lipid picture has probably obscured the more detailed thermodynamics of lipid-solute systems. Rejecting the simpler approaches a model based on regular solution theory is proposed and is used to simultaneously fit the phase behaviour (phase diagrams) and the transition enthalpies. Regular solution theory assumes that the entropy of mixing solute (here the peptide, 2) and solvent (here the lipid, 1) is ideal and that the enthalpy of mixing based on a quasi-lattice model is given by

$$\Delta H_m = x_1 x_2 \rho \tag{8}$$

where the interaction parameter ρ depends on the difference between solute/solvent interactions. The enthalpy change for the main transition on the basis of this theory is given by

$$\Delta H = n_1 \, \Delta H_1 \, (T_{c1}) + n_2 \, \Delta H_2 \, (T_{c2}) + \left(\frac{\rho^\ell - \rho^g}{\gamma} \right) \frac{n_1 n_2}{n_1 + \gamma n_2} \tag{9}$$

where n_1 and n_2 are the numbers of moles of lipid and solute respectively, ρ^ℓ and ρ^g are the interaction parameters for the liquid-crystalline and gel phases and γ is a lattice parameter. In equation (9) ΔH_2 is the enthalpy of transfer of the solute from the gel to the liquid-crystalline phase, while ΔH_1 is the observed enthalpy of chain melting of the pure lipid. The major consequence of this approach is that it is possible to reproduce the phase diagram while also incorporating the concentration dependence of the transition enthalpy without recourse to distinguishing between lipid in contact with solute and bulk lipid.

The above discussion has concentrated on thermal properties associated with the lipid in protein-lipid systems in which the protein has been previously incorporated. Thermal effects associated with both the lipid and the protein are observed when the enthalpies of interaction between proteins and lipid vesicles are measured. Enthalpies of interaction between glucagon [103] α-lactalbumin [107-109] and high density apolipoprotein [104,105,117] and DMPC vesicles have been measured and found to be markedly dependent on temperature. For glucagon and HDL A-II the enthalpies of interaction in the temperature range 20-35°C show a qualitatively similar behaviour in that ΔH at the lower temperature is positive, passes through zero at ~24°C, through a minimum at ~25°C and then increases becoming close to zero again at higher temperatures above 30°C.

Glucagon is a polypeptide hormone of 29 amino acid residues having a sequence which could lead to the formation of an amphipathic helix, and in association with lipid the complex contains ~50 lipid molecules per glucagon molecule with a 30% helix content. Epand and Sturtevant [103] attribute the observed enthalpy changes to changes in state of the lipid induced by the glucagon. The DSC transition observed for mixtures of glucagon and DMPC shows two overlapping peaks; the major component (~80% of the enthalpy change) peaks at 26.1°C and the minor component peaks at 23°C. At, for example 25°C, when glucagon interacts with a multilamellar suspension of DMPC the glucagon induces a change in the state of the lipid from liquid-crystalline to gel state with an accompanying exothermic enthalpy. From the DSC peak areas it is deduced that at a lipid to glucagon ratio of 350 or greater 57% of the lipid is converted from the liquid-crystalline to the gel state on interaction with glucagon. Since there are 50 lipid molecules affected per mole of glucagon the enthalpy change associated with the interaction can be estimated by taking 57% of the liquid-crystalline to gel transiton enthalpy per mole of lipid times 50 i.e. $0.57[\Delta H_{l-g} = -24 kJmol^{-1}]x50 = -648$ kJ (mol glucagon)$^{-1}$. The experimentally observed enthalpy of interaction is -628 kJ (mol glucagon)$^{-1}$ which is within 10% of the estimate.

In the study of the glucagon-DMPC system the thermal effects could be largely ascribed to the lipid, in contrast to studies of the enthalpy of interaction between HDL A-II and DMPC [105] where a conformational change in the protein contributes to the enthalpy of interaction. HDL A-II is a major component of blood serum, it has two identical polypeptide chains of 77 amino acids (molecular weight 8690) linked by a disulphide bond. Three HDL A-II complexes with DMPC have been characterized [104] with molar ratios of 1:45, 1:75 and 1:240. The 1:45 complex is an integral part of the other two consisting of HDL A-II and 45 so-called "boundary" non-freezable DMPC molecules. Circular dichroism was used to show that the reaction of protein with lipid resulted in an increase in the helical content of the HDL A-II with a substantial release of heat (-12.13 kJ/1% helix). At a high temperature above the chain melting temperature of pure DMPC (T_c) and the melting of DMPC in the complex, the formation of helical structure on interaction with DMPC is the sole source of heat. At lower temperatures the heat of interaction includes contributions from both the random coil to helix transition and the liquid-crystalline to gel transition. Thus the temperature dependence of the formation of the 1:75 complex can be analysed as follows.

At $T > T_c$ (for pure DMPC) and T_c (complex) i.e. 30°C (Table 9)

$\Delta H = \Delta H$ (α-helix formation) x increase in helix content ($\Delta\% = 20\%$)

$= -12.13 \times 20 = $ -243 kJ (mol HDL A-II)$^{-1}$.

At $T > T_c$ (for pure DMPC) $< T_c$ (complex) i.e. 24.5°C

$\Delta H = \Delta H$ (α-helix formation) x $\Delta\%$ helix + $\Delta H_{l \to g}$ ($= -24$ kJ mol^{-1}) x

number of moles of "freezable" lipid

For the 1:75 complex there are (75-45)=30 "freezable" lipids per HDL A-II molecule and the $\Delta\%$ helix is 42%.

$\therefore \Delta H = -12.13 \times 42 + 30 \times (-24.0) = $ -1229 kJ (mol HDL A-II)$^{-1}$

At $T < T_c$ (for pure DMPC) and T_c (complex) i.e. 23.5°C

$\Delta H = \Delta H$ (α-helix formation) x $\Delta\%$ helix + $\Delta H_{g \to l}$ (= +24 kJ mol^{-1}) x

number of moles of "non-freezable" lipid.

$\therefore \Delta H = -12.13 \times 42 + 45 (+24.0) = $ 570 kJ (mol HDL A-II)$^{-1}$

The experimental values for the enthalpies of interaction are -259, -1087 and +377 kJ (mol HDL A-II)$^{-1}$. The above estimates deviate from the experimental values by 6%, 13% and 51% respectively, however, perhaps more important than the deviations is the fact that the dramatic changes in the direction and signs of the enthalpies of interaction over such a narrow temperature interval can be largely explained. The explanation requires the division of the lipid into complexed and "freezable", although such a division has been seriously questioned it is clearly useful in the above context.

5.6. PROTEIN - SURFACTANT INTERACTIONS

The widespread application of surfactants in the field of biochemistry particularly in the study of membranes has given impetus to fundamental studies on the nature of the interactions between surfactants and proteins. Surfactants are used extensively in the solubilization of intrinsic membrane proteins and glycoproteins. Solubilization is a pre-requisite to the characterisation of membrane macromolecules by techniques such as polyacrylamide gel electrophoresis in sodium n-dodecylsulphate (SDS) and hence interactions between proteins and SDS are of particular interest. Much of the early pioneering work in this area has been documented by Steinhardt and

Reynolds [126] and more recently by the author [127]. As discussed above (section 5) the interaction between membrane proteins and lipids is of fundamental importance in relation to the structure and function of biological membranes. The study of rather simpler well characterised systems consisting of a globular protein and an amphipathic ligand, such as SDS, can give basic information on the nature of the non-covalent forces between proteins and amphipaths which might then be translated to the situation in biological membranes between intrinsic proteins and lipids and between intrinsic proteins and surfactants during the solubilization process. In the latter context attention should be drawn to the classic review on the solubilization of membranes by detergents by Helenius and Simon [128].

The factors which are responsible for the stabilization of the secondary and tertiary structure of biological macromolecules can be to some degree understood from studies of interactions which lead to the breakdown of native structure. Ionic surfactants are unique in this context in that they denature proteins at concentrations of the order of millimolar in marked contrast to denaturants such as guanidinium chloride or urea which function only at molar concentrations. The industrial importance of surfactants (or detergents) should also be noted. Apart from being components of washing and bleaching agents surfactants are importance ingredients in many cosmetics, toiletries and pharmaceutical preparations. All these materials come into contact with at least the surface of living organisms and hence their mode of interaction with proteinacious tissue is of importance.

5.6.1 Experimental methods

A complete study of interactions between a protein and a surfactant ligand cannot be satisfactorily made without recourse to a range of physical methods of which equilibrium dialysis, to obtain binding isotherms and related Gibbs energy changes, and calorimetry are two of the most useful techniques with which to obtain thermodynamic data. The interpretation of both binding isotherms and interaction enthalpies as a function of surfactant concentration cannot be made without some structural data on the nature of the complexes formed between the protein and surfactant and if the protein dissociates into subunits methods of obtaining the extent of such dissociation. In the latter context sedimentation rate and/or equilibrium measurements are of great value. If such equipment is not available polyacrylamide gel electrophoresis is a reasonable alternative, although the control of pH, ionic strength and surfactant concentration cannot be achieved with as much flexibility as is possible using ultracentrifugation. A range of physical methods and the information that can be obtained from them is shown in Table 10.

TABLE 10.   Physical Techniques Applicable to Surfactant-Protein Interactions

TECHNIQUE	INFORMATION OBTAINED
Equilibrium Dialysis	Surfactant binding, $\Delta G_{binding}$
Titrimetry	Proton binding, $\Delta G_{binding}$
Calorimetry (microcalorimetry and differential scanning (DSC))	Enthalpy of interaction, $\Delta H_{binding}$, $\Delta H_{binding + unfolding}$
Ultracentrifugation (sedimentation rate and equilibrium)	Sedimentation coefficients of complexes, subunit dissociation and molecular weights
Viscometry	Hydrodynamic volume and shape factors of complexes
Static and dynamic light scattering	Molecular weights – Diffusion coefficients – Dimensions
UV difference spectroscopy	Conformation changes
Enzyme kinetics	Binding to active site, denaturation conditions

The extent of surfactant binding to a protein is most conveniently measured by quantitative equilibrium dialysis using a high quality dialysis tubing with a specific cut-off at a molecular weight well below that of the protein and its subunits. In principle, the dialysis bag containing a known volume of protein solution of known concentration is dialysed to equilibrium against a known volume of surfactant solution and the free (unbound) surfactant concentration assayed. As the total amounts of protein and surfactant in the system are known the amount of surfactant bound to the protein ($\bar{\nu}$ = average number of bound surfactant molecules per protein molecule) is easily determined from the total free surfactant. Provided that the protein concentration is low enough (~0.1% w/v), relative to the ionic strength of the buffer, Donnan effects are negligible. Although there are more sophisticated methods of determining free surfactant in equilibrium with protein-surfactant complexes, e.g. specific ion electrodes, and more elaborate methods of carrying out dialysis, e.g. using compartmentalised cells separated by dialysis tubing, the simple dialysis bag procedure is thermodynamically sound and free from any electrical or mechanical problem. It should however, be noted that the passage of surfactant through dialysis membranes is a relatively slow process [129] and it is crucial to ensure that the system has come to equilibrium. With equilibrating volumes of 2 cm^3 inside and outside the bag periods of upto more than 96 hours are required.

The enthalpy of interaction between proteins and surfactants is most conveniently measured with a twin-cell microcalorimeter. The heat of mixing of equal volumes of protein and surfactant in one cell is compared with the heat of dilution of surfactant by an equal volume of solvent in the other cell. Generally the heat of dilution of protein is negligible or amounts to only a small constant correction. Provided the final surfactant concentration after mixing is less than the critical micelle concentration (cmc) of the surfactant the reaction and reference vessels will be thermally balanced. If this is not the case then problems of surfactant demicellization will give rise to an enthalpy effect which requires further corrections, the details of which have been discussed [130,131]. The calorimetric enthalpies of interaction and Gibbs energies of interaction derived from binding isotherms (see below) can be used to calculate the entropies of interaction. The thermodynamic parameters so derived are most usefully expressed as a function of the extent of surfactant binding ($\bar{\nu}$).

5.6.2 Theoretical Methods

The binding isotherm of a surfactant to a protein is most conveniently displayed as a plot of $\bar{\nu}$ vs. the logarithm of the free (unbound) surfactant concentration (log $[S]_f$). Fig. 16 shows some typical binding isotherms for the lysozyme sodium n-dodecyl sulphate system. Such typical isotherms can be divided into at least three reasonably distinct regions. At very low free surfactant concentrations the initial sharp increase in $\bar{\nu}$ is generally associated with the specific binding of the surfactant to charged sites on the surface of the protein (e.g. anionic surfactant binding to cationic amino acid residues). The specific binding region can terminate in a plateau or a region where $\bar{\nu}$ changes more slowly with log $[S]_f$ which can be defined as a non-cooperative binding region in contrast to the following region in which $\bar{\nu}$ increases rapidly with log $[S]_f$ and is characteristic of cooperative binding. The cooperative binding region generally occurs in the vicinity of the critical micelle concentration of the surfactant and terminates in saturation of the protein with surfactant. It is important to note that not all protein-surfactant systems display three distinct binding regions, at some pHs the specific binding region is not present and a simpler curve showing only the cooperative binding region is found.

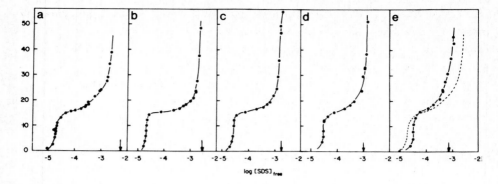

Fig. 16. Binding isotherms for sodium n-dodecyl sulphate (SDS) on interaction with lysozyme (concentration 1.25g l⁻¹) at 25°C, pH 3.2: a) ionic strength, I = 0.0119; b) I = 0.0269; c) I = 0.0554; d) I = 0.1119; e) I = 0.2119 (broken curve, I = 0.0119). The arrows denote the critical micelle concentrations (CMC). Reproduced from Jones, M.N., Manley, P. and Hold, A. International Journal of Biological Macromolecules, 1984, 6, 65-68, by permission of the publishers, Butterworth & Co. (Publishers) Ltd. ©.

The theoretical interpretation of a binding isotherm should take into account that the system, protein (P) plus surfactant (S) consists of a series of equilibria.

$$P \; + \; S \rightleftharpoons PS_1$$
$$PS \; + \; S \rightleftharpoons PS_2$$
$$PS_2 \; + \; S \rightleftharpoons PS_3$$
$$\vdots$$
$$PS_{n-1} \; + \; S \rightleftharpoons PS_n \tag{10}$$

However, even for the smallest of proteins the total number of binding sites can be as large as 50 so that a rigorous analysis to obtain all the binding constants is not practicable. There are however, several alternative approaches. The simplest of these which has found some application, particularly in the specific binding region, is the Scatchard equation [132].

$$\frac{\bar{\nu}}{[S]_f} = K(n - \bar{\nu}) \tag{11}$$

where K is an intrinsic binding constant. A plot of $\bar{\nu}/[s]_f$ vs. $\bar{\nu}$ should be linear and extrapolate to give the total number of binding sites n. The Scatchard equation is based on the assumption that the n binding sites are identical and independent, a condition which cannot rigorously hold for surfactant binding to different charged amino acid residues. Despite this however, a number of protein-SDS systems give superficially linear Scatchard plots in the specific binding region which extrapolate to values of n approximately equal to the number of cationic amino acid residues on the protein surface. At very low values of $\bar{\nu}$ the Scatchard plots pass through a maximum indicative of cooperativity between the specific binding sites [133] and demonstrating the limitations of the Scatchard equation.

Cooperative binding is better described in terms of the Hill equation [134].

$$\bar{\nu} = \frac{g\,(K\,[S]_f)^{n_H}}{1 + (K\,[S]_f)^{n_H}} \qquad (12)$$

where g is the total number of binding sites and n_H is the Hill coefficient, a measure of cooperativity (note the Scatchard equation is a special case of the Hill equation with $n_H=1$). Both the specific binding region and the cooperative binding region can be fitted using Hill equations, so that an entire isotherm can be fitted using a summation of "Hill terms" for each binding region. . In the simplest case two Hill terms suffice with six determinable parameters; binding constants for specific and cooperative binding, the corresponding Hill coefficients and the numbers of specific and cooperative binding sites.

The description of a complex multiple equilibrium in terms of overall binding constants is inevitably an approximation since the binding of every surfactant ligand must change the binding constant for subsequent ligands. A procedure which enables binding constants to be determined as binding proceeds based on the concept of a binding potential proposed by Wyman [135] has been used for a number of protein-surfactant systems [136-139]. Briefly the binding potential $\pi(P,T,\mu_1,\mu_2....)$ is related to the amount of surfactant bound, $\bar{\nu}_S$, and surfactant chemical potential, μ_S, by the equation

$$\bar{\nu}_S = \left(\frac{\partial\pi}{\partial\mu_S}\right)_{P,T} \qquad (13)$$

Assuming the chemical potential of the surfactant at concentration [S] is given by an ideal solution expression, $\mu_S=\mu_S^0+RTln[S]_f$, with a one molar standard state (μ_S^0), it follows that

$$\pi = RT \int_{\nu_S = 0}^{\nu_S = \nu} \nu_S \; d\ln[S]_f \qquad (14)$$

and hence the binding potential is simply RT times the area under the binding isotherm. Considering the formation of a particular protein-surfactant complex PS_n according to the equilibrium

$$P + nS \rightleftharpoons PS_n \qquad (15)$$

at a specific surfactant concentration, if it is assumed that the concentration of all species except P and PS_n are negligible then the overall stoichiometric binding constant is given by

$$K = \frac{[PS_n]}{[P][S]_f^n} \qquad (16)$$

and

$$\bar{\nu} = \frac{nK[S]_f^n}{1 + K[S]_f^n} \qquad (17)$$

Differentiating $\bar{\nu}$ with respect to $\ln[S]_f$ and substituting in equation (14) it can be shown [136] that

$$\pi = RT \ln (1 + K[S]_f^n) \qquad (18)$$

Since π can be determined from the binding isotherm at a given $[S]_f$ it follows that K can be determined as a function of $\bar{\nu}$. Furthermore at a given $[S]_f$ the predominant species present will be $PS_{\bar{\nu}}$ then $K[S]_f^n = K_{app}[S]_f^{\bar{\nu}}$. The apparent binding constant (K_{app}) can be related to an apparent Gibbs energy of binding ΔG_a at a given value of $\bar{\nu}$

$$\Delta G_a = -RT\ln K_{app} \qquad (19)$$

By this procedure it is possible to obtain plots of ΔG_a or the Gibbs energy of binding per surfactant molecule bound $\Delta G_a/\bar{\nu}$ defined as $\Delta G_{\bar{\nu}}$, as a function of $\bar{\nu}$. Such curves show clearly the changing strength of binding as increasing numbers of surfactant molecules (ions) bind to the protein.

5.6.3   Specific binding interactions

There is much evidence to support the view that the initial interaction between an anionic surfactant and a globular protein in acid or neutral solutions involves the binding of the anionic head group of the surfactant to the cationic amino acid residues of lysine, histidine and arginine.   Studies [140] on the enthalpy of interaction of n-alkyl sulphates with polylysine, polyhistidine and polyarginine have shown that these synthetic polypeptides bind n-alkyl sulphates stoichiometrically, each cationic residue binding one n-alkyl sulphate anion.   The enthalpy of the interaction is exothermic and for each polypeptide increases linearly with the chain length of the surfactant in the range C_8 - C_{12}.   That such interactions occur in proteins can be demonstrated by chemically modifying cationic amino acid residues and observing a decrease in specific binding sites and changes in the enthalpy of interaction.   For lysozyme, blocking the lysyl residues by acetylation reduces the number of specific binding sites for SDS, shifting the intercept of the Scatchard plot precisely in line with the number of groups acetylated [141]. Chemical modification of histidyl residues by photo-oxidation and arginyl residues by reaction with butane-2,3-dione also results in significant changes in the enthalpy of interaction of lysozyme with SDS [141].   Conversely the carboxyl groups of glutamyl and aspartyl residues inhibit specific interactions between anionic surfactants and proteins.   Thus the binding isotherm for the glucose oxidase-SDS system in acid solution (pH 3.65) shows a specific binding region, but at pH 6 only athermal cooperative binding is observed.   That both the loss of specific binding and the low energy of interaction is due to inhibition by carboxyl groups is demonstrated by amidation with glycine methyl ester.   Glycine methyl ester modified glucose oxidase interacts exothermically with SDS at pH 6.0, the enthalpy of interaction increasing with the extent of modification [142].   Similar studies have been made on chemically modified bovine catalase, for which it has been shown that the enthalpy of interaction with SDS is increased by amidation of carboxyl groups with glycine methyl ester or with ethylenediamine which replaces the negative carboxyl groups by positive sites.   Both these chemical modifications increase the susceptibility of catalase to dissociation into subunits by SDS [143-144].

Table 11 shows the thermodynamic parameters for specific binding of SDS to a number of globular proteins in acid or neutral solutions.   The number of specific binding sites were derived from Scatchard plots and the enthalpy changes were directly measured by microcalorimetry.   Although the table covers a diverse range of globular proteins the relatively crude Scatchard analysis gives numbers of specific binding sites which can be approximately correlated

Table 11   Thermodynamics parameters for the specific binding of sodium n-dodecyl sulphate to globular proteins in aqueous solution at 25°C

Protein (Mol wt) (ref)	No. of Cationic residues	pH	$K\,(dm^3\,mol^{-1})$	N	$\Delta\bar{G}_v$	$\Delta\bar{H}_v$ kJ mol^{-1}	$T\Delta\bar{S}_v$
Ribonuclease A (13682) [145]	18	7.0	$7.70.\times10^4$	19	-27.9	-1.27	26.6
Ribonuclease A (13682) [131]	18	3.2	15.2×10^4	25	-29.6	+1.26	30.9
Lysozyme (14306) [130]	18	3.2	3.93×10^4	18	-26.2	-8.66	17.5
Trypsin (23300) [146]	19	3.5	5.25×10^4	30	-26.9	-5.66	21.2
Ovalbumin (44000) [145]	42	7.0	1.77×10^5	37	-30.0	0	30.0
Bovine Serum Albumin (66000) [145]	98	7.0	2.17×10^4	57	-24.8	-6.95	17.9
Glucose Oxidase (147000) [142]	120	3.7	0.51×10^4	132	-21.2	-3.29	17.9
Catalase (245000) [138]	331	3.2	9.53×10^4	343	-28.4	-8.36	20.0

with the number of cationic amino acid residues in the proteins, with the exception of bovine serum albumin.

Bovine serum albumin is known to form specific complexes with a fixed stoichiometry BSA(SDS)$_{5-11}$, BSA(SDS)$_{38}$ and BSA(SDS)$_{76}$ [147,148]. The binding of the first 5-11 anions is thought to induce a conformation change in the protein which resembles the so-called N-F transition induced by acid or alkali denaturation [149]. Of the 98 potential cationic binding sites in BSA 58 are lysyl residues and of these, 40 are embedded within the cleft between the two globular units of the native structure. These lysyl residues become exposed in the N-F transition and it is believed that it is the 58 lysyl residues that are initially involved in complex formation [150] as the Scatchard analysis indicates (Table 11).

The Gibbs energies of surfactant binding lie in a relatively narrow range, approximately -20 to -30 kJ (mol SDS)$^{-1}$, and it is clear that it is the positive entropy change on binding which is the major contribution to the process. The enthalpy changes are small and differ markedly from protein to protein. For example despite the fact that both ribonuclease A and lysozyme have similar molecular weights and identical numbers of cationic sites the enthalpy of SDS interaction with ribonuclease A at pH 3.2 is endothermic whereas it is exothermic for lysozyme. Fig. 17 shows a comparison between the enthalpies of interaction for these two proteins as a function of SDS bound. These dramatic differences arise because lysozyme does not easily unfold. The maximum in the enthalpy curve suggests unfolding in the region of $\bar{v} \sim 40$ whereas ribonuclease A is readily unfolded at low binding levels and this is reflected in an endothermic contribution to the interaction enthalpy.

In general the enthalpy of interaction (ΔH_{int}) between a surfactant and a protein can be represented by the equation

$$\Delta H_{int} = \Delta H_c + \Delta H_d + \Delta H_B \qquad (20)$$

where ΔH_c, ΔH_d and ΔH_b are the enthalpy changes for specific binding, unfolding and apolar binding respectively. For anionic surfactants in acid and sometimes also in neutral solution ΔH_c and ΔH_d are the major contributions, the enthalpy of apolar binding is generally small. Ribonuclease A is perhaps the best example in which the enthalpy of interaction with SDS is dominated by the endothermic enthalpy of unfolding although curves of interaction enthalpy against SDS concentration showing the presence of an endothermic component due to unfolding have also been found for lysozyme [130,141], trypsin [146], BSA [145] and catalase [143]. It should also be noted that the

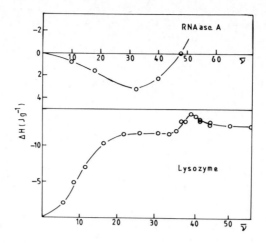

<u>Fig. 17</u>.   Enthalpy of interaction of lysozyme and ribonuclease A with sodium
<u>n</u>-dodecyl sulphate (SDS) in aqueous solution (pH 3.2, 25°C)   as a function of
the number of SDS molecules bound per protein molecule.

specific   complexes   formed   between   SDS   and   globular   proteins   are   often
insoluble   particularly   in   acid   solution.   One   might   expect   an   enthalpy
contribution   arising   from   precipitation,   however,   in   the   case   of   lysozyme   (Fig.
17) where the complexes are insoluble below $\bar{v} \sim 25$ and also for catalase [143]
the enthalpy curves show no "kinks" coincident with the solubility limits of the
complexes   and   it   would   appear   that   any   enthalpy   effects   arising   from
precipitation are relatively small.

        A   more   sophisticated   treatment   of   the   binding   isotherms   for   SDS   on
lysozyme   based   on   both   the   Hill   equation   and   binding   potentials   has   been
made   [139].   Fig.   16 (a-e) shows   the   binding   isotherms   which   cover   a   range   of
ionic strength from 0.01 to 0.21.   Fig. 16e compares the isotherms at extremes
of   the   ionic   strength   range.   From   this   it   can   be   seen   that   increasing   ionic
strength   shifts   the   specific   binding   region   to   higher   free   SDS   concentrations
while   the   hydrophobic   binding   region   is   shifted   to   lower   free   SDS
concentration.   Fitting   the   specific   binding   regions   of   the   curves   to   the   Hill
equation gives the thermodynamic parameters shown in Table 12.

        The   Hill   coefficients   lie   in   the   range   6-11   and   illustrate   that   there   is
considerable cooperativity in the specific binding domain.   Also given in Table
12   are   the   values   of $\Delta G_{\bar{r}}$ calculated   using   binding   potentials.   These   are   very
close to those determined from the Hill association constants.   Fig. 18 shows a
Scatchard   plot   for   the   lysozyme-SDS   system   with   a   maximum   characteristic   of

TABLE 12.
Thermodynamic and related parameters for the binding of sodium n-dodecyl sulphate
to lysozyme as a function of ionic strength at 25°C (pH 3.2)

Ionic strength	g	$K \times 10^{-4}$ (1 mol^{-1})	n_H	$-\Delta G_\nu^a$ (KJ mol^{-1})	$-\Delta G_\nu^b$ (kJ mol^{-1})	$-\Delta H_\nu^o$ (kJ mol^{-1})	$T\Delta S_\nu^o$ (kJ mol^{-1})
0.0119	16	5.13	6.78	26.9	26.8	8.41	18.5
0.0269	16	4.76	11.10	26.7	26.6	–	–
0.0554	16	3.66	7.19	26.1	25.9	10.7	15.3
0.1119	16	3.33	7.65	25.8	25.9	10.4	15.4
0.2119	17	2.93	6.51	25.5	25.7	10.9	14.6

[a]Calculated from K

[b]Calculated from the binding potential as described in the text.

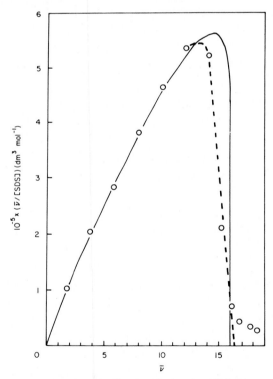

Fig. 18. Scatchard plot for sodium n-dodecyl sulphate on interaction with
lysozyme at 25°C, pH 3.2, I = 0.0269. The full curve was calculated from the
Hill equation. Reproduced from Jones, M.N., Manley, P. and Holt, A.
International Journal of Biological Macromolecules, 1984, 6, 65=68 by permission
of the publishers, Butterworth & co. (Publishers) Ltd. ©.

232

positive cooperativity. In the absence of data at low binding levels ($\nu < 10$) the apparent linearity of the plot could be erroneously interpreted in terms of a Scatchard model without too serious consequences for the determination of $\Delta G_{\bar{\nu}}$ or the number of specific binding sites as a comparison of the data in Table 11 and 12 shows.

5.6.4   Apolar binding interactions

After saturation of the specific binding sites which is either accompanied by protein unfolding or initiates unfolding, further binding is essentially of a hydrophobic nature and as the free detergent concentration approaches the critical micelle concentration (cmc) binding becomes increasingly cooperative. The Gibbs energy change for apolar cooperative binding is less than for specific binding.   The curve of $\Delta G_{\bar{\nu}}$ as a function of $\bar{\nu}$ for SDS binding to lysozyme is shown in Fig. 19 and illustrates the gradual decrease in binding energy, after saturation of the "high energy sites".   When the free SDS concentration reaches the cmc the Gibbs energy of binding and the Gibbs energy of micellization are in principle identical and this has been confirmed for the lysozyme-SDS system [139].

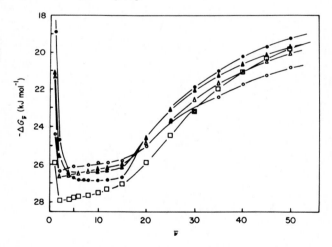

Fig. 19.   Gibbs energies of binding ($\Delta G_{\bar{\nu}}$) of sodium n-dodecyl sulphate to lysozyme as a function of amount bound ($\bar{\nu}$) at 25°C, pH 3.2: ☐ , I = 0.0119; ● , I = 0.0269; ▲ , I = 0.0554; △ , I = 0.119; ○ , I = 0.2119. Reproduced from Jones, M.N., Manley, P. and Holt, A. International Journal of Biological Macromolecules 1984, 6, 65-68 by permission of the publishers, Butterworths & Co. (Publishers) Ltd. ©.

Attempts can be made to isolate the contribution of the enthalpy of apolar binding (ΔH_b) of SDS from the total enthalpy of interaction by estimating the enthalpies of specific binding (ΔH_c) from data on SDS interaction with cationic polypeptides and allowing for the enthalpy of unfolding (H_d). Table 13 shows the results of such calculations for a number of globular proteins.

The data refers to interactions at a fixed SDS concentration (4mM). In estimating the values of ΔH_c the degree of ionisation of the cationic side chains at the appropriate pH was taken into account. ΔH_ν, the enthalpy of apolar binding per mole of SDS, is relatively high for lysozyme and trypsin where the proportion of apolar sites relative to the total number of binding sites at a total SDS concentration of 4mM is low. In these cases the ΔH_ν relates to the binding of SDS at the start of the apolar (cooperative) binding region. However, when larger numbers of SDS molecules are bound the values of ΔH_ν are small amounting to less than -10 kJ (mol SDS)$^{-1}$. At this level of apolar binding ΔG_ν is generally of the order of -20 kJ mol^{-1} so that $T\Delta S_\nu$ is positive and the dominating term, as expected for hydrophobic interactions.

Finally it should be noted that the binding of surfactant anions to proteins is accompanied by the binding of protons. This can be established from titration curves in the presence of surfactant [138,150]. The Gibbs energies of proton binding to the protein-surfactant complexes can be determined from the titration data and for both lsozyme-SDS complexes [150] and catalases-SDS complexes [138] becomes more negative with increasing surfactant binding.

Table 13 Contributions to enthalpies of interaction of globular proteins with sodium n-dodecyl sulphate (4mM) in aqueous solution at 25°C (140)

Protein (pH, Ionic strength, $\bar{\nu}$)	No. of cationic sites	No. of apolar sites	ΔH_{int} (kJ mol^{-1})	ΔH (kJ mol^{-1})	ΔH (kJ mol^{-1})	ΔH_b (kJ mol^{-1})	ΔH_v kJ(mol^{-1} SDS)
Ribonuclease A(7, 0.005, 68)	14.4	53.6	-17.1	-133	250	-134	-2.5
Lysozyme (3.6, 0.004, 25)	18	7	-168.8	-237	222	-154	-22
Trypsin (5.5, 0.01, 35)	18.3	14.7	-116.5	-147	∼280	-250	-17
β-Lactoglobulin (5.5, 0.01, 46)	6.6	39.4	-150	-84.3	∼190	-256	-6.5
Ovalbumin (7, 0.005, 97)	35.7	61.3	-220	-384	∼530	-366	-6.0
Bovine Serum Albumin (7.9, 0.005, 198)	82.7	115.3	-660	-758	∼790	-692	-6.0

REFERENCES

1. Danielli, J.F. in "Membranes and Transport" Vol. 1 Edited by A.N. Martonosi, Plenum Publishing Corporation 1982 page 3.

2. Singer, S.J. and Nicolson, G.L. Science 1972, 175, 720-731.

3. Deuvel, H.J. "The Lipids" Vol. 1 Interscience, New York 1951.

4. Quinn, P.J. "The Molecular Biology of Cell Membranes" Macmillan Press Ltd. London and Basingstoke 1976, Chapter 1.

5. Sundaralingam, M. Ann. N.Y. Acad. Sci. USA 1972, 195, 324-355.

6. Hauser, H., Pascher, I., Pearson, R.H. and Sundell, S. Biochim. Biophys. Acta. 1981, 650, 21-51.

7. Reid, D.S. J. Physics E Scientific Instruments 1976 9, 601-609.

8. Calvet, E. and Prut, H. "Recent Progress in Microcalorimetry" Pergamon, London 1963.

9. Benzinger,T.H. and Kitzinger, C. Methods Biochem. Anal. 1960, 8, 309-360.

10. Wadso, I. Acta. Chem. Scand. 1968, 22, 927-937.

11. Monk, P, and Wadso, I. Acta. Chem. Scand. 1968, 22,142-152.

12. Monk, P. and Wadso, I. Acta. Chem. Scand. 1969, 23, 29-36.

13. Chen, A. and Wadso, I. J. Biochem. Biophys. Meth. 192 6, 307-316.

14. Beezer, A.E., Hunter, W.H., Lipscombe, R.P., Newell, R.P. and Storey, D.E. Thermochim. Acta. 1982, 55, 345-349.

15. McKinnon, I.R., Fall, L., Parody-Morreale, A. and Gill, S.J. Anal. Biochem. 1984, 139, 134-139.

16. Pilcher, G. and Jones, M.N. unpublished work.

17. Privalov, P.L., Plotnikov, V.V. and Filimonov, V.V. J. Chem. Thermodynamics 1975, 7, 41-47.

18. Chapman, D. "The Structure of Lipids" Methuen and Co., Ltd, London 1965, Page 232.

19. Williams, R.M. and Chapman, D. Progr. Chem. Fats Other Lipids 1970, 11, 1-79.

20. Larson, K. Acta. Chem. Scand. 1966, 20, 2255-2260.

21. McClure, E.J. J. Cell Biol. 1976, 68, 90-100.

22. Luzzati, V. in "Biological Membranes" Ed. D. Chapman Academic Press, New York 1968, Page 71.

23. Philips, M.C., Williams, R.M. and Chapman, D. Chem. Phys. Lipids 1969, 3, 234-244.

24. Jones, M.N. "Biological Interfaces". Elsevier Publishing Co., Amsterdam 1975, Chapter 4.

25. Ruocco, M.J. and Shipley, G.G. Biochim. Biophys. Acta. 1982, 684, 59-66.

26. Ruocco, M.J. and Shipley, G.C. Biochim. Biophys. Acta. 1982, 691, 309-320.

236

27. Chen, S.C., Sturtevant, J.M. and Gaffney, J. Proc. Natl. Acad. Sci. USA 1980, 77, 5060-5063.

28. Fuldner, H.H. Biochemistry 1981, 20, 5707-5710.

29. Estep, T.N., Calhoun, W.I., Barenholz, Y., Biltonen, R.L., Shipley, G.G and Thompson, T.E. Biochemistry 1980, 19, 20-24.

30. Freire, E., Bach, D., Carrea-Freire, M., Miller, I. and Barenholz, Y. Biochemistry 1980, 19, 3662-3665.

31. Ruocco, M.J., Atkinson, D., Small, D.M., Oldfield, E., Skarjune, R.P. and Shipley, G.G. Biochemistry 1981 20, 5957-5966.

32. Janiak, M.J., Small, D.M. and Shipley, G.G. J. Biol. Chem. 1979, 254, 6068-6078.

33. Serrallach, E.N., Dijkman, R., de Haas, G.H. and Shipley, G.G. J. Mol. Biol. 1983, 170, 155-174.

34. Ranck, J.L., Keira, T. and Luzzati, V. Biochim. Biophys. Acta. 1977, 448, 432-441.

35. Hauser, H., Pascher, I. and Sundell, S.J. Mol. Biol. 1980, 137, 249-264.

36. McIntosh,T.J., Simon, S.A., Ellington, J.C. and Porter, N.A. Biochemistry 1984, 23, 4038-4044.

37. Seelig, J. and Seelig, A. Quart. Rev. Biophys. 1980, 13, 19-61.

38. Cameron, D.G., Gudgin, E.F. and Mantsch, H.H. Biochemistry 1981, 20, 4496-4500.

39. Yeagle, P.L. Acc. Chem. Research 1978, 11, 321-327.

40. Blume, A., Rice, D.M., Wittebort, R.J. and Griffin, R.G. Biochemistry 1982, 21, 6220-6230.

41. Pearson, R.H. and Pascher, I. Nature 1979, 281, 499-501.

42. Seelig, J., Dijkman, R. and de Haas, G.H. Biochemistry 1980, 19, 2215-2219.

43. Mabrey, S. and Sturtevant, J.M. Proc. Natl. Acad. Sci. USA 1976, 73, 3862-3866.

44. Hinz, H.J. and Sturtevant, J.M. J. Biol. Chem. 1972, 247, 3697-3700.

45. Phillips, M.C., Williams, R.M. and Chapman, D. Chem. Phys. Lipids 1969, 3, 234-244.

46. Mason, J.T., Huang, C. and Biltonen, R.L. Biochemsitry 1981, 20, 6086-6092.

47. Chen, S.C. and Sturtevant, J.M. Biochemistry 1981, 20, 713-718.

48. Stumpel, J., Nicksch, A. and Eibl, H. Biochemistry 1981, 20, 662-665.

49. Wilkinson, D.A. and Nagle, J.F. Biochemistry 1981, 20, 187-192.

50. Blume, A. Biochemistry 1983, 22, 5436-5442.

51. Blume, A. and Eible, H. Biochim. Biophys. Acta 1979, 558, 13-21.

52. Liao, M.J. and Prestegard, J.H. Biochim. Biophys. Acta 1981, 645, 149-156.

53. Cevc, G., Watts, A. and Marsh, D. Biochemistry 1981, 20, 4955-4965.

54. Mabrey-Gaud, S. in "Liposomes from Physical Structure to Therapeutic Applications" Ed. C.G. Knight, Elsevier/North Holland, Amsterdam 1981.

55. Kodama, M., Kuwabara, M. and Seki, S. Biochim. Biophys. Acta 1982, 689, 567-570.

56. Coolbear, K.P., Berde, C.B. and Keough, K.M.W. Biochem 1983, 22, 1466-1473.

57. Mason, J.T. and Huang, C. Lipids 1981, 16, 604-608.

58. Allgyer, T.T. and Wells, M.A. Biochemistry 1979, 18, 4354-4360.

59. Johnson, R.E., Wells, M.A. and Rupley, J.A. Biochemistry 1981, 20, 4239-4242.

60. Boggs, J.M. and Rangaraj, G. Biochemistry 1983, 22, 5425-5435.

61. Fleming, B.D. and Keough, K.M.W. Canad. J. Biochem. Cell. Biol. 1983, 61, 882-891.

62. Borle, F. and Seelig, J. Chem.Phys. Lipid 1985, 36, 263-283.

63. Rehfeld, S.J., Hansen, L.D., Lewis, E.A. and Eatough, D.J. Biochim. Biopys. Acta 1981, 691, 1-12.

64. Hauser, H. and Dawson, R.M.C. Eur. J. Biochem 1967, 1, 61-69.

65. Portis, A., Newton, C., Pangborn, W. and Papahadjopoulos, D. Biochemistry 1979, 18, 780-789.

66. Simon, S.A., Lis, L.J., Kauffman, T.W. and MacDonald, R.C. Biochim. Biophys. Acta 1975, 375, 317-326.

67. Chowdhry, B.Z., Lipka, G., Dalziel, A.W. and Sturtevant, J.M. Biophys. J. 1984, 45, 633-635.

68. Lee, A.G. Biochim. Biophys. Acta 1977, 472, 237-281.

69. Lee, A.G. Biochim. Biophys. Acta 1977, 472, 283-344.

70. Shimshick, E.J. and McConnell, H.M. Biochemistry 1973, 12, 2351-2360.

71. Davis, P.J., Coolbear, K.P. and Keough, K.M.W. Can. J. Biochem. 1980, 58, 851-858.

72. Luna, E.J. and McConnell, H.M. Biochim. Biophys. Acta 1977, 470, 303-316.

73. Wu, S.H.W. and McConnell, H.M. Biochem 1975, 14, 847-854.

74. Hammond, K., Lyle, I.G. and Jones, M.N. J. Coll. Int. Sci. 1984, 99, 294-296.

75. Lentz, B.R., Alford, D.R., Hoechli, M. and Dombrose, F.A. Biochemistry, 1982 21, 4212-4219.

76. Blume, A., Wittebort, R.J., Das Gupta, S.K. and Griffin, R.G. Biochemistry 1982, 21, 6243-6253.

77. Gaestel, M., Herrmann, A. and Hillerbrecht. B. Biochim. Biophys. Acta 1984, 769, 511-513.

78. Ohnishi, S. and Ito, T. Biochem 1974, 13, 881-887.

238

79. Ohki, K., Sekiya, T., Yamauchi, T. and Nozawa, Y. Biochim. Biophys. Acta 1981, 644, 165-174.

80. Demel, R.A. and de Kruijff, B. Biochim. Biophys. Acta 1976, 457, 109-132.

81. Ladbrooke, B.D., Williams, R.M. and Chapman, D. Biochim. Biophys. Acta 1968, 150, 333-340.

82. Davis, P.J. and Keough, K.M.W. Biochem 1983, 22, 6334-6340.

83. Blume, A. Biochem 1980, 19, 4908-4913.

84. Estep, T.N., Freire, E., Anthony, F., Barenholz, Y., Biltonen, R.L. and Thompson, T.E. Biochem. 1981, 20, 7115-7118.

85. Bach, D., Miller, I.R. and Sela, B-A. Biochim. Biophys. Acta 1982, 686, 233-239.

86. Merajver, S.D., Sheridan, J.P. and Siguel, E.N. J. Theor. Bio. 1981, 93, 737-755.

87. Jacobs, R.E., Hudson, B.S. and Anderson, H.C. Biochemistry 1977, 16, 4349-4359.

88. Scott, H.L. and Cheng, W-H. Biophys. J. 1979 28, 117-132.

89. Tinker, D.O. and Low, R. Can. J. Biochem 1982, 60, 538-548.

90. Mondat, M., Georgallas, A., Pink, D.A. and Zuckerman, M.J. Canad. J. Biochem. Cell. Biol. 1984, 62, 796-802.

91. Chapman, D. and Hayward, J.A. Biochem. J. 1985, 228, 281-295.

92. Devaux, P.F. and Seigneuret, M. Biochim. Biophys. Acta 1985, 822, 63-125.

93. Brandts, J.F., Erickson, L., Lysko, K., Schwartz, A.T. and Taverna, R.D. Biochemistry 1977, 16, 3450-3454

94. Brasitus, T.A., Tall, A.R. and Schachter, D. Biochemsitry 1980, 19, 1256-1261.

95. Blazyk, J.F. and Newman, J.L. Biochim. Biophys. Acta 1980, 600, 1007-1011.

96. Papahadjopoulos, D., Moscarello, M., Eylar, E.H. and Isac, T. Biochim. Biophys. Acta 1975, 401, 317-335.

97. Epand, R.M. and Surewicz, W.K., Canad. J. Biochem. Cell Biol. 1984, 62, 1167-1173.

98. Boggs, J.M. in "Membrane Fluidity in Biology" Ed. R.C. Aloia Academic Press 1983 Vol. 2 89-130.

99. Surewicz, W.K. and Epand, R.M. Biochem. 1985, 24, 3135-3144.

100. Epand, R.M. and Sturtevant, J.M. Biophys. Chem. 1984, 19, 355-362.

101. Huschilt, J.C., Hughes, R.S. and Davies, J.H. Biochem. 1985, 24, 1377-1386.

102. Morrow, M.R., Huschilt, J.C. and Davies, J.H. Biochem. 1985, 24, 5396-5406.

103. Epand, R.M. and Sturtevant, J.M. Biochem. 1981. 20, 4603-4606.

104. Massay, J.B., Rohde, M.F., Van Winkle, W.B., Gotto, A.M. and Pownall, H.J. Biochem. 1981, 20, 1569-1574.

105. Massey, J.B., Gotto, A.M. and Pownall, H.J. Biochem. 1981, 20, 1575-1584.

106. Bendzko, P. and Pfeil, W. Biochim. Biophys. Acta 1983, 742, 669-676.

107. Hanssens, I., Houthuys, C., Herreman, W. and Van Cauwelaert, F. Biochim. Biopys. Acta 190, 602, 539-557.

108. Van Cauwelaert, F., Hanssens, I., Herreman, W., Van Ceunebroeck, J-C, Baert, J. and Berghmans, H. Biochim. Biophys. Acta 193, 727, 273-284.

109. Hanssens, I., Herreman, W., Van Ceunebroeck, J-C, Dangreau,H., Gielens, C., Preaux,G. and Van Cauwelaert, F. Biochim. Biophys. Acta 1983, 728, 293-304.

110. Heyn, M.P., Blume, A., Rehorek, M. and Decher, N.A. Biochem. 1981, 20, 7109-7115.

111. Heyn, M.P., Cherry, R.J. and Dencher, N.A. Biochem. 1981, 20, 840-649.

112. Rehorek, M., Dencher, N.A. and Heyn, M.P. Biochem. 1985, 24, 5980-5988.

113. Goodwin, G.C. and Jones, M.N. Biochem. Soc. Trans. 1980, 8, 323-324.

114. Goodwin, G.C., Hammond, K., Lyle, I.G. and Jones, M.N. Biochim, Biophys. Acta 1982, 689, 80-88.

115. Petri, .W.A., Estep, T.N., Pal, R., Thompson, T.E., Biltonen, R.L. and Wagner, R.R. Biochem. 1980, 19, 3088-3091.

116. Wiener, J.R., Wagner, R.R. and Freire, E. Biochem. 1983, 22, 6117-6123.

117. Rosseneu, M., Soetewey, F., Blaton, V., Lievers, J. and Peeters, H. Chem. Phys. Lipids 1976, 17, 38-56.

118. Rigell, C.W., Saussure, C de and Freire, E. Biochem. 1985, 24, 5638-5646.

119. Chen, C-W. and Guard-Friar, D. Biopolymers 1985 24, 883-895.

120. Yu, C.A., Gwak, S.H. and Yu, L. Biochim. Biophys. Acta 1985, 812, 656-664.

121. Pink, D.A. and Chapman, D. Proc. Natl. Acad. Sci. USA. 1979, 76, 1542-1546.

122. Owicki, J.C. and McConnell, H.M. Proc. Natl. Acad. Sci. USA. 1979, 76, 4750-4754.

123. Pink, D.A., Georgallas, A. and Chapman, D. Biochem. 1981, 20, 7152-7157.

124. Pink, D.A. Canad. J. Biochem. Cell Biol. 1984, 62, 760-777.

125. Paddy, M.R. and Dahlquist, F.W. Biophys. J. 1982, 37, 110-112.

126. Steinhardt, J. and Reynolds, J.A. "Multiple Equilibrium in Proteins" 1969 Academic Press, New York.

127. Jones, M.N. "Biological Interfaces" Elsevier, Amsterdam, 1975, Chapter 5 pp. 101-134.

128. Helenius, A. and Simon,K. Biochim. Biophys. Acta 1975, 415, 29-79.

129. Abu-Hamdiyyah, M. and Mysels, K.J. 1967, J. Phys. Chem. 71, 418-426.

130. Jones, M.N. and Manley, P. in "Solution Behaviour of Surfactants" Ed. K.L. Mittal 1983 Plenum Publishing Corporation Vol. 1, 1403-1406.

131.Paz Andrade, M.I., Boitard, E., Saghal, M.A., Manley, P., Jones, M.N. and Skinner, H.A. J. Chem. Soc., Faraday Trans. I 1981, 77, 2939-248.

132. Scatchard, G. Ann. N.Y. Acad. Sci. 1949, 51, 660-672.

133. Schwarz, G. Biophys. Struct. Mechanism 1976, 2, 1-12.

134. Hill, A.V. J. Physiol 1910, 40, 10P.

135. Wyman, J. J. Mol. Biol. 1965, 11, 631-644.

136. Jones, M.N. and Manley, P. J. Chem. Soc. Faraday Trans I. 1979, 75, 1736-1744.

137. Jones, M.N. and Manley, P. J. Chem. Soc., Faraday Trans I 1981, 77, 827-83.

138. Jones, M.N. and Manley, P. Int. J. Biol. Macromol 1982, 4, 201-206.

139. Jones, M.N., Manley, P. and Holt, A. Int. J. Biol. Macromol 1984, 6, 65-68.

140. Paz-Andrade, M.I., Jones, M.N. and Skinner, H.A. J. Chem. Soc. Faraday Trans I 1978, 74, 2923-2929.

141. Jones, M.N. and Manley, P. J. Chem. Soc. Faraday Trans I, 1980, 76, 654-664.

142. Jones, M.N., Manley, P. and Wilkinson, A.E. Biochem. J. 1982, 203, 285-291.

143. Finn, A., Jones, M.N., Manley, P., Paz-Andrade, M.I. and Regueira, L.N. Int. J. Biol. Macromol. 1984, 6, 284-290.

144. Jones, M.N., Wilkinson, A.E. and Finn, A. Int. J. Biol. Macromol 1985, 7, 33-38.

145. Tipping, E., Jones, M.N. and Skinner, H.A. J. Chem. Soc. Faraday Trans. I 1974, 70, 1306-1315.

146. Jones, M.N. Biochim. Biophys. Acta 1977, 491, 120-128.

147. Decker, R.V. and Foster, J.F. Biochemistry 1966, 44, 1242-1254.

148. Reboiras, M.D. and Jones, M.N. Electrophoresis 1982, 3, 317-321.

149. Leonard, W.J. and Foster, J.F. J. Biol. Chem. 1961, 236, PC73.

150. Vijai, K.K. and Foster, J.F. Biochemistry 1967, 6, 1152-1159.

CHAPTER 6

THERMOCHEMISTRY OF LIVING CELL SYSTEMS

INGEMAR WADSO

6.1   INTRODUCTION

The present treatise will concentrate on recent calorimetric work
conducted on microorganisms and animal cells. Work on plant materials
will also be treated but very little calorimetric work has been done
in that field. Some investigations on pieces of human tissue, like
muscle fibres, will also be discussed.

Work in these areas reported prior to 1978 or 1979 was discussed
in different chapters of the monographs edited by Jones [1] and Beezer
[2]. The monograph edited by James [3] contains several reviews and
discussions dealing with the thermochemistry and the calorimetric
methods used in work on living cells. In the present discussion,
references will frequently be made to these articles and to other
reports listed in Table 1.

In thermochemical experiments with living cells, it is usually
the rate of heat production, i. e., the thermal power, P, which is
determined as a function of time. Usually the measurements are performed by use of micro reaction calorimeters operated under essentially isothermal conditions. For the determination of important
auxiliary data in studies of growth of microorganisms, traditional
combustion calorimetry is of significant importance, but such measurements are only performed in a few laboratories at the present time.
Temperature-scanning methods (DSC), being very important in current

Table 1. Reviews and discussions on calorimetric work on living
cellular systems

Area	Authors
Microorganisms:	
Bacteria, general aspects	Kresheck [4], Belaich [5], James [6]
Yeasts, general aspects	Lamprecht [7], Miles, Beezer, Perry [8]
	Perry, Miles, Beezer [9]
Identification and charac-terization of microorganisms	Newell [10]
Medical and pharmaceutical investigations	Bettelheim and Shaw [11], Beezer and Chowdhry [12]
Animal cell systems:	
Blood cells	Monti and Wadsö [13], Levin [14], Monti [15]
Other animal cells	Kemp [16], Kemp [17]
Instrumentation and measure-ent problems:	Spink and Wadsö [18], Martin and Marini [19], Spink [20], Wadsö [21,22],

biochemical thermochemistry, have so far only been of limited interest
in work on living cells.

Calorimetric investigations on living systems are conducted in
some cases as strict thermodynamic experiments. Typically, then, the
goal is to get information about chemical and biological mechanisms of

importance for material-energy balances. In other cases, the calorimeters are primarily used as general monitors for the cellular processes, i.e. as kinetic or analytical instruments. For complex
reaction systems, the use of calorimeters as quantitative but non-
specific process monitors has several advantages. One is that unknown
or unexpected phenomena are more likely to be recorded than if a more
specific analytical technique were used. In such work, which often is
conducted in applied areas such as biotechnology, clinical analysis
and ecology, the relationship between the observed thermal power and
the investigated process is usually not well-understood on the molecular level. It is then common that no use is made per se of the fact
that a thermodynamic property of basic importance, the thermal power,
is determined. It is sometimes difficult to determine relevant analytical quantities which would make it possible to analyze the power
signal in molecular-thermochemical terms: rates of consumption and
formation, respectively, of substrate(s), metabolites and cellular
mass and variations in the sizes of important material pools, like
that of ATP. Therefore, a large part of these calorimetric-analytical
experiments are conducted on an empirical "finger-print" level. But
undoubtedly   this promising analytical technique will be more
useful in basic as well as in applied work when the thermochemistry of
cellular processes is better understood.

It is quite common that the calorimetric power signal of a cell
system is taken as a measure for the poorly-defined property
"biological activity". One must then be aware that no simple and
general relationship exists between the thermal power and expressions
of biological "activity" like "growth rate", "rate of substrate consumption" or "rate of product formation". However, there is a general
and lasting value in the determination of well-documented thermodynamic quantities for biological systems, even if at the present time

they cannot be interpreted in great detail. Therefore, even when calorimeters are primarily used as general process monitors, it is important that they are adequately calibrated, that details of the experimental procedure are well-documented, and that the results are reported using units and terminology internationally recommended for thermochemical work. A brief guideline on these matters was prepared by the Interunion Commission on Biothermodynamics (IUPAC, IUB, IUPAB) [23]. Gnaiger in particular has recently suggested several new terms and expressions in connection with bioenergetics - thermochemistry of living systems, see e.g. [24], but it remains to be seen if they will gain a wider acceptance. One such term is "ergobolic" reactions, used for processes leading to net changes of concentrations of compounds with high phosphoryl transfer potential, e.g. ATP and creatinphosphate.

6.2   INSTRUMENTATION

It is usually microcalorimeters which are employed in work on cellular systems. The term "microcalorimeter" is not a well-defined concept, but it usually implies that the instrument is very sensitive and that only small quantities of material are required. For instance, a micro reaction calorimeter can normally be used for precise determination of millijoule quantities of heat and the power sensitivity is characteristically on the order of a microwatt or higher. However, typical microcalorimeters are often used at a much reduced sensitivity, e.g. in work on small animals or concentrated microbial suspensions, where the thermal power evolved may be several orders of magnitude larger than the sensitivity of the instrument. Ordinary (macro) solution or reaction calorimeters can also be useful for such work.

Calorimetric instrumentation and some methodological problems

connected with work on living cell systems were recently discussed in
some detail [22], cf. [21]. From the point of view of measurement principles, calorimeters form two main groups: adiabatic or nearly adiabatic calorimeters and heat conduction calorimeters (also called heat
leakage, heat flow or heat flux calorimeters). Microcalorimeters
which presently are used in work on living cells are nearly always of
the heat conduction type and this measurement principle will therefore
be briefly described.

In the ideal heat conduction calorimeter, heat released is quantitatively transferred from the reaction vessel to a surrounding heat
sink, which normally consists of a metal block. Usually the heat flow
is recorded by positioning a "thermopile wall" between the calorimetric vessel and the heat sink. The temperature difference over the
thermopile (usually consisting of semi-conducting thermoelectric
plates, "Peltier effect plates") will give rise to a voltage signal.
The instrument signal is thus the thermopile voltage, V, which in the
ideal case is related by the Tian equation (eqn (1)) to the thermal
power produced in the vessel,

$$P = \varepsilon(V + \tau\frac{dV}{dt}),$$

(1)

where ε is a constant determined by calibration experiment, τ is the
time constant for the instrument and $\frac{dV}{dt}$ is the time derivative of the
voltage signal. The value for τ is proportional to the heat capacity
of the reaction vessel and inversely proportional to the thermal
conductivity of the thermopile. In accurate kinetic work, it is
sometimes desirable to use more than one time constant, see e.g. [25].

If a constant power is produced, V will reach a steady-state value, V_s. The power released in the vessel is then quantitatively balanced by the heat leakage to the surroundings and equation (1) will be reduced to

$$P = V_s \qquad\qquad (2)$$

i.e., the power is directly proportional to the voltage signal from the thermopile. To a good approximation, the simple eqn. (2) can also be applied to processes where the power changes slowly compared to the time constant. This is typically the case for processes involving living cells, but if the thermal power changes rapidly with time, the recorded voltage-time curve will be distorted compared to the true kinetic curve. To obtain precise kinetic information under such conditions, it is necessary to reconstruct the power-time curve from the thermopile output data, see e.g. [25]. Such corrections can now be performed automatically by use of a computer connected on-line to the calorimeter and such units have also become commercially available. But as yet it is very rare to find reports where the signals from thermopile conduction calorimeters are corrected to true kinetic curves.

Integration of eqn. 1 or 2 will give

$$Q = \int_{t_1}^{t_2} V dt \qquad\qquad (3)$$

or

$$Q = \mathcal{E} \cdot A \qquad\qquad (4)$$

i.e. the heat quantity evolved is proportional to the surface area under the voltage-time curve (A). If the integration is extended to the time where the voltage signal has returned to the baseline, it is thus not necessary to know the time constants in order to determine Q.

Most microcalorimeters used in cell biology are designed as twin, or differential, instruments, where one calorimetric vessel usually contains the reaction system and the other an inactive material. If the differential signal is recorded, contributions from disturbances are expected to cancel. This property is of special importance in cases where the experiments are conducted over long periods of time.

The reaction vessel containing the cell material is sometimes made up of a simple static ampoule, or it can have a more complex design, e.g. a stirred vessel or vessels with two compartments where reagents are mixed by rotation or rocking of the calorimeter. Sometimes the vessels are fitted with titration inlet tubes. Flow calorimetric vessels typically consist of spiralized tubes or ampoules through which reagents or medium are pumped via heat exchangers. If an object, such as a piece of tissue or cells anchored to a support, is kept in a calorimetric vessel through which liquid or gas can flow, the instrument is often called a perfusion calorimeter.

The instruments which are the most widely used in work on living cells are the thermopile heat conduction microcalorimeters manufactured by Setaram (Caluire, France) and LKB Produkter (Bromma, Sweden). One of the best known types of heat conduction calorimeters is the Calvet microcalorimeter (Setaram), developed by E. Calvet in Marseille and based on the early design by Tian, see e.g. Calvet and Prat [25]. This instrument is suitable for measurements on the microwatt level and has a good long-term stability. A large aluminium block serves as the heat sink for two (or four) calorimetric units, which normally are used as a differential system(s). In each unit, a thin-walled metal cylinder defines the space for the calorimetric vessels. They can consist of simple tubes with stoppers, reaction volume 15-100 ml, but many different vessels for special purposes (e.g., injection, stirring, introduction of light) have been described, see e.g. [18]. Several specialized vessels are commercially available.

More recently, Setaram has introduced a new multipurpose calori-metric system,"Bio-DSC", which is also based on the heat conduction principle. This instrument is designed for temperature scanning ex-periments as well as for isothermal batch and flow calorimetry, see [26]. Vessel volume is 1 ml and the power sensitivity is stated to be 0.2 µW when used in an isothermal mode.

In the LKB 2107 microcalorimetric system, the calorimetric vessels or vessel holders are shaped like boxes or thin plates. Semi-conducting thermocouple plates are interspersed between the vessels and metal blocks forming parts of the heat sink. In the batch reaction calorimeter [27], reagents in the two-compartmented reaction vessel can be mixed by rotation of the calorimetric block. The flow instrument [28] has been used in many studies involving cells in suspension, in particular with microorganisms. Different flow-through and mixing vessels have been described, see e.g. [22]. Flow calori-meters used in work on cell suspensions are sometimes connected in series with other instruments such as oxygen electrodes, meters for optical density, fraction collectors or valve systems for sequential sample collection or introduction, see e.g. [18, 22, 29].

The prototype of the LKB 2107 ampoule microcalorimeter [18, 30] has been used extensively in work with blood cells and other animal cells, [13, 15], and with microbial systems including soil [21]. This instrument uses simple closed ampoules (volume 1-10 ml) or a perfusion vessel as insertion reaction vessels. A related instrument, the LKB "sorption" microcalorimeter [31] has, for example, been used in studies on cells anchored to beads placed in its adsorption column [16, 32].

The new LKB microcalorimetric system [33] (called "BioActivity Monitor", or the more recent model, "Thermal Activity Monitor") has a thermostated water bath which can hold up to four independently operated twin calorimeters ("channels"). The temperature stability of the bath is $\pm$ 1·10^{-4} ^{o}C over long periods of time (days), provided that the room temperature does not fluctuate more than $\pm$ 1 ^{o}C. The long-time baseline drift and the maximum short-term deviation from the baseline for channels using simple closed vessels are typically less than 0.1 µW. The twin calorimetric unit can also be fitted with two permanently installed flow vessels. Recently a stirred "titration and perfusion vessel" was developed [34]. It can be used with cells in suspension or attached to a support such as polystyrene film or micro-carriers [34, 35]. It can also be used with small pieces of tissue, e.g. muscle fibres [36], in which case the biological material is placed in a cage attached to the stirrer shaft. Experiments can be performed with gas phase present or with the vessel completely filled with medium when liquid is perfused through the vessel. Small quantities of reagents can be added to the sample compartment through injection needle(s) during the measurement and stepwise titrations can be per-formed by an automatic procedure. The baseline stability is typically $\pm$ 0.3 µW over 10 h (stirring rate 50 r.p.m., perfusion rate $\lesssim$ 20 ml/h). Gas can also be pumped through the system, with some increase of the noise level [34, 37].

The "Bioflux" microcalorimeter (Thermanalyse, Grenoble, France) is a rather simple twin heat conduction calorimeter which has been used in several studies on living cells, e.g. skin cells [38]. It is used with insertion vessels with a maximal volume of 25 ml. Short-time baseline stability is stated to be $\pm$ 2 µW.

Lovrien and coworkers have reported the design of several heat conduction calorimeters used for work with cells and other biological objects. These instruments include rotating batch calorimeters holding two [39], three [40], or even five [41] mixing vessels.

A 2-channel heat conduction calorimeter manufactured by Rhesca, Tokyo, has been used in work on suspensions of tumour cells [47]. The reaction vessel can be stirred and reagents can be injected but the sensitivity of this instrument is rather low.

The Picker type of flow calorimeter [43] (Sodev Inc., Sherbrooke, Canada, and Setaram) is not a heat conduction calorimeter. Essential-ly it is a differential flow calorimeter with counter-current heat exchangers and a detection device requiring a single thermistor sensor. Versions of the instrument have been used, e.g. in studies of biodegradation processes of waste water sytems. In such connections the power detection limit is reported as 1-2 μW [44].

Fujita and coworkers have reported the design of a rotating batch calorimeter which was equipped with vessels specially designed for studies of aerobic microbial growth. A recent version of this instru-ment [45], which can be used in either batch or flow mode, is produced by Tokyo Riko Co., Tokyo.

The instrument reported by Ishikawa et al. [46] was specially designed to handle the problem of efficient oxygen supply to microbial cultures. The reaction vessels, which resemble small conventional fermentors, have a rather large volume, 30 ml. Instead of applying vigorous agitation of the gas-liquid system during aerobic microbial experiments, Dermoun et al. [47] used shallow reaction vessels, where the medium has a large contact area with the gas phase. The vessels, which rest on thermocouple plates, are equipped with stirrers. Inoculum can be injected after the instrument has equilibrated and humidified gas can be introduced to the vessels during the experiments.

Kawabata et al. [48] described a 6-channel heat conduction calorimeter for work on soil microbial systems. The calorimetric vessels, volume 30 ml, were placed in a hexagonal arrangement with thermocouples interspersed between their bottoms and a large aluminium block used as a common heat sink. A reference calorimetric unit was placed in the center of the hexagon.

Practically all physical and chemical processes are accompanied by heat effects. This property makes reaction calorimeters useful as general process monitors, but it will also make calorimetric techniques vulnerable to systematic errors. Such problems are particularly important to watch out for in microcalorimetric work where a small power signal can be much affected by various experimental parameters, which thus must be under strict control: evaporation and condensation effects, friction and other mechanical effects, sedimentation and adhesion of the cells in flow lines and in reaction vessels, and oxygen supply. Problems of this sort were discussed in some detail in [21, 22].

Reaction vessels for work with living cells must be designed with many practical factors in mind. It is therefore common that the electrical calibration of the vessel cannot be conducted under conditions which are ideal from a calorimetric point of view. In order to control electrical calibration values for calorimetric vessels, not the least flow vessels, chemical test processes are frequently needed. It was recently shown that the hydrolysis of triacetin in imidazol-

acetic acid buffers can be useful in this context [49]. Different
mixtures giving power values in the range 7-90 μW per ml (25, 37 $^{\circ}$C)
were described. Accurate power values for the mixtures ($\pm$ 0.5%) can
be calculated as a function of time during extended reaction periods,
20 h or more.

6.3   INVESTIGATIONS ON DIFFERENT CELL SYSTEMS

A major part of the calorimetric work on cellular systems has
been performed with microorganisms. In most cases the studies have
been made on pure cultures of bacteria or yeast but poorly defined
mixtures such as those found in natural systems have also been
investigated. The work on pure cultures has to a large extent been
conducted on a thermodynamic level and the thermochemistry of
microbial growth and maintenance is much better understood than
corresponding work on animal cells. However, it appears that work in
the latter area is presently making rapid progress.

6.3.1   Bacteria

General studies of growth and metabolism. Work reported before
the late 1970s was discussed in the review by Belaich [5]. James has
covered the same field in his very recent review [6] and the treatment
here will therefore be quite summary and limited to recent work.

Dermoun and Belaich [50, 51] have reported from two important
investigations on the thermochemistry of aerobic growth of E. coli in
a minimal medium using succinic acid as substrate. The microcalorimetric measurements, performed by use of a newly designed vessel for
aerobic growth [47], were combined with results of combustion calorimetric measurements of the formed biomass. The experimentally determined heat production during growth ($\Delta_{met}H$) was found to be significantly lower than the value for the enthalpy of combustion for

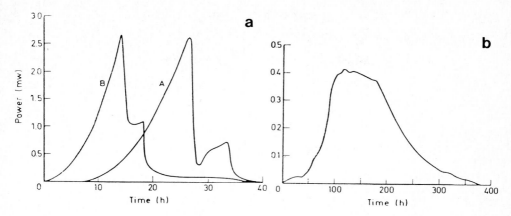

Fig. 1a   Power time curves of <u>Cellulomonas</u> sp. 21399 growing

aerobically in glucose, 2 g/l (A), and amorphous cellulose

(B).

Fig. 1b   Power time curve of <u>Cellulomonas</u> sp. 21399 growing

aerobically on crystalline cellulose. Reprinted from

Dermoun et al. [54]

succinic acid, corrected for the fraction of substrate which was

transformed to biomass.

Ishikawa et al. [52] have studied the aerobic growth of E. coli K-

12 under glucose-, nitrogen-, or oxygen-limited conditions. A twin

heat conduction calorimeter [46] resembling a conventional fermentor was

used. Enthalpies of combustion and elementary composition of the

cells were determined and a detailed thermochemical analysis was

performed.

Gonda et al. [53] studied the initial reaction phases when solu-

tions of glucose and other sugars were added to a suspension of washed

E. coli cells kept in phosphate buffer. The calorimetric measurements

were made by use of a Tronac Model 450 isoperibolic calorimeter fitted

with a 25 ml vessel. Three different reaction steps were observed

which were interpreted as connected with the uptake of the substrate and with two steps in the glycolytic pathway, respectively.

In two papers Dermoun et al. [54, 55] compared the thermochemistry of growth of Cellulomonas on glucose, cellobiose and amorphous and crystalline cellulose. The P-t curves and the thermochemical proper- ties were found to be similar for glucose, cellobiose and amorphous cellulose whereas the results for crystalline cellulose were different, Fig. 1. It was concluded that the cellulolytic enzymes readily degrade amorphous region of cellulose but become less active when the crystallinity increases.

The energetics of growth of Desulfovibrio vulgaris was investigated by Traore et al [56] under strictly anaerobic conditions using a Tian Calvet microcalorimeter. Lactate or pyruvate was used as energy substrate together with a high concentration of sulfate. The enthalpy of combustion of the obtained cellular mass was also determined. The derived $\Delta_{met}H$ values for the two substrates were in good agreement with the values based on product formation (which were significantly different from earlier results reported for this bacterial system).

James and his coworkers have conducted extensive thermochemical investigations on the growth of a strain of Klebsiella aerogenes (NCTC418) by use of the LKB 10700-1 flow calorimeter. In a first series of papers [57-60], the energy changes during growth in simple salt-glucose media were investigated. The influence on the growth curves from basic experimental parameters such as the state of the inoculum, aeration in the fermentor and in the calorimeter, flow rate, and the glucose concentration were studied. The organisms were grown in a fermentor where pH, $p(O_2)$ and optical density were monitored. A flow type oxygen electrode was inserted into the flow line immediately after the calorimeter.

From results of the microcalorimetric measurement and measurements of the elementary composition and the enthalpy of combustion of the cellular mass, the thermochemical quantities for the growth processes were calculated. P-t curves indicated that the metabolic pattern varied with the temperature even if aerobic conditions prevailed but values for stored and waste energy were constant below 40 ^{o}C [59]. At higher temperatures a larger part of the available energy went to waste.

Bowden and James correlated thermochemical growth parameters and the size of the ATP pool [61] and the influence by limited supply of phosphate [62]. For glucose-limited medium, it was found that the ATP

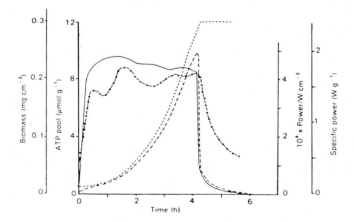

Fig. 2   Power, specific power and ATP profiles for cells of
Klebsiella Aerogenes growing aerobically in glucose limited
medium. -------, Biomass; — — — —, power; ————, specific
power; — . —. — ., ATP pool. Reprinted from Bowden and
James [61].

pool and the specific power increased markedly during the early part
of the exponential growth. During later stages of the growth these
parameters were essentially constant but oscillations around the
mean values were observed for each parameter, Fig. 2. It could not be
shown that the oscillations were due to synchronous growth which,
possibly, could be initiated through the cryogenic treatment of the
inoculum. When the growth took place under phosphate limited conditions three different growth phases could be identified. Mass and
energy balances were established for each phase.

James and Djavan investigated the heat production for the same
strain of Klebsiella aerogenes during aerobic growth in chemostat
culture under different limiting conditions [63, 64]. The effect of
uncouplers was also studied [65].

Gordon and Millero [32] have measured the heat production from
Vibrio alginolyticus metabolizing on glucose media. The cells were
allowed to attach to polyacrylamide beads which were placed in the
perfusion vessel of an LKB sorption microcalorimeter (2107). The heat
production per cell was found to be much lower than earlier found for
these cells in free suspension [66].

Lovrien and coworkers used a batch heat conductor calorimeter for
metabolic studies involving simple sugars and aromatic compounds [39].
E. coli was used with eight different sugars under aerobic and
anerobic conditions and Pseudomonas multivarians and an Acinetobacter
strain were used for aerobic metabolism of benzene, toluene,
naphtalene and 2-methylnaphtalene. Comparison was made between the
calorimeter results and the values for the total combustion of the
substrates, giving a measure of the efficiencies by which the

substrates were used by the organisms.

Monk has reported from a series of flow microcalorimetric studies
of lactic acid bacteria (cf. Belaich [5], Newell [10]). In [67] the growth
of Str. thermophilus and L. Bulgaricus were investigated in single and
in mixed cultures. Anaerobic and aerobic catabolism of glucose at
different pH values by Str. agalactia were reported in [68]. The effect
of stimulation of glucose metabolism by addition of the uncoupler
2,4 dinitrophenol was also investigated.

Bezjak [69] has reported from a flow calorimetric study on E. coli
grown in two different broth media. The increase in viable cell
numbers was not accompanied by a proportional power output but showed
a stepwise increase.

Marison and von Stockar [71] have modified a commercial bench scale
calorimeter (Ciba-Geigy, Basel, Switzerland, model BSC-81; reaction
volume 2 1, primarily designed for chemical process development) for
microbiological work. The instrument has been used for studies of
aerobic growth of Klyveromyces fragiles on lactose [71, 72] and other
sugar substrates [72]. The results were used to optimize the medium
composition. Relationships between heat evolution, biomass
concentration, rate of oxygen consumption, and rate of carbon dioxide
formation were determined. It was suggested that calorimetry may
serve as a valuable on-line measurement technique for determinations
of parameters needed for the design and control of microbial processes
on a technical scale.

Cordier et al [73] have discussed the relationships between
elemental composition and heat of combustion of microbial biomass. An
improved sample preparation technique for use in heat of combustion

work was developed. Heats of combustion of several strains of
bacteria and yeasts, grown on different substrates, were determined.
Results were compared with values calculated from different models
from the literature.

Miles et al. [74] have studied the heat resistance of different
vegetative bacteria by use of a Perkin Elmer differential scanning
calorimeter (DSC-2C). Complex thermograms were obtained showing four
major peaks. Thermograms of heat-sensitive organisms showed features
which were absent in the more heat-resistant types. Cell death and
irreversible denaturation of cell components occurred within the same
temperature range. Equations were derived to calculate the thermal
death rates of bacteria when the temperature is increased at a
constant rate.

Characterization of bacteria. The identification of bacteria by
use of their characteristic and very complex P-t curves when grown on
a complex medium appeared to be a very promising technique when it was
first demonstrated by Boling et al. in 1973 [75]. The potential has
since been verified in several reports, see e.g. Fujita et al. [76] and
Herman et al. [77]. However, it has also been shown that several
experimental parameters, in particular the inoculum history, must be
strictly controlled, which restricts the procedure's general
usefulness for identification. It seems to be in the narrower field
of characterization of microorganisms that calorimetry is most
attractive, for instance in quality control for assessing performance
of microorganisms from a standard origin or for evaluating media
batches. For a critical account of the technique see Newell [10].

Antibiotic tests. When the metabolism of microorganisms is disturbed, it will be reflected by the power-time curve. This has been used in several microcalorimetric studies involving the kinetics of antibiotic action. Work in this field is discussed by Beezer and Chowdhry [12], who concluded that microcalorimetric methods can be very useful for (a) quantitative and qualitative assay of drugs, (b) investigations of the mode of action of the drug on the molecular level and (c) for the design and screening tests of new drugs. Several examples of such investigations were reviewed by Beezer and Chowdhry [12]. More recently, Seminitz et al. [78] have reported from an extensive study where the action of several antibiotics on E. coli was observed by a flow microcalorimetric technique (LKB 2107), turbidometry, phase contrast spectroscopy and electron microscopy. A good correlation was found between the results for the four different methods. Chernykh and Firsov [79] have studied the in vitro kinetics of antimicrobial action of sisomicin on E. coli by microcalorimetry. They concluded that the calorimetric method was more accurate than the count of colony-forming units (CFU) and more sensitive and selective than turbidometry.

Bowden and James [80] have recently studied, on a basic thermochemical   level, the action of naldixic acid on Klebsiella aerogenes. The enthalpy of production of cellular material and the enthalpy of metabolism of glucose became more negative with increasing inhibition.

Perry et al. [81] have evaluated the use of flow microcalorimetry as a "bioactivity" screening procedure. The effect from a series of novel Schiff bases on an E. coli test strain was investigated and it was concluded that the technique was simple, rapid and discriminating.

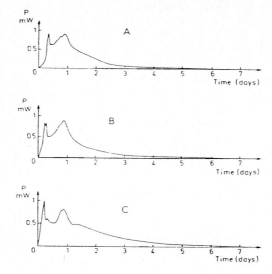

Fig. 3   Reproducibility of P-t curves of straw fermentation by a
          mixed culture. A. Initial experiment. B. Experiment A
          was repeated after one month of maintenance of the culture
          on straw. C. Corresponding experiment after two months.
          The amount of straw was 1.5 g. Reprinted from Fardeau et
          al. [82].

Some degradation processes. Fardeau et al. [82], cf. [83, 84] used a
Calvet microcalorimeter for an investigation where finely divided
straw from barley was degraded under anaerobic conditions by a mixed
bacterial culture. Despite the complexity of the system, the
reproducibility of the P-t curves was high, Fig. 3. The combined
results of calorimetry and chemical analysis showed that the limiting
factor in biodegradability of the straw was the celluolytic activity
of the bacterial culture, cf. [54, 55]. The enthalpy change associated
with fermentation of each cellulose unit ($C_6H_{10}O_5$) was found to be
similar to values obtained earlier for hexose fermentation by pure
cultures.

Glanser et al. [85] used an LKB flow calorimeter to follow changes in a mixed culture during bio-oxidation of molasses slops containing betain. The calorimetric results were used to optimise the regulation of the process.

The potential use of microcalorimetry as a rapid method for estimation of bacterial levels in ground meat [86] and in fish [87] has been explored by Gram and Søgaard. Correlations were obtained between values for the thermal power and traditional plate count methods.

The use of microcalorimetry in connection with microbial degradation of minerals has been explored in some model experiments. Goodman and Ralph [88] used an LKB flow calorimeter to measure the growth of two strains of Thiobacillus using thiosulphate as a substrate. Soljanto and Tuovinen [89] used the same type of instrumentation to monitor the action of Thibacillus ferroxidens on $FeSO_4$ and $FeSO_4$ together with a suspension of UO_2

Microbial systems of ecological interest. Several laboratories have been concerned with development of calorimetric methods for the characterization of the "biological activity" of ecological systems such as soil, sediment, waste water masses and other natural microbial systems. The microorganisms are normally not well-defined, although it is usually assumed that the dominating biological activity is due to bacteria. An early systematic study on bacterial activity in soil was performed by Ljungholm et al [90]. In order to avoid significant changes of the gas phase composition during long-term experiments, a special ampoule technique was developed where O_2/CO_2 exchange could take place through silicone rubber membranes, cf. [21]. Ljungholm et al. also studied the effect on the thermal power from acidification of soil with or without addition of extra energy substrates (cellulose powder) to the soil [91, 92].

Sparling [93-95] has investigated the heat production from a large number of different soils by use of a Calvet microcalorimeter. The calorimetric values correlated well with rates of respiration and, to a lesser extent, with the biomass and ATP content [93]. The heat production rate was found to be nearly the same for a wide range of freshly collected soils [95].

Takahashi and coworkers have used a 6-channel heat conduction calorimeter [48] in their studies of soil. Experiments were made on soil to which typical chemical pollutants [48] or different sugars [96] had been added. In [97] Kimura and Takahashi reported from a study on the degradation of glucose by soil bacteria. On the basis of the results the authors proposed a method to evaluate the rate of microbial degradation of organic substances in soil. In two papers, Yamano and Takahashi [98, 99] developed methods for computer simulation of microbial growth curves. In [99] the P-t curves from experiments with soil were analyzed.

Wojcik [100] has studied the microbial activity of different forest soils by use of direct calorimetry in combination with respirometry. A clear decrease in activity and in respiration with soil profile depth was seen. The activity and the proportion anaerobic processes depended on the type of habitat and the soil moisture.

Pamatmat et al. [101] investigated the thermal power, ATP concentration and the activity of the electron transport systems (ETS) in samples of sea sediments. These workers concluded that direct metabolic rate measurements, preferably by calorimetry, are essential in addition to ATP and ETS activity measurements for an understanding of the differences and changes in sediment metabolism.

Lock and Ford have reported the design of a simple flow calorimeter for studies of attached and sedimentary aquatic microorganisms [101]. Results from measurements on river epilithon, river sand and marine sand were reported.

Gustafsson and Gustafsson have discussed the use of flow microcalorimetry in ecotoxicity tests and compared it with other techniques [103]. Qualitatively good agreement was obtained between heat production and respiratory activity, as well as between different types of biomass determined as ATP-pool and number of microorganisms. The effect of a quaternary amine (aliquate 336) on the bacterial degradation of cellulose and of river sedements was investigated in some detail by K. Gustafsson [104, 105]. Gustafsson and Gustafsson [37] have also reported from an evaluation of LKB's stirred titration and perfusion vessel [34] for studies of the biological activity of natural sediments.

Lasserre and Tournié [106, 107] have developed a microcalorimeter procedure where metabolic processes occurring in the interface between sea water and sediment were measured. An LKB flow calorimeter was used. From experimental microcosms the interface sea water was pumped continuously through the calorimeter and an oxygen electrode connected in series with the flow calorimeter. The seasonal adaptive metabolic changes were studied [106].

Jolicoeur, Beaubien and coworkers have used a Picker-type flow microcalorimeter [43] in several studies on biodegradation processes in waste water systems [44, 108, 109]. A comparison was made with respiratory measurements. Toxicity effects from heavy metal salts [108], aliphatic alcohols [109], and surfactants [108] were investigated. In [110] their calorimetric technique was tested in model experiments and in on-site experiments at an industrial waste water plant. A certain amount of waste-water substrate was added to a reservoir with microorganisms from a waste-water treatment plant. The mixture was pumped from the reservoir through the flow calorimetric vessel. The resulting P-t growth curve was analyzed starting from Monod's kinetic model. It was found that t_{max} (the time during which the power was maximal) was linearly correlated to the biodegradable content of the waste water. The effect on t_{max} of the addition of $HgCl_2$ was also investigated.

Weppen and Schuller [111] measured the metabolic heat production from a chemostat culture of <u>Acinetobacter</u> <u>calcoaceticus</u> and studied the toxic influence of potential environmental chemicals. Ecological aspects were discussed, cf. also Weppen's work with E. coli in chemostat culture [112].

Redl and Tiefenbrunner [113] used an LKB flow calorimeter to characterize the microbial hydrolytic activity of different substrates in waste water systems. The same type of instrument was used in a direct practical application by Harjo-Jeanty [114], who assessed the activity of microorganisms causing slime problems in papermaking waters. The technique was used to assess the minimum inhibitory concentration of a biocide. The formation of resistant strains was also studied.

Lamprecht and his colleagues have studied heat evolution from microbial degradation of solid natural substrates like manure [115]. They employed an adiabatic calorimeter useful for 25 g samples. The possible applications for energy production from microbial degradation of waste materials were discussed, cf. also discussion in [116].

6.3.2 Mycoplasmatales

The first calorimetric measurements of the growth of organisms belonging to the order of Mycoplasmatales (Acholeplasma, Mycoplasma and Ureaplasma) were reported by Ljungholm et al. [117]. These workers, who used a static ampoule method, suggested the possibility that calorimeters could be used for identification and numeration of these organisms. More recently Miles et al. [118, 119] have studied the growth of Mycoplasma mycoides in complex media containing different sugars and sugar derivatives. Measurements were made using a proto-type of the LKB 4-channel microcalorimeter [33]. A continuous flow method or a static ampoule method was used. Enthalpy changes asso-ciated with metabolism of the energy source were in good agreement with theoretical values. Values for the Michaelis constant (K_m) and maximum specific rates of substrate utilization were calculated from the calorimetric results. P-t curves were reproducible and showed details of growth which were not observed by conventional microbiological techniques. In [120] glucose-negative and pyruvate-negative mutants were characterized by the microcalorimetric technique.

6.3.3 Yeast

Calorimeter work on growth and metabolism of yeast reported before 1977 or 1978 was thoroughly covered in Lamprecht's review [7]. Some special aspects were treated by Beezer and Chowdhry (interaction with drugs) [12] and by Newell (identification or characterization) [10]. Both fundamental and industrial aspects have more recently been dis-cussed in some detail by Beezer and his colleagues [8, 9]. The treat-ment here will therefore be kept very brief and will concentrate on work which was reported after Lamprecht's review [7].

General studies on growth and metabolism. Nunomura and Fujita
used a rotating batch calorimeter in a series of studies with different yeasts under nongrowing conditions [121-123]. A considerable
amount of heat was evolved when a washed suspension of yeast cells was
diluted with buffer [121]. Oxygen was consumed during the process,
suggesting that the observed heat effect was due to the endogenous
metabolism of the cells, cf. corresponding "heat of dilution" experiments performed with bacteria by Forrest and Walker [124]. When glucose
was added to the washed cell suspension under aerobic conditions, the
P-t curve showed three distinct phases which could be related to
glucose uptake, metabolism of formed product (ethanol) and finally a
phase of endogenous metabolism. Heat effects associated with the
glucose transport were also investigated [123].

Lamprecht's group has continued its fundamental metabolic studies
of yeast grown under different conditions. Brettel et al. [125-128]
have reported from work made with Saccaromyces cerivisiae (bakers'
yeast) in batch or chemostat culture. The thermal power was measured
by use of an LKB flow-through microcalorimetric vessel. Aerobic
conditions were maintained by use of a mixed flow of cell suspension
and air. A correction to the power value was applied in order to
account for the flow time of cells between the fermentor and the
calorimetric vessel. With glucose as energy substrate and under
optimal aerobic conditions a triphasic growth was observed [125]. Thermochemical and thermokinetic data for the growth processes were reported [126, 127]. With glucose media and with batch cultures on ethanol the energy balances were completely established. With chemostat
cultures using ethanol as a substrate, a lack of about 25% of the
energy input occurred which was not converted to biomass or dissipated
heat [128]. This group of workers used the same calorimetric technique
in their work with synchronous cultures of bakers' yeast [129, 130].
Both in batch and in continuous cultures the cell count and the heat

production showed an oscillatory pattern.

Louriero-Dias and Arrabaca [131] used an LKB flow calorimeter in an investigation of respiration-deficient mutant of S. cerevisiae under aerobic and anaerobic conditions.

Lamprecht and Schaarschmidt [132]have reported from an exploratory investigation on the simultaneous effect from light and certain dye-stuffs ("photodynamic effect") on the growth and metabolism of several strains of S. cerevisiae. Different microcalorimeters were used.

Calorimetric growth pattern and yeast characterization. Schaarschmidt et al. [133] used an LKB flow microcalorimeter to

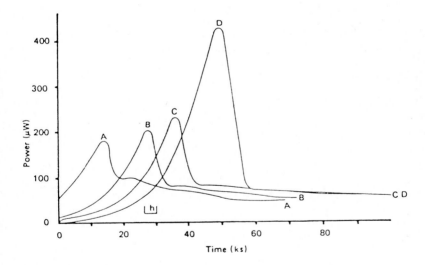

Fig. 4   P-t curves showing the effects of innoculum density on the growth of S. cerevisiae D1. Calorimetric incubations: 10 mM glucose at inoculum densities of 5 x 10^8 cfu/150 ml (A); 1 x 10^8 cfu/150 ml (B); 5 x 10^7 cfu/150 ml (C); 1 x 10^7cfu/150 ml (D). Reprinted from Perry et al. [138].

investigate the effect of the antibiotic nystatin on yeast, cf. work by
Beezer and coworkers reviewed in [12]. Beezer and Chowdhry [134] used
flow microcalorimetry to compare the activity of nystatin and its N-
acetyl derivative. The responding organism was S. cerevisae. The
activity was found to be significantly lower for the N-acetyl compound
than for the apparent antibiotics. In another flow calorimetric
study, Beezer and Chowdhry [135] investigated the effects of pH, metal
ions (Ca^{2+}, Mg^{2+}) and sterols (cholesterol and ergosterol) on the
interaction of polyene antibiotics with respiring S. cerevisiae cells.
Different effects were observed for the different antibiotics tested.

Perry et al. [136] have investigated the different conditions which
will influence the growth curves of yeast (S. cerevisiae). An LKB
flow microcalorimeter was used. It was shown that the shape of the P-
t curves were significantly influenced by factors like inoculum variation (e.g. size, history, phase of growth) or glucose concentration,
and these workers were very sceptical about the possibility to use
microcalorimetry for identification of yeast strains, cf. Newells
review [10]. In a later study Perry et al. [137] observed marked differences in P-t curves for different brewing, baking and distilling
strains of S. cerevisiae and S. uravum grown on defined media. The
results led these workers to conclude that their flow microcalorimetric procedures could be of practical use as a rapid method of characterization of commercial yeast strains. The influence from inoculum
density and glucose concentration on the growth curves were further
investigated for several commercial yeast strains [138]. Fig. 4 shows
the obtained P-t curves of S. cerevisae. It is seen that with increasing innoculum density both P_{max} and the area below the principal
peak declined. An anomalous, large and unexplained power following
exhaustion of glucose was observed. The shape of the P-t curves was

269

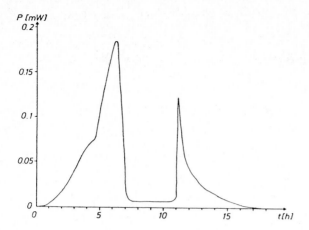

Fig. 5   P-t curve of an aerated culture of S. cerevisiae RXII
          growing in a complex medium containing a low concentration
          of glucose (2.5 mM). The substrate consumed in the sharp
          second peak, following a 4 h period of nearly zero heat
          production, was believed to be acetate. Reprinted from
          Lamprecht et al. [141].

similar to those observed for the other strains of S. cerevisae.   In
another study [139], Perry et al. investigated P-t curves for a number
of commercial strains of S. cerevisiae grown on defined media containing glucose and maltose. Contrary to earler results reported by
Schaarschmidt and Lamprecht [140], Perry et al. found in all cases
monophasic P-t curves.

     Lamprecht et al. [141] investigated the occurrence of a heat
production peak late after the termination of growth of a strain of
bakers' yeast in low concentration glucose media, cf. Beezer et al.
[142]. The substrate reponsible for the effect was not identified but
acetate was considered a likely candidate, Fig. 5.

     Itoh and Takahashi [143] have reported from a thermodynamic and kinetic study of the growth of S. cerevisiae with glucose as energy substrate. A twin heat-conduction calorimeter with very large reaction vessels (300 ml) was used. Measurements were performed at 7 temperatures in the range 16-39 $^\circ$C. The calorimetric results were analyzed using Monod's growth model. Arrhenius plot of the maximum growth rate constant showed a sharp discontinuity at 33 $^\circ$C. Thermodynamic parameters were derived for the glucose uptake by the cells at 25 $^\circ$C.

Owusu and Finch [144] used the LKB sorption microcalorimeter for a basic investigation of S. cerevisae cells entrapped in calcium alginate. Rod-shaped pieces (2 x 4 mm, 150-200 mg) of the immobilized yeast cells were charged into the (flow) adsorption vessel. When the cells were exposed to nutrients, the resulting power increased exponentially with a doubling time of 2.2 h. The half-life of the cells under non-growing conditions with sucrose as substrate was 84 h. The kinetics of the transformation of a series of sugars was characterized, leading to values for Michaelis' constant and maximum power values.

6.3.4   Protozoa

Nässberger and Monti have reported from two microcalorimetric investigations of Tetrahymena-fed Amoeba proteus. In [145] a static 1 ml ampoule [18, 30] was used as reaction vessel for cells which had been starved 3-5 days. The duration of the starvation period did not affect the metabolic activity. With 1500-4000 cells in the ampoule, good steady-state P-t curves were obtained, P = 0.8 nW/cell. When higher cell concentrations were measured, the steady-state value was lower ("crowding effect") and was preceeded by a peak. In [146] a non-stirred perfusion vessel [18, 30] was used in a study of pinocytosis processes induced by 67 mM Na^+, 75 mM K^+ or 0.43 mg/ml lysozyme.

Corresponding increases in the power values were about 130, 325 and 100% respectively.

6.3.5 Animal cells

Since the early part of 1970 a significant number of investigations on mammalian cells have been reported. In most cases these studies have had the character of methodological or exploratory work and have been concerned with human cells. Not surprisingly, much attention has been given to the possibility of applying microcalorimetric methods in clinical analysis and the work has often been performed on cells prepared from patients.

So far there have been relative few reports where the calorimetric measurements have been combined with results of extensive biochemical analysis performed on the cellular systems. Thus much work remains before the thermochemistry of mammalian cells is well-known on a molecular thermochemical level.

In calorimetric experiments on animal cells the power evolved is usually much lower than in work on microorganisms and often only very small quantities of cell material are avilable. Batch calorimetric methods employing insertion vessels ("ampoules") are much more commonly used than flow methods, which tend to consume more materials. Furthermore, these latter methods often cause problems with cells which tend to adhere to the flow lines. Some mammalian cells, such as macrophages and epithelial cells and many cultured cell lines are studied while attached to a solid support.

During the last few years there have been significant developments in the design of calorimeters suitable for work with mammalian cells in suspension or adhered to a support [22]. It is now possible to perform measurements on the sub-microwatt level using less than 1 ml of continuously stirred suspension. Injections or reagents can be made during measurements and for cells adhered to a solid support, such as a film or a suspension of microcarrier beads, medium can be perfused through the vessel.

Some years ago Krescheck [4] and Kemp [16] reviewed work on mammalian cells other than blood cells. Kemp has brought his discussion up to date in [17]. Monti and Wadsö [13] and Levin [14] reviewed earlier work on blood cells. In Monti's recent review [15] on calorimetric work on human blood cells, which is strongly oriented towards clinical analysis, a summary of preparation procedures used for these cells is also given.

Although the number of reports on calorimetric work on mammalian cell systems is quite large, there are so far very few groups involved in such work. This is particularly true for blood cells. Extensive methodological work has now been performed on the major cell fractions of human blood and values have been reported for the basal heat production under different experimental conditions: cell preparation methods, suspension media (glucose containing buffer, plasma, serum), calorimetric techniques, pH, temperature and storage time. Most of the experiments have been conducted under static conditions using small ampoules (1 ml), but flow methods have also been used. Besides the purely methodological studies, reported work on blood cells consists mainly of comparisons between cells prepared from healthy subjects and patients. In some cases, the possibility of using calorimetry as a clinical diagnostic tool has been explored; in other cases, the goal has been to characterize a suspected metabolic deficiency by power measurements.

Blood cells do not grow under normal experimental conditions and
their heat production rate under basal conditions is usually kept
constant. The power values are influenced by many experimental parameters, but if "standard conditions" are employed, characteristic and
reproducible power values are found. However, the "biological variation" between populations of the same type of blood cells from different healthy individuals is large and will also change significantly
with time. This fact undoubtedly decreases the value of calorimetry
as a direct clinical diagnostic tool, cf. e.g. fig. [8]. Further work
seems warranted to investigate the cause of the "normal" variations.
The very large spread in power values for cells prepared from patients
seems to suggest the possibility of using the power value as a means
of grading or subdividing the diseased condition.

Antibiotics are frequently used in experiments with animal cell
cultures in order to prevent bacterial growth. Its possible effect on
the heat production rate in blood cells has been investigated by
Nässberger and Monti [147]. At concentrations corresponding to therapeutic levels, no effect was observed for erythrocytes and lymphocytes, whereas a slight increase in power values was found for granulocytes. At high concentrations gentamicin caused an increased heat
production in erythrocytes.

Whole blood. In an early study, Bandman et al. [148] measured the
heat production from whole blood (P = 63 ½W/ml) and found a good
agreement with the sum of the values for the main cell fractions.
Monti et al. [149] mixed whole blood with defined immunocomplexes, using
an LKB batch microcalorimeter [27], and found a very large increase in
the heat effect, with maximal power values 4-5 times higher than the
basal value. Subsequently it was shown [150] that by far most of the
activation effect was due to interaction between the immunocomplexes

and the granulocytes. Small contributions also came from the mononu-
clear fraction, but no activation was observed for erythrocytes,
lymphocytes, platelets or cell-free plasma. It was shown that F_c
receptors for IgG and C3b receptors were involved in the activation
process. When corresponding experiments were performed with blood
from patients with acute leukemia in remission, a significantly lower
response was found (remission duration < 6 months) [151] than for healthy subjects.

Erythrocytes. Several studies have been conducted on the basal
heat production of human erythrocytes under different experimental
conditions, see [13-15]. About half of the heat production in whole
blood stems from the erythrocyte fraction. Therefore suspensions of
erythrocytes, in their normal concentrations, produce rather high
power values and the very highest sensitivity of instrumentation may
not be required. However, reported values from normal human erythrocytes show a rather large spread and a number of factors have been
identified, including method of preparation and suspension medium [13,
15], which do influence the power value. The cells sediment rapidly
and are not easy to keep in a uniform suspension in a microcalorimetric vessel. Most studies of erythrocytes have been performed by use
of static vessels where the cells have been packed at the bottom.
This leaves some uncertainty concerning possible contributions from
"crowding effects", including a poorly defined pH in the cell sediment. The power value for erythrocytes is particularly sensitive to
variations of pH. At the normal blood pH value of 7.40, the power
will decrease by 1.5% when the pH decreases by 0.01 pH units. This
value was determined for cells uniformly suspended in a buffer.[152].
Earlier, a slightly lower value (1.2%) was found for cells sedimented
from plasma suspension [153].

More than 90% of the substrate (glucose) is metabolized by the
glycolytic pathway. The rest passes the hexose monophosphate shunt.
Several investigators have tried to sort out quantitatively the thermochemistry of erythrocyte metabolism, but in most cases the calorimetric results have been combined with rather meager analytical data.
In a recent study [152], it was found that the power value for erythrocytes in hepes buffer containing glucose could be accounted for quantitatively by the production of lactate and CO_2. These measurements
were performed on a stirred suspension with a low hematocrit (10%).

Monti and co-workers used LKB-type static ampoule calorimeters [30]
cf, e.g. [15], in a series of clinical studies involving erythrocytes.
Cells from hyperthyroid patients were investigated before and after
treatment [154, 155]. Inhibition of the heat production rate by addition of ouabain (Na^+K^+-ATPase inhibition) did not change as a result
of treatment with thyroid hormones [154]. Similarly, inhibition by
ouabain gave the same results for cells from healthy subjects and from
obese patients [156]. This finding contradicts the hypothesis that
deficient Na^+K^+-pump activity is of importance in the development of
obesity. It was found [155] that the metabolic rates along the glycolytic pathway, as well as along the hexose monophosphate pathyway, is
stimulated by thyroid hormones.

Platelets (thrombocytes). As for erythrocytes, several methodological calorimetric studies have been reported for human platelets
and the influence from several experimental parameters is well-known
[13-15]. Earlier studies were all cases mainly directed towards clinical goals and were not accompanied by any detailed biochemical analysis. However, Nanri and Minakami [157] have performed a rather detailed
analysis of the thermochemistry of pig platelets suspended in plasma.

The rates of heat production, oxygen consumption and formation of metabolites were measured simultaneously. Heat production associated with the glycolytic pathway was estimated from the decrease in thermal power caused by addition of NaF. An LKB flow calorimeter was used. Results are summarized in figure 6 , which shows the total power and its main components as a function of pH. It is seen that about half of the heat production could be ascribed to glycolysis, whereas smaller fractions were due to respiration and an unidentified residual power term. The total power at pH = 7.40 (45 fW/cell) is signifi- cantly lower than corresponding values reported for human platelets, 59-68 fW/cell [13-15].

Monti and coworkers have recently reported results from several studies of platelets oriented towards the clinical field. In all cases static ampoule calorimeters [30] were used. Fagher and Monti [158] investigated platelets from patients with prosthetic heart valves. The possible relationship between the power value for platelets and thromboembolism in patients with rheumatic heart disease was dis- cussed. Valdemarsson et al. [159] studied the heat production in plate- lets from 10 hyperthyroid patients before and after therapy with L- thyroxine, cf. also work on adipocytes, p. 41. Before the treatment, the value was low, $P = 52 \pm 2$ fW/cell, but after 3 months of therapy it had increased to $P = 57 \pm 2$ fW/cell. This value is close to the normal value for the same experimental conditions.

Monti and coworkers have also investigated the effect of several xantine derivatives on platelet metabolism. Administration of 100-200 mg coffein to healthy subjects (corresponding to the intake of 1-2 cups of coffee) caused a significant increase in the power value for their platelets [160]. The value returned to normal 1 h after admini- stration. When coffein, theophyllin or enprofylline were added to

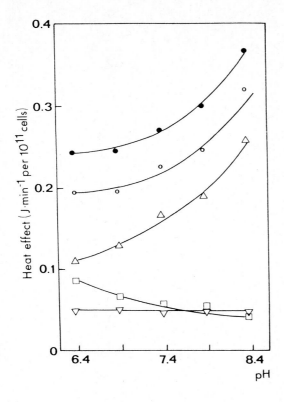

Fig. 6   Heat production of pig platelets as a function of pH.
● total heat production; o sum of the heat production
associated with the lactate formation and oxygen
consumption; Δ heat production associated with the lactate
formation; □ heat production associated with oxygen
consumption; ▽ heat production associated with neither
respiration nor glycolysis (the residual heat effect).
Total heat production was obtained experimentally.
(1 J·min^{-1} per 10^{11} cells equals 0.167 pW·cell^{-1}).
Reproduced from Nanri and Minakami [157].

platelets in vitro, their heat production rates were significantly
decreased [161], cf. [162]. This line of work was continued with an

investigation of the effect on platelet metabolism of adenosine and
adenosine derivatives [163].

Lymphocytes. A systematic study of the heat production from
peripheral blood lymphocytes free from phagocytizing cells was reported by Ikomi-Kumm et al. [164]. Measurements were performed using a
static ampoule as a reaction vessel. The effect of the cell-isolation
method, the presence of other leucocytes, lymphocyte concentration,
temperature, pH and suspension medium (plasma, RPMI-buffer) were evaluated. For cell concentrations in the range of $(0.5-5.0)\cdot 10^6$
cells/ml good steady-state P-t curves were usually found. For preparations containing phagocytizing cells and for pure lymphocyte preparations at concentrations $>6\cdot 10^6$ cells/ml, the P-t curves were de-
clining, Fig. 7. Measurements were performed at 25, 32, 37 and 42 $^{\circ}$C.
Arrhenius plot was linear for the three lower temperatures, leading to
an apparant activation energy of 63 kJ·mol.

The same calorimetric technique was used by Brandt et al. [165] in a
clinical study where heat production from lymphocytes prepared from
patients with chronic lymphatic leukemia was compared with values for
cells obtained from healthy subjects. The values for the healthy group
(P = 2.6 $\pm$ 0.5 pW/cell) were found to be slightly higher than the value
for the patients (P = 1.8 $\pm$ 0.5 pW/cell). Results are summarized in
Fig. 8. The figure illustrates the large spread in power values per
cell typically obtained in this type of work. However, this spread is
mainly due to biological variation and to errors in cell count. The
calorimetric uncertainty is small, about 0.1 pW/cell [164].

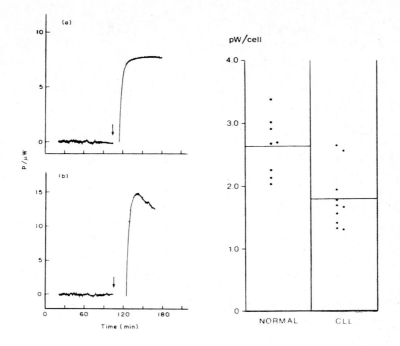

Fig. 7   (a) Steady state power-time curve typically obtained for
         lymphocyte suspensions in the concentration range (0.5-5.0) x
         10^6 cells/ml. (b) Decreasing power-time curve usually
         obtained for lymphocytes containing phagocytic cells and for
         pure lymphocytes at concentrations greater than 6.0 x 10^6
         cells /ml. ↓ Sample ampoule inserted into the calorimeter.
         Reprinted from Ikomi-Kumm et al. [164].

Fig. 8   Power values for lymphocytes from normal subjects and from
         patients with chronic lymphocytic leukemia. Reprinted from
         Brandt et al. [165].

Granulocytes (polymorphonuclear neutrophiles, PMNs). The basal
heat production was measured by Bandman et al. [148] using a static
vessel. The influence of experimental parameters on the power value
was discussed in [13]. Very significant differences in power values can
be obtained with different calorimetric techniques and using different

media. For instance, using the same calorimetric method, Bandman et al. [148] found the values P = 3.5 pW/cell and P = 1.3 pW/cell for cells suspended in plasma and in phosphate (glucose) buffer, respectively. More recently, Eftimiadi and Rialdi [166], using an LKB flow calorimeter, found the value P = 2.5 pW/cell for cells suspended in Gey's saline solution. From results of metabolite analyses, these workers concluded that the rate of lactic acid production corresponded to a power contribution of about 1.8 pW/cell, whereas about 1 pW/cell could be accounted for by aerobic metabolism. Extensive measurements of granulocytes activated by phorbol-12-myristate-13-acetate (PMA) were also performed.

Eftimiadi and Rialdi have also carried out a large number of flow calorimetric measurements of granulocytes activated by phagocytosis of bacteria [167] and protein A-IgG complexes [168]. The calorimetric determinations were accompanied by determination of metabolites, leading to quantitative thermochemical discussions of the metabolic bursts following the stimulation processes. It was concluded that the heat burst caused by PMA was related to two different non-mitochondrial oxygen reduction pathways. The metabolite analyses suggested that both the hexos monophosphate shunt and glycolysis were involved. In the work with phagocytosis of bacteria, several strains of staphylococcus were investigated. An inverse relationship was found between the degree of phagocytosis and the heat burst (and corresponding oxygen consumption). The heat bursts caused by $\underline{S}$. $\underline{aureus}$ strains were about twice as large as the effect from other staphylococci strains tested. The oxycalorific coefficient was the same for all bacterial strains, $Q/O_2 = 479 \pm 5$ kJ/mol. This value is very close to the Q/O_2 value determined for the PMA-stimulation, 488 ± 6 kJ/mol, which is the predicted value for aerobic catabolism of glucose. In a separate study with $\underline{S}$. $\underline{aureus}$, a batch calorimetric technique using a Calvet calorimeter, was used [169]. The results agreed with those obtained with the flow technique.

Eftimiadi and Rialdi have also conducted activation experiments with granulocytes in the presence of erythrocytes [170]. The red cells which cause an increased heat effect are believed to act as scavengers of some toxic metabolites produced by the granulocytes. In still another study on PMA activation of neutrophiles, Eftimiadi and Rialdi [171] compared results from calorimetric and chemiluminiscence measurements. It was found that the chemoluminiscence signal could be related with a reaction step which accounts for only a small fraction of the total heat burst.

Fäldt et al. [172] compared the heat production in granulocytes from normal subjects and from patients with acute myelogenous leukemia (AML). Increased power values were found for most AML patients at diagnosis and during the first 6 months of remission. Normal cells suspended in leukemic sera usually had increased power values, possibly due to the presence of immune complexes. Ankerst et al. [151] studied the activation of granulocytes in samples of whole blood from healthy subjects and from acute leukemia in remission, see p. 274.

Monocytes. Takahashi and Hirsh [174] have performed a low-temperature DSC study of human monocyte suspensions with and without the presence of dimethylsulfoxide (a cryoprotectant). A Perkin Elmer DSC-4 instrument was used. The mortality of the cells could be correlated with several phase transitions observed under a variety of cooling and warming conditions.

6.3.6   Macrophages

Thorén et al. [175] measured the basal heat production from rabbit alveolar macrophages using a static ampoule technique. The cells were allowed to adhere in a monolayer to polystyrene discs which were suspended in a 5 ml ampoule of an LKB microcalorimeter. Only about 10^5 cells were used in each experiment, leading to power values in the range of 2-3 $\frac{1}{2}$W (24 $\overset{+}{-}$ 3 pW/cell, 37 ^{o}C, pH 7.40, MEM + 20% serum). In subsequent investigations, Thorén et al. [176] have used this technique to study the toxic effects on macrophages from particles of materials which form important air pollutants, such as quartz and metal oxides.

Loike et al. [177] used a Privalov differential scanning calorimeter [178] to study the heat production from mouse macrophages in the temperature range 10-37 ^{o}C. Scanning rate was 1 ^{o}C per min. A comparatively large quantity of cells was used (5·10^6-60·10^6); they formed a sediment in the calorimetric vessel. The power value for the cells increased according to a complex pattern from nearly zero at 10 ^{o}C to about 10 pW/cell at 37 ^{o}C. Concentrations of lactate, CO_2, ATP and creatine phosphate were determined before and after simulated temperature scan experiments and a detailed thermochemical analysis was carried out. However, results obtained in temperature scanning experiments performed on sediments of macrophages seem to be of a limited physiological interest.

Skin cells. Following the early work by Hellgren et al. [179] and by Anders et al. [180], there have been several studies reported by Pätel, Schaarschmidt and Reichert on human fibroblasts grown in monolayers on a film support under static conditions [38, 180, 182-186]. A Bioflex calorimeter (Thermoanalyse, Grenoble, France) was used. Their work, which has a clear direction toward pharmaceutical applications, has recently been summarized [186] and will therefore be discussed only very briefly here. Fibroblasts from human foreskin were seeded on a 5 x 7 plastic foil (Steriline No. 337, Teddington, UK) [182]. When the cells had attached, the foil was bent and put into the calorimetric ampoule so that the cell-free side covered the inner wall of the vessel. Through this method the growth of the cells could be recorded until the essentially constant power during confluency was reached. The effects of antraline and related compounds on cell growth and metabolism were investigated [184, 185].

Using a modified technique, Reichert and Schaarschmidt have also studied the heat production of human epidermal keratinocytes [38]. In work with these cells, Petriperm dishes (Heraeus, Orsay, France) were used as support for the cell culture. The flexible (oxygen permeable) bottoms of the dishes were coated with acid-soluble collagen to facilitate attachment of the cells. After pre-cultivation of the keratinocytes, the membrane was removed from the dish and was transferred to the calorimetric ampoule. Following initial fluctuations in the power, essentially steady-state signals were observed. Transformed keratinocytes were also studied.

Adipocytes. Extensive measurements on isolated human adipocytes have been performed by the groups in Lund [159, 187-192]. The cells were prepared by collagenase treatment of fat tissue obtained during operations. The calorimetric measurements were performed by use of static ampoules and the adipocytes thus formed a layer floating on top of the aqueous medium in contact with the air phase. In a methodological investigation [188], the effects on the power value from variations in pH, temperature, concentrations of glucose and insulin were also investigated. The apparent activation energy was found to be low, 44 $kJ \cdot mol^{-1}$ (25-40 ^{o}C).

In other studies differences between cells obtained from normal subjects and from obese patients were investigated [187, 189, 191, 192]. Cells from obese patients were found to produce significantly less heat than cells from normal subjects. A moderate weight reduction (15%) of obese patients through gastroplasty led to a significant increase in the adipocyte power value, but the value remained lower than for lean subjects. A rapid and more substantial weight reduction (30%) by means of gastroplasty surgery caused a substantial (20%) increase in the power value per cell [191]. Clinical analytical determinations were performed in parallel with the calorimetric work.

Valdemarsson et al. [159] have studied the heat production in adipocytes from hyperthyroid patients [cf. work on platelets, p. 35], and found a very low power value before treatment, $P = 19 \pm 2$ pW/cell. After therapy the value was significantly higher, $P = 32 \pm 3$ pW/cell, but did not reach the level for healthy subjects under the same experimental conditions, $P = 49 \pm 4$ pW/cell.

Nilsson-Ehle and Nordin [190] characterized the effect from addition of glucose, insulin and norepinephrine to rat adipocytes. The variations in power with cell size and with the region from which the fat tissue was obtained were also studied. Experimental obesity induced by brain lesion caused up to a 5-fold increase in the power value per cell.

Kuroshima et al. [193] used a rotating batch calorimeter (Rhesca Co. Ltd., Tokyo) to study the effect from glucogen and noradrenaline on isolated adipocytes from rat epidymal fat pads. In all cases the hormones caused an increase in the thermal power. The results were significantly influenced by the temperature at which the animals had been acclimatized.

In an early study, Nedergaard et al. [194] investigated the heat production of cells isolated from brown adipose tissue from hamster. Recently Clark et al. [195] have reported from an extensive investigation on brown adipocytes from rat. Measurements were performed by use of an LKB rotating batch calorimeter which had been slightly modified [196] in order to decrease the equilibration time and to improve the mixing of the cell suspension. The cells were prepared by a collagenase procedure. The effects from sequential addition of glucose, noradrenaline, propanolol and oleic acid on the rates of O_2 consumption and heat production were compared. Measurements were performed on cells prepared from "cafeteria-fed" animals and from a control group.

Calorimetric measurements on adipose tissue fat pads from obese and from lean mice have been reported by Hansen and Knudsen [195]. These workers used an LKB ampoule microcalorimeter equipped with a perfusion ampoule of the type described in [30].

Hepatocytes. Jarret and Clark and their associates have reported from extensive calorimetric studies of rat hepatocytes prepared by a collagenase procedure. A slightly modified LKB rotating batch microcalorimeter was used [196]. The work was summarized in some detail in Kemp's recent review [12].

Jarret et al. [196] made a comparison between ratios for O_2 consumption and thermal power for cells prepared from starved and from fed rats. The experiments were performed in different media and·the effect of addition of different reagents was observed. Addition of fructose or dihydroxyacetone caused a marked increase in heat production in cells from starved animals. No corresponding increase in the O_2 consumption was observed, which led to the conclusion that the increased heat production was partly due to wasteful metabolism ("futile cycle"). Clark et al. [198] followed up this study with a detailed investigation of the effect of fructose concentration on the carbohydrate metabolism of rat hepatocytes.

In another study, Clark et al. [199] studied the thermogenic role of (Na^+,K^+)-dependent ATPase in hepatocytes from hyperthyroid rats. The ATPase activity, the heat production and the oxygen consumption were increased by 59%, 61% and 75%, respectively, for hepatocytes from hyperthyroid rats compared to controls. Oxygen uptake and thermal power were decreased by ouabain both for cells from hyperthyroid animals and from controls. The decreases were larger for cells from the hyperthyroid rats, but the differences were not considered large enough to support the view that (Na^+,K^+)-ATPase plays a major role in the increased heat production.

Nässberger et al. [200] have reported from a methodological work on rat hepatocytes. A static ampoule calorimetric technique was used. The cells were allowed to adhere to small plates (6 x 12 mm) stamped out of Petri dishes and layered with acid-soluble collagen. It was claimed that the technique allows long-term as well as repeated measurement on the same cell population. A significant "crowding effect" was observed. The results suggested that this effect was not only due to "contact crowding" as suggested by previous studies.

Tumor cells. It is apparent that much attention is currently focused on the potential use of microcalorimetry as a clinical analytical tool and many studies on cells prepared from patients, including a few on tumor cells, have been reported [15]. Monti et al. have investigated the heat production from lymphoid cells from patients with non-Hodgkins lymphoma (NHL) [201, 202]. The cells were prepared from tumor masses obtained by fine-needle punctures and measurements were made under static conditions. A comparison was made between power values determined before treatment and those from an evaluation of the malignancy two years after treatment was started. A significant correlation was found between high power values and degree of malignancy. The results suggested that the power value (before treatment) is a more important prognostic parameter than age, clinical stage and morphological grade of the malignancy according to the Rappaport classification [202].

Fäldt and Ankerst [203] investigated the heat production from leukemic blood cells and their interaction with immunoglobulin-coated latex particles. A comparison was made with granulocytes from patients in remission and from healthy blood donors.

Ito et al. [42] have measured the heat production from Ehrlich ascites carcinoma cells from mice. The measurements were performed by use of a Rhesca twin microcalorimeter (Tokyo) employing stirred reaction vessels. The effect of additions of several metabolic inhibitors was investigated. The power level for starving cells was almost constant. The amount of heat produced by addition of glucose was equivalent to 40% of the expected value for glycolysis.

Established cell lines. Cultured cells from established cell lines form an important research tool in cell biology, but until recently few calorimetric studies have been conducted with such cells. The early work in this field was summarized by Kemp [16], who also discussed results from this area in his recent review [17].

The early work included investigations on mouse leukemia cells in suspension (L1210) by Krescheck [204] and mouse fibroblasts (L929) in monolayer by Nicolić and Nešcović [205]. KB and HeLa cells were studied by Cerretti et al. [206] and HeLa cells by Ljungholm et al. [207]. The latter study included results from exploratory experiments with adenovirus-infected cells.

More recently Loesberg et al. [208] studied Reuber H35 rat hepatoma cells at different temperatures in the range 35-43 $^{\circ}$C. These workers used a simple isoperibol reaction calorimeter (Dewar vessel fitted with thermistor and stirrer) containing a large quantity of cells ($2 \cdot 10^8$ cells suspended in 100 ml of medium). Comparatively low power values were found, ca. 15 pW/cell, which increased slightly with the temperature in the range 35-40 $^{\circ}$C. At 42 $^{\circ}$C and 43 $^{\circ}$C, increasingly lower values were found, but these hyperthermal conditions did not lead to cell death during the course of the calorimetric measurements.

Hoffner et al. [29] used a modified LKB flow calorimeter, together with a semi-automated pump system which allowed sequential registration of heat production from cell suspensions treated with different inhibitors and from control samples. The cells investigated were LS cells, a suspension-adapted derivative of the mouse fibroblast cell line L919. In experiments conducted in parallel to the calorimetric measurements, the oxygen consumption was measured. The power value for untreated growing cells increased slightly with time and was 36 ± 3 pW/cell after 7 h. The corresponding value for suspensions containing 1 mM 2.4 DNP was 66 ± 4 pW/cell and for suspensions with 0.1 mM cyanide 21 ± 4 pW/cell. The oxycalorific equivalent was high, 969 kJ/mol O_2 (after 7 h), suggesting contributions from an inefficient anoxic metabolic mechanism.

Ankerst et al. [209] used human melanoma cells (SK-MEL 28) in their studies of complement-dependent cytoxicity. The calorimetric experiments were conducted with the LKB 4-channel microcalorimeter (2277), using static reaction ampoules. Measurements were performed with trypsinized cells suspended in medium containing fetal calf serum, monoclonal antibodies and rabbit serum with high complement

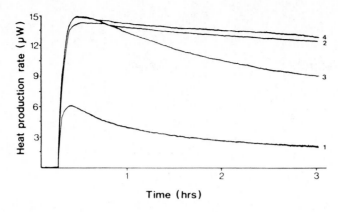

Fig. 9   Typical power time curves obtained with human melanoma cells

(SK-MEL28) incubated with a monoclonal antibody in the

presence of active or inactive complement. 1. antibody

diluted 1:500 in the presence of active complement; 2.

antibody diluted 1:500 in the presence of inactive comple-

ment; 3. antibody diluted 1:62,500 in the presence of active

complement; 4. antibody diluted 1:62,500 in the presence of

inactive complement. Reprinted from Ankerst et al. [209].

activity but containing a minimum of heterophilic antibodies against

the melanoma cells. Only 10^5 cells were used in each experiment.

Typical power-time curves are shown in Fig. 9 . It is seen that the

power values obtained with high and low antibody concentrations in the

presence of inactive complement are closely similar and decrease only

slowly with time. At low antibody concentration in the presence of

active complement, the power value declined more rapidly with time,

whereas at high antibody concentration and with active complement

present, the initial power value was low and it decreased to a very

low value after about 3 h.

In a subsequent study Ankerst et al. [210] used the same

calorimetric technique to study the kinetics of antibody-dependent

cellular toxicity (ADCC) against melanoma cells. Human peripheral lymphocytes were used as effector cells. It was concluded that microcalorimetry is a sensitive and particularly suitable method for the analysis of cytotoxic kinetics.

In the author's laboratory a series of investigations has been performed on cultured tissue cells in suspension and with the cells anchored to a solid support. In all cases the recently developed titration-perfusion vessel [34] was used, together with the LKB 4-channel microcalorimeter. In [34] human T-lymphoma cells (CCRF-CEM) and human melanoma cells (H1477) were studied as part of the methodological work on the new reaction vessel. The lymphoma cells were measured in a stirred suspension under a gas phase, whereas the melanoma cells were attached to a polystyrene film, cf. Schaarschmidt and Reichert [180]. The film, which was shaped into a tube with the cells facing inward, was inserted into a 3 ml sample container. During the measurement the medium was perfused through the vessel.

In a following study with the lymphoma cells, performed under non-growing conditions, Schön and Wadsö [211] combined results of calorimetric measurements with results of determinations of rates of formation of CO_2 and lactate. Changes in concentrations of ATP and pyruvate were found to be insignificant. The power signal was found to be constant for several hours. The power value at pH 7.40 was 7.8 ± 0.6 pW/cell, which may be compared with the much lower value for human peripheral lymphocytes [164], about 2 pW/cell. Notably the variation in thermal power with pH was found to be much lower for lymphoma cells than for lymphocytes, 21% and 78% per pH unit respectively. Results showed that 75% of the measured power could be accounted for through the formation of lactate and bicarbonate from glucose and from the protonation of the buffer. It was suggested that the residual 25% of the power was due to breakdown of cellular materials which were not accounted for by the ^{12}C analyses.

The same calorimetric technique was used by Borrebaeck and Schön [212] in a recent study of the inhibitor effects of PHA isolectins on lymphoma cells. Large differences were found between different lectins.

Using the same instrumentation, Schön and Wadsö [35] measured the heat production from green-monkey kidney fibroblasts (Vero cells) attached to microcarriers (Cytodex, Pharmacia) under non-growing conditions. The microcarriers, to which about 10^6 cells were attached, were kept in suspension during the calorimetric measurements. The measurements were conducted without perfusion of medium, but the stirring arrangement allows perfusion ($\lesssim$ 20 ml/h) of medium through the vessel during the measurement without losing any microcarriers [36]. The power values recorded were essentially constant with time for at least 4 h. The thermochemical analysis led to a residual power value as high as 67 $\pm$ 4%. As for the lymphoma cells, it was suggested that this was due to a degradation of biomass, presumably under the formation of (unlabelled) CO_2.

Eggs and sperm cells. Microcalorimetric methods developed for work with cells ought to be well-suited for studies of metabolism of small eggs and sperm, fertilization processes and the development of the embryo. But so far few reports from this area have appeared.

Elia et al. [213] have investigated the binding of spermatozoa to the receptor sites on vitelline coat (VC) of glycerol-treated eggs ("ghost eggs") of the Ascidian Ciona intestinalie. Glycerol treatment will not affect the binding of spermatozoa to VC and will not interfere with the metabolism of the egg, but the fertilization process

will not continue. Calorimetric experiments of the sperm-egg binding
process were performed using a rotating LKB batch calorimeter. The
experiments led to the conclusion that only monomeric spermatozoa
induce metabolic activation of the egg.

Grudnitskii [214] has conducted a microcalorimetric study on hi-
bernating silkworm eggs and larvae immediately after hatching. The
rate of oxygen consumption was also determined. Lovrien and coworkers
[215] have studied the heat evolution from fertilized eggs of the flour
beetle Tribolum confuscam. Measurements were performed by use of a
twin batch calorimeter allowing exchange of air in the reaction vessels. Parallel measurements of the oxygen consumption were made.

Innskeep and Hammerstedt [41] and Innskeep et al. [216] have used
another batch calorimeter of Lovrien's design [cf 40] in extensive
measurements on bovine sperm. The calorimeter holds five reaction
vessels. Each vessel holds two compartments, the contents of which
can be mixed by rotation of the instrument block. In [41] the basal
metabolism of ejaculated sperm was studied using glucose as external
energy source, with or without the presence of electron transport
inhibitors (antimycin, rotenone). Metabolite determinations (lactate,
acetate, acetylcarnitine, CO_2) were also performed. From the thermochemical analysis it was concluded that about 10% of the heat evolution was due to endogenous metabolism. In [216] a comparison was made
between ejaculated sperm (EJS) and sperm recovered after surgical
removal of candia epididymidis (CES). It is known that during ejaculation sperm will interact with sex-gland fluids, leading to an activation of the sperm cell metabolism. In experiments parallel to the
calorimetric measurements, the rates of the O_2 consumption and the
sperm mobility were measured. From the results it was concluded that
ejaculation is associated with a large increase in the degradation of
glucose. No difference in percentage of motile cells was observed,

but the mean velocity was lower for CES than for EJS.

Muscle tissues. In the past a large number of fundamental
calorimetric investigations have been performed on activated and
resting muscles, see review article by Woledge [217]. Work in
this field is largely considered to be outside the present report.
Nevertheless, the most recent calorimetric studies on muscles will be
mentioned below and some recent microcalorimetric work on small
samples of human muscle fibres will be discussed briefly.

Schlipf and Höhne [218] used a thermopile heat conduction calorimeter for measurements of frog muscles. Results of preliminary experiments were reported. Prudnikov and Bakarev [219] have also reported
results from calorimetric measurements on frog muscles.   Lörenczi
[220] has summarized and discussed results from calorimetric studies on
frog striated muscles. Ponce-Hornos et al. [221] have described a new
calorimeter for studies of arterially perfused rabbit interventricular
septum and results of measurements were reported. Coulsen [222] has
performed calorimetic measurements on perfused hearts from rabbit.

A series of investigations on the heat production from bundles of
resting human muscle fibres has been reported by Fagher et al. In a
methodological study [36], three different microcalorimetric vessels
were used: (a) 1 ml static vessel [18, 30]; (b) 1 ml non-stirred
perfusion vessel [18, 30]; and (c) a 3 ml stirred perfusion vessel [34].
In the two non-stirred vessels the fibers rested on the bottom of the
vessels. With the perfusion techniques the flow rate of medium
through the vessels was 5 ml/h. With the stirred perfusion vessel,
which fits into the LKB 4-channel microcalorimeter, the fibers were
contained in a rotating cage. In Fig. 10   typical results from
measurements with the three methods (a, b and c, respectively) are

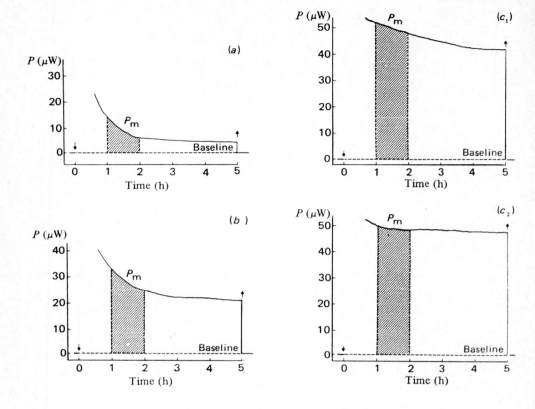

Fig. 10   Power-time curves for human skeletal muscles obtained
with a static ampoule as reaction vessel (a) and with a non-
stirred perfusion vessel (b). A stirred perfusion vessel
was used in (c_1) and (c_2), with the samples prepared at 4 ^{O}C
and 37 ^{O}C, respectively. The power mean value during the
second hour of registration, P_m, was used to compare values
for fibres from different muscle. Reprinted from Fagher et
al. [36].

compared. With all techniques the power-time curve initially showed a
fast decline if the samples had been chilled prior to the measurements
taking place (at 37 ^{O}C). When the stirred perfusion vessel was used
with samples which had not been chilled, a high, near steady-state

value was obtained after about 1.5 h after the sample was introduced
into the instrument. The results illustrate the importance of an
adequate contact between fresh medium and the tissue and it also
illustrates that the handling of the biological sample can be very
critical.

The non-stirred perfusion vessel [21, 30] has been used in several
clinical investigations reported by Fagher et al. In [223-225] fibres from
human skeletal muscle obtained from hemodialysis patients were
investigated. Power values were found to be lower than
normal and were significantly correlated with clinical values for the
thyroid function. In [226], biopsy samples from vastus lateralis muscle
were taken from a group of volunteers to which β-blockers had been
administered. It was found that the heat production became
significantly reduced by a non-selective β-adrenoreceptor blockade
(propranolol) but not by β_1-selective blockade.

In an attempt to develop a diagnostic test for malignant hyperpy-
rexia (MH), Ranklev et al. [162] measured the heat produced by muscle
fibres obtained from a group of individuals suspected of being MH-
susceptible, and from a control group. The non-stirred perfusion
vessel [30] was used. During the calorimetric measurements, the normal
perfusing medium was changed to one which had been equilibrated with
air containing 4% halothane. A significant exothermic effect followed
the change in medium but no differences were found between the two
groups of subjects.

6.4   PLANT CELLS

Very few calorimetric investigations of living plant materials
have been reported. Bogie et al. [227] studied the heat effects con-

nected with elongation of caleoptile segments from oat (<u>Avena</u> <u>sativa</u>) with and without presence of the auxine indol-3-acetic acid. In a more recent, largely methodological study, Anderson and Lovrien [228] investigated the influence of the same auxine on the metabolism of coleoplic tissue from corn (<u>Zea</u> <u>mays</u>). Measurements were conducted on 250 mg samples in a microcalorimeter which allowed perfusion of media (30. ml/h) through the sample vessel during the measurement.

Heytler and Hardy [229] have reported from an extensive calorimetric study on the heat production from isolated soybean nodules infected by Rhizobium japonicum. A new adiabatic calorimeter employing a miniaturized Dewar vessel (volume 2.8 ml) as reaction chamber was used. The goal of the study was to investigate the metabolic cost associated with the symbiotic N_2-fixation system. Results confirmed the view that the biological N_2 fixation is an energy inefficient process. The heat production connected with the maintenance metabolism of the nodules was estimated to be about 30% of the maximal heat evolution from fresh nodules taking part in the N_2-fixation process.

Breidenbach et al.[230] used a Tronac isothermal calorimeter (Provo, Utah, USA) in a study of short-term metabolic responses to thermal stress of cultured cells and tissue segments of tomato and carrot. The temporary cycling of the temperature range 0-38 $^{\circ}C$ did not permanently affect the cultures. Apparent activation energies were found to be different for the two species.

6.5   CONCLUSIONS

Microcalorimetric instruments and methods have now been developed to a stage where they can be used routinely on different types of microorganisms and animal cells. It is apparent that there has been notable methodological progress in these fields during recent years, in particular with respect to work on animal cells. For microorganisms, the work reported has involved a significant number of basic thermodynamic and kinetic investigations in addition to the practically oriented studies in which the instruments have merely been used as general process monitors. In the work on animal cells there have so far been few basic thermodynamic studies. It is judged that progress in this area - which mainly has been concerned with human cells and has a direction towards clinical analysis - now would gain from a more basic approach. It is striking that very little calorimetric work has been performed on cellular materials from plants.

REFERENCES

1. Jones, M. N. (ed)., (1979) Biochemical Thermodynamics, Elsevier, Amsterdam.

2. Beezer, A. E. (ed.), (1980), Biological Microcalorimetry, Academic Press, London.

3. James, A. (ed) (1987) Thermal and Energetic Studies of Cellular Biological Systems, Wright, Bristol

4. Kresheck, G. C. (1979) in reference [1], 281-307.

5. Belaich, J.-P. (1980) in reference [2], 1-42.

6. James, A. (1987) in reference [3], 68-105.

7. Lamprecht, I. (1980) in reference [2], 43-112.

8. Miles, R. J., Beezer, A. E., Perry, B. F. (1987) in reference [3], 106-130.

9. Perry, B. F., Miles, R. F. and Beezer, A. E. (1987) (Spencer, J. F. T. and Spencer, D. M.) Springer Verlag, London, in press

10. Newell, R. D. (1980) in reference [2], 163-186.

11. Bettelheim, K. A. and Shaw, E. J., in reference [2], 187-194.

12. Beezer, A. E. and Chowdhry, B. Z. in reference [2], 195-246.

13. Monti, M. and Wadsö, I. (1979) in reference [1], 256-280.

14. Levin, K. (1980) in reference [2], 131-143.

15. Monti, M. (1987) in reference [3].

16. Kemp, R. B. (1980) in reference [2], 113-130.

17. Kemp, R. (1987) in reference [3], 147-181.

18. Spink, C. and Wadsö, I. (1976) in Methods of Biochem. Analysis (D. Glick, ed.). Wiley & Sons, Inc., 23, 1-159.

19. Martin, C. J. and Marini, M. A. (1979) in CRC Crit. Rev. Anal. Chem. 8:3, 221-283.

20. Spink, C. (1980) in CRC Crit. Rev. Anal. Chem. 9:1, 1-95.

21. Wadsö, I. (1980) in reference [2], 247-274.

22. Wadsö, I. (1987) in reference [3], 34-67.

23. IUPAC-IUPAB-IUB Interunion Commission on Biothermodynamics (1977), Biochim. Biophys. Acta, 461, 1-14.

24. Gnaiger, E. (1983) J. Exp.Zool. 228, 471-490.

25. Randzio, S. and Suurkuusk, J. (1980) in reference [2], 311-341.

26. Benoist, L. and Pithon, F. (1984). Thermochim. Acta 72, 189-194.

27. Wadsö, I. (1968) Acta Chem. Scand. 22, 927-937.

28. Monk, P. and Wadsö, I. Acta Chem. Scand. 22, 1842-1852.

29. Hoffner, S. E. S., Meredith, R. W. J. and Kemp, R. B. (1985), Cytobios 42, 71-80.

30. Wadsö, I. (1974) Science Tools 21, 18-21.

31. Kusano, K., Nelander, B. and Wadsö, I. (1971) Chem. Scripta 1, 211-216.

32. Gordon, A. S. and Millero, F. J. (1983) J. Microbiol. Meth. 1, 291-296.

33. Suurkuusk, J. and Wadsö, I. (1982) Chem. Scripta 20, 155-163.

34. Görman, N. M., Laynez, J., Schön, A., Suurkuusk, J. and Wadsö, I. (1984) J. Biochem. Biophys. Meth. 10, 198-202.

35. Schön, A. and Wadsö, I. (1986) J. Biochem. Biophys. Meth. 13, 135-143.

36. Fagher, B., Monti, M. and Wadsö, I. (1986) Clin. Sci. 70, 63-72.

37. Gustafsson, K. and Gustafsson, L. (1985) J. Microbiol. Meth. 4, 103-112.

38. Reichert, U. and Schaarschmidt, B. (1986) Experienta 42, 173-174.

39. Lovrien, R. E., Jorgensson, G., Ma, M. K. and Lund, W. E. (1980) Biotechnol. Bioeng. 22, 1249-1269.

40. Hammerstedt, R. H. and Lovrien, R. E. (1983) J. Exp. Zool. 228, 459-469.

41. Innskeep, P. B. and Hammerstedt, R. H. (1983) J. Biochem. Biophys. Meth. 7, 199-210.

42. Ito, E., Sakihama, H., Toyama, K. and Matsui, K. (1984) Cancer Res. 44, 1985-1990.

43. Picker, P., Jolicoeur, C. and Desnoyers, J. E. (1969) J. Chem. Thermodyn. 1, 469-483.

44. Fortier, J. L., Reboul, B., Philip, P., Simard, M. A., Picker, P. and Jolicoeur, C. (1980) J. Water Pollut. Control. Fed. 52 89-97.

45. Kobayashi, K., Fujita, T. and Hagiwara, S. (1985) Thermochim. Acta 88, 329-338.

46. Ishikawa, Y., Shoda, M. and Maruyama, H. (1981) Biotechnol. Bioeng. 23, 2629-2640.

47. Dermoun, Z., Boussand, R., Cotten, D. and Belaich, J. P. (1985) Biotechn. Bioeng. 27, 996-1004.

48. Kawabata, T., Yamano, H. and Takahashi, K. (1983) Agric. Biol. Chem. 47, 1281-1288.

49. Chen, A-t. and Wadsö, I. (1982) J. Biochem. Biophys. Meth. 6, 297-306.

50. Dermoun, Z. and Belaich, J. P. (1979) J. Bacteriol. 140, 377-380.

51. Dermoun, Z. and Belaich, J. P. (1980) J. Bacteriol. 143, 742-746.

52. Ishikawa, Y., Nonoyama, Y. and Shoda, M. (1981) Biotechnol. Bioeng. 23, 2825-2836.

53. Gonda, K., Wada, H. and Murase, N. (1985) Thermochim. Acta 88, 339-344.

54. Dermoun, Z., Gaudin, C. and Belaich, J. P. (1984) Appl. Microbiol. Biotechnol. 19, 281-287.

55. Dermoun, Z. and Belaich, J. P. (1985) Biotechnol. Bioeng. 27, 1005-1011.

56. Traore, A. S., Hatchikian, C. E., Belaich, J. P. and LeGall, J. (1981) J. Bact. 145, 191-199.

57. Nichols, S. C., Pritchard, E. F. and James, A. M. (1979) Microbios 25, 187-203.

58. Nichols, S. C. and James, A. M. (1980) Microbios 29, 95-104.

59. Nichols, S. C. and James, A. M. (1980) Microbios 31, 153-160.

60. Nichols, S. C. and James, A. M. (1983) Microbios 38, 51-63.

61. Bowden, C. P. P. and James, A. M. (1985) Microbios 43, 93-105.

62. Bowden, C. P. P. and James, A. M. (1985) Microbios 44, 75-85.

63. James, A. M. and Djavan, A. (1980) Microbios 29, 171-183.

64. James, A. M. and Djavan, A. (1981) Microbios 30, 163-170.

65. James, A. M. and Djavan, A. (1982) Microbios 34, 17-30.

66. Gordon, A. S., Millero, F. J. and Gerchakov, S. M. (1982) Appl. Environ. Microbiol. 44, 1102-1109.

67. Monk, P. (1979) J. Dairy Res. 46, 485-496.

68. Monk, P. (1984) J. Gen. Appl. Microbiol. 30, 329-336.

69. Bezjak, V. (1984) Microbios 41, 167-176.

70. Marison, I. W. and von Stockar, U. (1985) Thermochim. Acta 85, 493-496.

71. Marison, I. W. and von Stockar, U. (1985) Thermochim. Acta 85, 497-500.

72. Marison, I. W. and von Stockar, U. (1987) Enz. Microb. Tech. 9, 33-43.

73. Cordier, J.-L., Butsch, B. M., Birou, B. and von Stockar, U. (1987) Appl. Microbiol. Biotechnol. (in press).

74. Miles, C. A., Mackey, B. M. and Parsons, S. E. (1986) J. Gen. Microbiol. 132, 939-952.

75. Boling, E. A., Blanchard, G. C. and Russel, W. J. (1973) Nature, London, 241, 472-473.

76. Fujita, T., Monk, P. and Wadsö, I. (1977) J. Dairy Res. 45, 457-463.

77. Herman, J. P. M., Jacubzak, E., Izard, D. and Leclerc, H. (1980) Can. J. Microbiol. 26, 413-419.

78. Seminitz, E., Casey, P. A., Pfaller, W. and Gstraunthaler, G. (1982) Chemotherapy 29, 192-207.

79. Chernykh, V. M. and Firsov, A. A. (1985) Antibiotics and Med. Biotechnol (USSR) 30, 498-503.

80. Bowden, C. P. P. and James, A. M. (1985) Microbios 44, 201-216.

81. Perry, B. F., Beezer, A. E., Miles, R. J., Smith, B. V., Miller, J. and da Graza Nascimento, M. (1986) Microbios 45, 181-191.

82. Fardeau, M.-L., Plasse, F. and Belaich, J. P. (1980) European J. Appl. Microbiol. Biotechnol 10, 133-143.

83. Plasse, F., Fardeau, M.-L. and Belaich, J. P. (1980 Biotechnol. Lett. 2, 11-16.

84. Partos, J. and Fardeau, M.-L. (1980) Biotechnol. Lett. 2, 505-508.

85. Glanser, M., Ban, S. and Kniewald, J. (1979) Process Biochem. February, 17-25.

86. Gram, L. and Søgaard, H. (1985) J. Food Prot. 48, 341-345.

87. Gram, L. and Søgaard, H. (1985) Thermochim. Acta 95, 373-381.

88. Goodman, A. and Ralph, B. J. (1980) Proc. 4th ISEB and Leaching Conference, Austr. Academy of Science, Canberra, 1980, 477-483.

89. Soljanto, P. and Tuovinen, O. H. Proc. 4th ISEB and Leaching Conference, Austr. Academy of Science, Canberra, 1980, 469-475.

302

90. Ljungholm, K., Norén, B., Sköld, R. and Wadsö, I. (1979) Oikos 33, 15-23.

91. Ljungholm, K., Norén, B. and Wadsö, I. (1979) Oikos 33, 24-30.

92. Ljungholm, K., Norén, B. and Odham, G. (1980) Oikos 34, 98-102.

93. Sparling, G. P. (1981) Soil Biol. Biochem. 13, 93-98.

94. Sparling, G. P. (1981) Soil Biol. and Biochem. 13, 373-376.

95. Sparling, G. P. (1983) J. Soil Science 34, 381-390.

96. Yamano, H. and Takahashi, K. (1983) Agric. Biol Chem. 47, 1493-1499.

97. Kimura, T. and Takahashi, K. (1985) J. Gen. Microbiol. 131, 3083-3089.

98. Yamano, H. and Takahashi, K. (1986) Agric. Biol. Chem. 50, 3145-3155.

99. Yamano, H. and Takahashi, K. (1986) Agric. Biol. Chem. 50, 3157-3164.

100. Wojcic, W. (1982) Ekol. Pol. 30, 29-80.

101. Pamatmat, M. M., Graf, G., Bengtsson, W. and Novak, C. S. (1981) Mar. Ecol. Prog. Ser. 4, 135-144.

102. Lock, M. A. and Ford, T. E. (1983) Appl. Environ. Microbiol. 46, 463-467.

103. Gustafsson, K., Gustafsson, L. (1983) Oikos 41, 64-72.

104. Gustafsson, K. (1984) in Dickson, L. and Dutka, B. J., eds., Toxicity Screening Procedures Using Bacterial Systems. Marcel Dekker, Inc., New York, p. 251-259.

105. Gustafsson, K. (1985) Ecol. Bull. (Stockholm) 36, 130-134.

106. Tournié, T. and Laserre, P. (1984) J. Exp. Mar. Biol. Ecol. 74, 111-121.

107. Lasserre, P. and Tournié,T. (1984) J. Exp. Mar. Biol. Ecol. 74, 123-139.

108. Beaubien, A. and Jolicoeur, C. (1984) Drug Chem. Toxicol. 1, 261-281.

109. Beaubien, A., Simard, M. A., Lavallee, J. F., Desrosiers, O. and
      Jolicoeur, C. (1985) Water Res. 19, 747-755.

110. Beaubien, A. and Jolicoeur, C. (1985) J. Water Pollut. Control
      Fed. 57, 95-100.

111. Weppen, P. and Schuller, D. (1984) Thermochim. Acta 72, 95-102.

112. Weppen, P. (1985) Thermochim. Acta 94, 139-149.

113. Redl, B. and Tiefenbrunner, F. (1981) Water Res. 15, 87-90.

114. Harju-Jeanty, P. (1982) Appita 36, 26-31.

115. Bolouri, H., Lamprecht, I. and Schaarschmidt, B. (1980) Thermal
      Analysis (ICTA 80), Birhauser Verlag, Basel, p. 577-582.

116. Lamprecht, I. and Schaarschmidt, B. (1980) Thermochim. Acta 40,
      157-166.

117. Ljungholm, K., Wadsö, I. and Mårdh, P.-A. (1976) J. Gen.
      Microbiol. 96, 283-288.

118. Miles, R. J., Beezer, A. E. and Lee, D. H. (1985) J. Gen
      Microbiol. 131, 1845-1852.

119. Miles, R. J., Beezer, A. E. and Lee, D. H. (1986) Microbios 45,
      7-19.

120. Lee, D. H., Miles, R. J. and Beezer, A. E. (1986) FEMS
      Microbiol. Lett. 34, 283-286.

121. Nunomura, K. and Fujita, T. (1981) J. Gen. Appl. Microbiol. 27,
      357-364.

122. Nunomura, K. and Fujita, T. (1982) J. Gen. Appl. Microbiol.
      28,479-490.

123. Nunomura, K. and Fujita, T. (1983) J. Gen. Appl. Microbiol. 29,
      233-238.

124. Forrest, W. W. and Walker, D. J. (1963) Biochem. Biophys. Res.
      Commun. 13, 217-222.

125. Brettel, R., Lamprecht, I. and Schaarschmidt, B. (1980) Radiat.
      Environ. Biophys. 18, 301-309.

126. Brettel, R., Lamprecht, I. and Schaarschmidt, B. (1981)

304

Eur. J. Appl. Microbiol. Biotechnol. 11, 205-211.

127. Brettel, R., Lamprecht, I. and Schaarschmidt. B. (1981) Eur. J.
      Appl. Microbiol. Biotechnol. 11, 212-215.

128. Brettel, R., Lamprecht, I. and Schaarschmidt, B. (1981)
      Thermochim. Acta 49, 53-61.

129. Sayadi, P., Lamprecht, I. and Schaarschmidt, B. (1980)
      Thermochim. Acta 40, 63-71.

130. Brettel, R., Lamprecht, I. and Schaarschmidt, B. (1980) Thermal
      Analysis (ICTA 80), Birkhaeuser Verlag, Basel, p. 553-557.

131. Loureiro-Dias, M. C. and Arrabaca, J. D. (1982) Z. Allg.
      Microbiol. 22, 119-122.

132. Lamprecht, I. and Schaarschmidt, B. (1979) Experienta, Suppl
      37, 166-175.

133. Schaarschmidt, B., Lamprecht, I. and Simonis, M. (1979)
      Experienta, Suppl. 37, 176-183

134. Beezer, A. E. and Chowdhry, B. Z. (1980) Talenta 27, 1-5.

135. Beezer, A. E. and Chowdhry, B. Z. (1980) Microbios 28, 107-121.

136. Perry, B. F., Beezer, A. E. and Miles, R. J. (1979) J. Appl.
      Bact. 47, 527-537.

137. Perry, B. F., Beezer, A. E. and Miles, R. J. (1983) J. Appl.
      Bact 54, 183-189.

138. Perry, B. F., Beezer, A. E. and Miles, R. J. (1983) Microbios
      37, 117-129.

139. Perry, B. F., Beezer, A. E and Miles, R. J. (1983) Microbios
      Lett. 24, 121-127.

140. Schaarschmidt, B. and Lamprecht, I. (1978) Thermochim. Acta 22,
      333-338.

141. Lamprecht, I., Schaarschmidt, B. and Siemens, J. (1985)
      Thermochim. Acta 94, 129-138.

142. Beezer, A. E., Newell, R. D. and Tyrell, H. J. V. (1978)
      Microbios 22, 73-85.

143. Itoh, S. and Takahashi, K. (1984) Agric. Biol. Chem. 48, 271-
      275.

144. Owusu, R.K. and Finch, A. (1985) J. Gen. Appl. Microbiol. 31,
      221-230.

145. Nässberger, L. and Monti, M. (1984) Protoplasma 123, 135-139.

146. Nässberger, L. and Monti, M. (1987) J. Protozool. (in press).

147. Nässberger, L. and Monti, M. (1987) Human Toxicol. (in press).

148. Bandman, U., Monti, M. and Wadsö, I. (1975) Scand. J. clin. Lab.
      Invest. 35, 121-127.

149. Monti, M., Fäldt, R., Ankerst, J. and Wadsö, I. (1980) J.
      Immunol. Meth. 37, 29-37.

150. Fäldt, R., Ankerst, J., Monti, M. and Wadsö, I. (1982) Immunology
      46, 189-197.

151. Ankerst, J., Fäldt, R. and Monti, M. (1984) Leukemia Res. 8,
      997-1002.

152. Bäckman, P. and Wadsö, I. (1987) to be published.

153. Monti, M. and Wadsö, I. (1976) Scand. J. Clin. Lab. Invest. 36,
      565-572.

154. Monti, M., Hedner, P., Ikomi-Kumm, J. and Valdemarsson, S.
      (1987) Metabolism 36, 155-159.

155. Monti, M., Hedner, P., Ikomi-Kumm, J. and Valdemarsson, S.
      (1987) Acta Endocrinol. 115, 87-90.

156. Monti, M. and Ikomi-Kumm, J. (1985) Metabolism 34, 183-187.

157. Nanri, H. and Minakami, S. (1983) Biomed. Biochim. Acta 42, 315-
      321.

158. Fagher, B. and Monti, M. (1984) Europ. Heart J. 5, 55-61.

159. Valdemarsson, S., Fagher, B., Hedner, P., Monti, M. and Nilsson-
      Ehle, P. (1985) Acta Endocrinol. 108, 361-366.

160. Ammaturo, V. and Monti, M. (1986) Acta Med. Scand. 220, 181-184.

161. Monti, M., Edvinsson, L., Ranklev, E. and Fletcher, R. (1986)
      Acta Med. Scand. 220, 185-188.

306

162. Ranklev, E., Monti, M. and Fletcher, R. (1985) Br. J. Anaesth. 57, 991-993.

163. Edvinsson, L., Ikomi-Kumm, J. and Monti, M. (1986) Br. J. Clin. Pharmac. 22, 685-689.

164. Ikomi-Kumm, J., Monti, M. and Wadsö, I. (1984) Scand. J. clin. Lab. Invest. 44, 745-752.

165. Brandt, L., Ikomi-Kumm, J., Monti, M. and Wadsö, I. (1979) Scand. J. Haematol. 22, 141-144.

166. Eftimiadi, C. and Rialdi, G. (1982) Cell Biophys. 4, 231-244.

167. Eftimiadi, C. and Rialdi, G. (1984) J. Infect. Diseases 150, 366-371.

168. Eftimiadi, C. and Rialdi, G. (1985) Microbiologica 8, 255-261.

169. Eftimiadi, C. and Rialdi, G. (1985) Microbiologica 8, 297-301.

170. Eftimiadi, C. and Rialdi, G. (1985) Studia Biophys. 110, 83-88.

171. Eftimiadi, C. and Rialdi, G. (1985) Thermochim. Acta 85, 489-492.

172. Fäldt, R., Ankerst, J. and Monti, M. (1984) Br. J. Haematol. 58, 671-678.

173. Ankerst, J., Fäldt, R. and Monti, M. (1984) Leukemia Res. 8, 997-1002.

174. Takahashi, T. and Hirsh, A. (1985) Biophys. J. 47, 373-380.

175. Thorén, S., Holma, B., Monti, M. and Wadsö, I. (1984) Thermochim. Acta 72, 117-122.

176. Thorén, S. et al., to be published.

177. Loike, J. D., Silverstein, S. C. and Sturtevant, J. M. (1981) Proc. Natl. Acad. Sci. USA 78, 5958-5962.

178. Privalov, P. L. (1980) Pure Appl. Chem. 52, 479-497.

179. Hellgren, L., Larsson, K. and Vincent, J. (1977) Arch. Derm. Res. 258, 295-303.

180. Anders, A., Schaefer, H., Schaarschmidt, B., and Lamprecht, I. (1979) Arch. Derm. Res. 265, 173-180.

181. Pätel, P., Reichert, U. and Schaarschmidt, B. (1980) in Thermal Analysis (Ed. Hemminger, W.) Birkhäuser Verlag, Basel, Vol. II, 559-564.

182. Schaarschmidt, B. and Reichert, U. (1981) Exp. Cell. Res. 131, 480-483.

183. Pätel, M., Schaarschmidt, B. and Reichert, U. (1981) Br. J. Dermatol. 105, Suppl. 20, 60-61.

184. Pätel, M. (1981) Thermochim. Acta 49, 123-129.

185. Pätel, M., Schaarschmidt, B. and Lamprecht, I. (1982) in Thermal Analysis (Ed. Miller, B), J. Wiley, Chichester, Vol. II, 857-862.

186. Schaarschmidt, B. and Reichert, U. (1986) Thermochim. Acta 94, 123-128.

187. Sörbris, R., Nilsson-Ehle, P., Monti, M. and Wadsö, I. (1979) FEBS Lett. 101, 411-414.

188. Monti, M., Nilsson-Ehle, P., Sörbris, R. and Wadsö, I. (1980) Scan. J. clin. Lab. Invest. 40, 581-587.

189. Sörbris, R., Nilsson-Ehle, P., Monti, M. and Wadsö,I. (1982) Metabolism 31, 973-978.

190. Nilsson-Ehle, P. and Nordin, G. (1985) Int. J. Obesity 9, Suppl. 1, 169-172.

191. Olsson, S.-Å., Monti, M., Sörbris, R. and Nilsson-Ehle, P. (1986) Int. J. Obesity 10, 99-105.

192. Sörbris, R., Olsson, S.-Å., Monti, M. and Nilsson-Ehle, P. (1986) Clin. Nutr. 195-197.

193. Kuroshima, A., Kurahashi, M. and Yahata, T. (1979) Pflügers Arch. 381, 113-117.

194. Nedergaard, J., Cannon, B. and Lindberg, O. (1977) Nature 267, 518-520.

195. Clark, D. G., Brinkman, M. and Neville, S. (1986) Biochem. J. 235, 337-342.

196. Jarret, J. G., Clark, D. G., Filsell, O. H., Harvey, J. W. and

308

Clark, M. G. (1979) Biochem. J. 180, 631-638.

197. Hansen, E. S. and Knudsen, J. (1982) Biochim. Biophys. Acta 721, 418-424.

198. Clark, D. G., Filsell, O. H. and Topping, D. L. (1979) Biochem. J. 184, 501-507.

199. Clark, D. G., Brinkman, M., Filsell, O. H., Lewis, S. J. and Berry, M. N. (1982) Biochem. J. 202, 661-665.

200. Nässberger, L., Jensen, E., Monti, M. and Florén, C.-H. (1986) Biochem. Biophys. Acta 882, 353-358.

201. Monti, M., Brandt, L., Ikomi-Kumm, J., Olsson, H. and Wadsö, I. (1981) Scand. J. Haematol. 27, 305-310.

202. Monti, M., Brandt, L., Ikomi-Kumm, J. and Olsson, H. (1986) Scand. J. Haematol. 36, 353-357.

203. Fäldt, R. and Ankerst, J. (1986) Leukemia Res. 10, 1147-1150.

204. Kresheck, G. C. (1974) Cancer Biochem. Biophys. 1, 39-41.

205. Nicolic, D. and Nescovic, B. (1976) Jugoslav. Physiol. Pharmacol. Acta 12, 191-197.

206. Cerretti, D. P., Dorsey, J. K. and Bolen, D. W. (1977) Biochim. Biophys. Acta 462, 748-758.

207. Ljungholm, K., Wadsö, I. and Kjellén, L. (1978) Acta Pathol. Microbiol. Scand. Sect. B 86, 121-124.

208. Loesberg, C., van Miltenburg, J. C. and van Wijk, R. (1982) J. therm. Biol. 7, 87-90.

209. Ankerst, J., Fäldt, R. and Sjögren, H. O. (1985) J. Immun. Meth. 77, 267-274.

210. Ankerst, J., Sjögren, H. O. and Fäldt, R. (1986) J. Immunol. Meth. 88, 259-264.

211. Schön, A. and Wadsö, I. (1986) Cytobios 48, 195-205.

212. Borrebaeck, C. and Schön, A. (1987) Cancer Res. in press.

213. Elia, V., Rosati, F., Barone, G., Monroy, A. and Liquori, A. M. (1983) EMBO J. 2, 2053-2058.

214. Grundnitskii, V. A. (1983) Biofizika 28, 485-488.

215. Dunkei, F., Wensman, C. and Lovrien, R. (1979) Comp. Biochem. Physiol. 62A, 1021-1029.

216. Innskeep, P. B., Magargee, S. F. and Hammerstedt, R. H. (1985) Arch. Biochem. Biophys. 241, 1-9.

217. Woledge, R. C. (1980) in reference [2], 145-162.

218. Schlipf, M. and Höhne, G. W. H. (1981) Thermochim. Acta 49,1-9.

219. Prudnikov, V. M. and Bakarev, A. M. (1982) Fiziol. Zh. SSSR 67, 313-319.

220. Lörenczi, D. (1982) Studia Biophys. 89, 81-97.

221. Ponce-Hornos, J. E., Ricchiuti, N. V. and Langer, G. A. (1982) Am. J. Physiol. 243, H289-H295.

222. Coulson, R. L. (1982) Fed. Proc. 41, 199-203.

223. Fagher, B., Thysell, H., Nilsson-Ehle, P., Monti, M., Olsson,L., Eriksson, M., Lindstedt, S. and Lindholm, T. (1982) Acta Med. Scand. 212, 115-120.

224. Fagher, B., Cederblad, G., Eriksson, M., Monti, M., Moritz, U., Nilsson-Ehle, P. and Thysell, H. (1985) Scand. J. clin. Lab. Invest. 45, 169-178.

225. Fagher, B., Monti, M., Nilsson-Ehle, P. and Thysell, H. (1987) Scand. J. clin. Lab. Invest. 47, 91-97.

226. Fagher, B., Liedholm, H., Monti, M. and Moritz, U. (1986) Clin. Sci. 70, 435-441.

227. Bogie, H. E., Kresheck, G. C. and Harment, K. H. (1976) Plant· Physiol. 57, 842-845.

228. Andersson, P. C. and Lovrien, R. E. (1979) Anal. Biochem. 100, 77-86.

229. Heytler, P. G. and Hardy, R. W. F. (1984) Plant Physiol. 75, 304-310.

230. Breidenbach, R. W., Criddle, R. S., Lewis, E. A., Eatough, D. J. and Hansen, L. D. (1986) Plant Physiol. 80 (4 Suppl.), 122

CHAPTER 7

CONFORMATIONAL CHANGES IN PROTEINS IN RELATION TO ENERGY TRANSDUCTION

A.G. Lowe and A.R. Walmsley

7.1 INTRODUCTION

Proteins can exist in many states ranging from the "native" conformation to unfolded forms, and native states often include substrate-bound forms of the "induced fit" type and allosteric forms (usually rearrangements of subunits resulting from the binding of regulator-molecules). Furthermore, the relative abundancies of the different conformational states varies with the physical conditions prevailing with temperature, pH, ionic strength and (especially for membrane-bound enzymes) the hydrophobicity of the environment playing important parts in determining which particular state is predominant. For the majority of proteins evolutionary forces have presumably led to conformational states with catalytic activities and responsiveness to metabolic conditions which are optimal in relation to the metabolic needs of the çell. However, in a few instances, notably myosin/actin in muscle and active transport proteins in biological membranes, an additional function has evolved: the proteins can undergo conformational changes which are driven unidirectionally through a cycle by an energy-yielding degradative reaction, typically the hydrolysis of ATP. In passing round the cycle such proteins effect useful work: muscular contraction or the movement of a solute across a membrane against its electrochemical gradient.

The molecular nature of the conformational changes responsible for energy transduction remains relatively poorly understood for most systems, but this situation will certainly improve as the amino acid sequences and three-dimensional structures of energy transducing proteins are elucidated. There is also relatively little information available about the thermodynamics of energy transducing systems since it is rarely possible to make direct measurements of the heat changes associated with protein conformational changes because of the small quantities of proteins available in purified form.

However, in at least some systems estimates of enthalpy and entropy changes can be obtained indirectly through measurements of the temperature dependence of the equilibria between pairs of protein conformers. Even though detailed thermodynamic and structural information about individual systems is often unavailable it is possible to build up partly intuitive pictures of energy-transducing conformational changes of proteins by first considering examples of non-energy-transducing systems about which there is more information, then extrapolating the lessons learned from these to energy-transducing proteins. This article is not intended to be an exhaustive analysis of the problem of energy transduction in proteins and for more extensive treatment the reader is referred to books and reviews by Jencks (1), Hill (2), Tanford (3), Knowles (4) and Kodama (5).

7.2 MEMBRANE TRANSPORT SYSTEMS

7.2.1 Facilitated diffusion of glucose

At first glance membrane carrier systems seem unlikely to be particularly informative as models for the thermodynamic characteristics of protein conformational changes because knowledge of the folded structure of membrane proteins is largely restricted to interpretation of amino acid sequences in terms of the number of α-helical coils large enough to span the membrane, and the complex kinetic properties of transport systems have led to the proposal of many different models for even the simplest systems such as the glucose carrier of the human red blood cell. Furthermore, membrane carrier proteins can usually be obtained only in very limited quantity so that it is normally impossible to make direct measurements of enthalpy changes associated with conformational transitions. Despite this apparently unpromising scenario, recent studies of the glucose carrier of the human erythrocyte (6) have shown that it is possible to exploit the unusual and apparently complex kinetic properties of glucose transport in a way which ultimately yields information about the enthalpy and entropy changes associated with glucose transport and conformational changes of the carrier molecule. Furthermore, when thermodynamic information is combined with knowledge of the structure of the carrier derived from the sequencing of the cDNA for the glucose carrier (7) it is possible to suggest a more detailed mechanism for the red cell glucose carrier than has previously been attempted.

In fact the experimental approach used to solve the kinetic mechanism of glucose transport constitutes a general method that could be applied to the analysis of many other membrane carriers and possibly also enzymes not involved in membrane transport. In view of this generality it seems appropriate to describe the use of the method in the analysis of red cell glucose transport in some detail.

Early work on red cell glucose transport identified a very active catalytic system capable of equilibrating glucose between the cytoplasm of the red cell and the surrounding medium within a few seconds at physiological temperatures. In early studies of glucose transport it was expected that, since there was no input of metabolic energy, the carrier kinetics would be entirely symmetrical with equal affinities for glucose at the two sides of the membrane and equal rates of transition of the carrier between its inward and outward-facing conformations. Thus if we represent transport by the kinetic model shown in Fig. 1, it was assumed (implicitly if not explicitly) that $b/a = e/f$ and that $c = d = g = h$. Subsequent work (e.g. (8)), made it increasingly

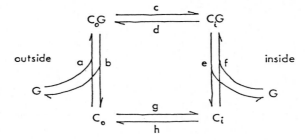

Fig 1.

The conventional 4-state carrier model. C_o and C_i represents the outward- and inward-facing carrier conformations, G is glucose and a-h are rate constants.

evident that this was not in fact the case since (at 20^oC) the maximum rate of influx into glucose free cells (V_{oi}^{zt}, governed chiefly by rate constants h and c) was substantially slower than the rate of exchage of extracellular glucose with glucose present in preloaded cells (V^{ee}, governed chiefly by rate constants c and d) or the rate of efflux of intracellular glucose into a glucose-free medium (V_{io}^{zt}, governed chiefly by rate constants d and g). Superficially this result is surprising but a thermodynamically acceptable model demands simply that $aceh = bdfg$ or, as shown by Haldane (9), that

$$\frac{V^{ee}}{K^{ee}} = \frac{V_{oi}^{zt}}{K_{oi}^{zt}} = \frac{V_{io}^{zt}}{K_{io}^{zt}}$$

In fact Lowe & Walmsley (6) demonstrated that the Haldane relationship does in fact hold for red cell glucose transport at both 20° and 0°C, despite the fact that V^{ee} exceeds V^{zt}_{oi} by two orders of magnitude at the latter temperature.

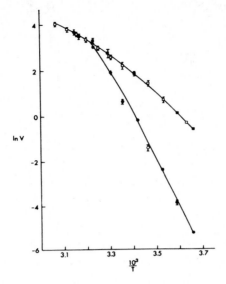

Fig 2.

Arrhenius plots of the temperature dependences of V_{max} for influx of D-(^{14}C)-glucose into human red cells under zero trans (●,○) and equilibrium exchange (■,□) conditions.

Fig. 2 illustrates the temperature dependences of V^{ee} and V^{zt}_{oi} and it is immediately apparent, firstly that the difference between the two V_{max} decreases with increasing temperature and, secondly, that both Arrhenius plots are non-linear.   One simple explanation of this non-linearity is that two or more rate constants (with differing activation energies) are significant determinants of V^{ee} and V^{zt}_{oi}.   Using the model in Fig. 1 it can readily be shown that, provided that the rate of dissociation of glucose from the carrier is fast relative to the rates of the conformational change of the carrier (in fact this must be the case at temperatures where V^{ee} exceeds V^{zt}_{oi} and V^{zt}_{io} and NMR studies support this view (10)), then

$$V^{ee} = \frac{[C]}{(1/c + 1/d)}, \quad V^{zt}_{oi} = \frac{[C]}{(1/c + 1/h)}, \quad \text{and } V^{zt}_{io} \; \frac{[C]}{(1/d + 1/g)},$$

where [C] is the concentration of carrier (per litre cell water) and c, d, g and h are the rate constants governing changes in carrier conformation. Furthermore, the temperature dependencies of the various rate constants can be expected to be described by Arrhenius equations of the types

$$h = A_h e^{-E_h/RT}, \qquad g = A_g e^{-E_h/RT} \quad etc.$$

and it is reasonable to assume, at least as a first approximation, that the pre-exponential terms (A_h etc.) are independent of temperature in the relatively narrow range investigated.

Given this model it has been possible to use the data in Fig. 2 together with additional data for "zero trans" efflux of glucose, to obtain definitive values for rate constants c, d, g and h by non-linear least squares regression. Initial estimates of rate constants h and c (and their activation energies) can be obtained by inspection of the shape of the Arrhenius plot of V_{ci}^{zt}, then refined by least squares analysis of the data under this experimental conditon. The value of c estimated in this way can then be used as an initial estimate in the analysis of the temperature dependence of V^{ee}, which then yields a new estimate of c together with an estimate of d and the associated activation energies. An estimate of g can be calculated from data for V_{io}^{zt} (which is dependent of g and d), and then the values of all 4 rate constants (c, d, g and h) and their activation energies can be used as initial estimates in a concerted optimisation of these parameters to the combined data measured under conditions of "zero trans" influx and efflux and equilibrium exchange of glucose throughout the whole temperature range studied.

The results of this analysis are shown in Table 1 and it was these fitted parameters that were used to draw the lines describing the temperature dependence of glucose transport in human red cells in Fig. 2. Some caution is necessary when considering this analysis since it would not be surprising if, with a system of 4 temperature-dependent constants, there was at least one alternative solution. In fact another statistically less satisfactory fit can be obtained with the values of c and d reversed at 0°C. However, this fit can be rejected as physically unacceptable since if c and d are reversed the fit demands that d has a negative activation energy and that the affinity of the carrier for glucose on opposite sites of the membrane differs by two orders of magnitude. The results of the above analysis are broadly consistent with a wide range of data on glucose transport in human red cells collected by a number of different research laboratories (8,11-13), with the concept of an unequal transmembrane distribution of the carrier molecule and its glucose complexes (14), and with numerous studies of effects of inhibitors on glucose transport (e.g. 15,16). This comprehensive kinetic description of the glucose transport system was derived from measurements of initial rates of glucose transport, but it should be noted that results conflicting with this concept of a simple carrier model for glucose transport have been obtained (17-19) after derivation of kinetic parameters from prolonged time courses of glucose movement into or out of red cells using the integrated rate equation.

<u>Table 1</u>

Fitted values of the rate constants governing conformational changes of the glucose carrier, with associated activation energies.

Rate constant	Glucose carrier	
	$k[C](\text{mmol} \cdot 1^{-1} \cdot s^{-1})$	$E_A \ (\text{kJ} \cdot \text{mol}^{-1})$
c	7.42 ± 3.29	31.7 ± 5.11
d	0.602 ± 0.0231	88.0 ± 6.17
g	0.0809 ± 0.00654	127 ± 4.78
h	0.00484 ± 0.000332	173 ± 3.10

7.2.2 Thermodynamics of the glucose carrier system

The preceding kinetic description of glucose transport in human red cells makes possible an appraisal of the thermodynamics of the transport system (20). The rate constant ratios, c/d and g/h define the equilibria between the inward- and outward-facing forms of the glucose-loaded and unloaded forms of

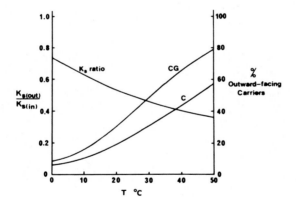

Fig. 3.

The asymmetry of affinities of the outward- and inward-facing glucose carriers (Ks(out)/Ks(in)) and the proportions of outward-facing carriers (C) and carrier-glucose complexes (CG) as a function of temperature. Values plotted are calculated from the fitted values of rate constants, c, d, g and h given in Table 1.

the carrier molecule. As shown in Fig. 3 the carriers are mostly in the inward-facing conformation at 0°C, but become progressively more outward-facing as the temperature is increased. The van't Hoff relationship makes it possible to calculate the enthalpy changes associated with carrier reorientations, and using the relationships,

$$\Delta G = -RT \ln K$$
$$\text{and} \quad \Delta G = \Delta H - T\Delta S,$$

the free energy changes and entropy changes associated with carrier reorientation can also be calculated (Table 2) (20). The magnitudes of ΔH and ΔS are consistent with a relatively modest conformational change in the carrier since formation of one hydrogen bond is associated with an energy of 12-38 kJ.mol^{-1} and release of water with an entropy change of about 40 J.K^{-1}mol^{-1} (30).

Using the model in Fig. 2 it is also possible to calculate the true affinity of the carrier for glucose from the relationships

$$Ks(out) = \frac{b(1+(g/h))}{a(1+(c/h))} \quad \text{and} \quad Ks(in) = \frac{e(1+(h/g))}{f(1+(d/g))}$$

where Ks(out) and Ks(in) are the dissociation constants of the carrier glucose complexes on the outside and inside of the cell. The standard free energy of binding of glucose to the inward-facing conformation turns out to be about -12 kJ/mol and to be almost independent of temperature with ΔH (+5.5 kJ/mol) balanced by ΔS (+58 J.K^{-1}mol^{-1} (Table 2). The positive entropy change for glucose binding presumably reflects the fact that the increase in order associated with binding of glucose to the carrier molecule is more than counterbalanced by increases in entropy associated with dissociation of water from both glucose itself and from the glucose binding site of the unloaded carrier. This finding is consistent with the work of Levine et al. (21) on the temperature-dependence of the inhibitory effect of glucose on transport of

Table 2

Thermodynamic parameters for the transport system for glucose in human red blood cells (20). Standard enthalpies and entropies were calculated as described in the text. Basic Gibbs free energies were calculated (according to (2)) for a glucose concentration of 5mM. Ci and Co are the inward- and outward-facing carrier states, while G represents glucose.

Glucose carrier	$\Delta H°$ (kJ·mol^{-1})	$\Delta S°$ (J·K^{-1}·mol^{-1})	Basic Gibbs free energy change (kJ·mol^{-1})	
			0°C	37°C
Substrate binding to the				
outward-facing carrier	5.51 ± 3.39	58.1 ± 24.0	1.7	1.2
inward-facing carrier	− 4.95 ± 3.73	17.3 ± 21.9	2.3	3.3
Carrier reorientation from inside to outside				
unloaded carrier	46.0 ± 5.69	145 ± 21.0	6.4	1.05
carrier-glucose complex	56.3 ± 8.01	185 ± 30.7	5.8	−1.05

L-sorbose by the red cell glucose transport system. These workers concluded that binding of glucose to the carrier required loss of about 3 or 4 molecules of water bound to glucose in free solution, and that the number of glucose-bound water molecules appeared to decrease with increasing temperature. In aqueous solution glucose is hydrated by about 4 molecules of water at 20°C and about 2.5 molecules water at 37°C (22).

<u>The intermediate transition state of the carrier</u>. Transitions of the glucose carrier between inward- and outward-facing conformations can be considered as equivalent to chemical reactions whose rate is limited by the need to attain sufficient energy to reach an intermediate transition state. The enthalpies, entropies and free energies associated with this intermediate state can be calculated from the Boltzmann equation

$$ k = \frac{K_B T e^{\Delta S/R}. e^{-\Delta H/RT}}{h} $$

(where k can be rate constant c, d, g or h, E_A is the corresponding activation energy, k_B is the Boltzmann constant, T is the absolute temperature, h is Planck's constant, R is the gas constant and ΔG, ΔS and ΔH are the Gibbs free energy, entropy and enthalpy changes for formation of the intermediate transition state) and the relationships

$$ \Delta H = E_A + RT \text{ and} $$
$$ \Delta G = \Delta H - T\Delta S $$

Taking the concentration of carriers in the red cells to be 6.66 μmole.(litre cell water)$^{-1}$, it is possible to derive the energy diagram for glucose transport via the carrier in the human red cell shown in Fig. 4 and this can be interpreted in terms of a possible model for glucose transport. As has already been discussed, the relatively small change in entropy associated with glucose binding (particularly to the outward-facing carrier) probably derives from the fact that the (anticipated) decrease of entropy of binding is compensated for by dissociation of bound water from both glucose and the binding site of the carrier, which is presumably made up of an array of hydrogen-bonding amino acid side chains known to be present in the α-helical domain of the carrier within the membrane (7). Carrier-bound water may also be important in determining the relative entropies of the inward- and outward-facing conformations and the transition state of the carrier.

In the absence of glucose the large increases in entropy observed for the

318

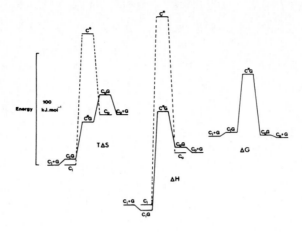

Fig. 4

Gibbs free energy, enthalpy and entropy diagrams associated with transport of glucose via the glucose carrier at 37°C. Standard Gibbs free energies are shown except for glucose binding, for which basic Gibbs free energies (2) with glucose at 5mM on both sides of the membrane are given.

conversions of the inward- and outward-facing carriers to the transition state could well involve loss of water form both the glucose-binding sites and, in the case of the inward-facing carrier, loss of additional water from elsewhere in the carrier molecule. Thus the lower entropy of the inward-facing carrier, relative to the outward-facing conformation, would be due to binding of this additional water, perhaps to the globular domain. The very large endothermic enthalpy changes associated with attainment of the transition state of the carrier in the absence of glucose are also consistent with the breakage of hydrogen bonds involved in water binding at the substrate binding site. Thus, in this model (Fig 5) the transition state represents an arrangement of the carrier's membrane-bound α-helices with no water between the helices.

The situation is somewhat different for the carrier-glucose complexes. Arrival at the transition state from the outward-facing complex involves a small decrease in entropy, presumably indicating that there is little or no loss of bound water, and instead some loss of internal freedom of the macromolecular glucose complex. This could arise from a strengthening of carrier-glucose hydrogen bonds in the transition state relative to the initial Michaelian complex. However, the overall enthalpy change is endothermic so that the enthalpy associated with the strengthening of any carrier-glucose bonds in the transition state is apparently outweighed by breakage of other bonds, possibly in the globular domain. Conversion of the inward-facing

319

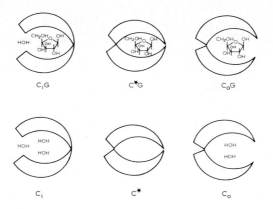

C_iG C^*G C_oG

C_i C^* C_o

Fig.5

Diagramatic representation of states of the carrier and carrier-glucose complexes based on the involvement of intermediate states (C^* and C^*G) in glucose transport.

carrier-glucose complex to the transition state involves an increase entropy. In this case there may be, in addition to the strengthening of hydrogen bonds between the protein and glucose in the transition state, a change in the globular domain of the carrier (perhaps loss of water on breakage of salt linkage(s)) leading to increased degrees of freedom (see below).

One important feature of this model is that the transition states of the carrier and its glucose complexes are dehydrated species which could be expected to be stabilised by dehydrating conditions such as high concentrations of ammonium sulphate. There is in fact evidence that this stabilisation takes place since Kahlenberg et al. (23) showed that in the presence of concentrated ammonium sulphate the glucose carrier in human red cell ghosts binds [14C]glucose and similar sugars with much higher affinity than is found with the carrier under normal physiological ionic conditions. Furthermore Kahlenberg & Dolansky (24) found that the specificity of hydrogen bonding between different sugars and the carrier was very similar at high ionic strength to the hydrogen bonding pattern found at normal ionic strength by Barnett et al. (25,26). The rate of formation of the glucose carrier complex was very slow in the presence of ammonium sulphate (23) and this is consistent with the idea that the carrier mainly as its transition state at high ionic strength. Indeed, the transition state of the carrier-glucose complex can be regarded as an occluded species analogous to the occluded K^+ (or Rb^+) complex which can be detected under certain conditions with the (Na,K)ATPase (27).

The origin of the dissociable water present complexed to the inward-facing carrier at location(s) other than the glucose binding site could possibly be the highly charged globular part of the carrier rather than the hydrophobic α-helices in the membrane. If so, dissociation of this water could be accompanied by, or be the trigger for, other changes in the globular domain such as rearrangement of intra-carrier salt bridges, or possibly salt bridges between the globular domain and the head-groups of the membrane lipids. It is also possible that rearrangements in the globular domain have consequential effects on the α-helices in the membrane and hence dictate whether the glucose binding site adopts the inward- or outward-facing conformation. In this connection it is interesting to note that when the purified glucose carrier is incorporated into phospholipid vesicles of defined lipid composition glucose transport kinetics vary with the nature of the lipid head group (28), and that in membrane vesicles of phosphatidyl choline transport kinetics at 37°C show marked asymmetry and resemble those found at 0°C in intact red blood cells (29). These effects presumably reflect interactions between the carrier's globular region and the lipid head groups. This potential for a modulating effect of the globular domain on the transport process may appear to have little obvious importance to the mode of action of non-energy-transducing systems such as the glucose carrier. However, as will be seen later, changes in the globular domain may well be of crucial importance in active transport systems.

The system discussed above conforms with the requirements given by Fersht (30) for a perfectly evolved biological catalyst: binding of glucose to the carrier is not particularly strong so that the activation energy for transport is not increased by the binding process. Furthermore Ks for glucose binding at both sites of the membrane is of the same order as, but higher than, the physiologically normal concentration of glucose (about 5mM) so that the rate of transport will be sensitive to small changes in glucose concentration *in vivo*. At physiological temperature the carrier properties are also optimal for equilibrating glucose across the membrane because the two loaded and unloaded forms of the carrier are more or less equally distributed between the two sides of the membrane, but at low temperatures the transport process is less efficient because of the much greater stability of the inward-facing conformation. In fact at low temperature glucose effectively acts as a catalyst for the interchange between carrier conformations possibly by providing a pathway of hydrogen bonds between the two conformations.

7.2.3   Active transport systems

In contrast to the downhill and exchange transport processes catalysed by the red cell glucose carrier, active transport systems such as the Na,K-pump drive movements of the carried ions against their electrochemical gradients. A typical physiological setting for operation of the Na,K-pump is cytoplasmic Na^+ 10mM cytoplasmic K^+ 150mM, extracellular Na^+ 145mM, extracellular K^+ 5mM, and on this basis and with the pump exporting 3 Na^+ and importing 2 K^+ ions per cycle (31), and a membrane potential of -60 mV (inside negative) the free energy involved can be calculated as

$$3RTln([Na]_o/[Na]_i) + 2RTln([K]_i/[K]_o) + 0.06F \;\;=\;\; 40 \text{ kJ/mol pump cycle.}$$

Since the pump hydrolyses 1 molecule of ATP per cycle the energy available to drive these cation movements is the free energy of hydrolysis of ATP under conditions prevailing in the cell (e.g. 5mM ATP, 1mM ADP, 5mM inorganic phosphate) or

$$\Delta G \;=\; \Delta G^o + RTln \; [(ADP)(P_i)]/(ATP) \;=\; -58 \text{ kJ/mol.}$$

There is therefore ample energy available from ATP hydrolysis and the question of interest is how this free energy of ATP hydrolysis is translated into cation movements through the mechanism of action of the Na,K-pump.

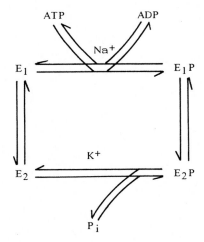

Fig. 6

The mode of operation of the Na,K-pump. Here the pump is represented as phosphorylated and non-phosphorylated forms with Na^+ necessary for phosphorylation by ATP, and K for dephosphorylation of E_2P to E_2. Transport of Na^+ occurs during conversion of E_1P to E_2P and transport of K^+ during conversion of E_2 to E_1.

Many years of intensive study have led to the following generally accepted model (simplified from 32,33) for the mode of action ATPase activity of the Na,K-pump (Fig. 6). The pump is made up of α- and β-subunits (possibly arranged as a tetramer) which have been sequenced (34-36) and appear to be membrane-spanning proteins with both α-helical membrane domains and, in the case of the α-subunit, a globular domain containing the ATP-binding site. The system can exist in at least two different conformations (E_1 and E_2) which each have phosphorylated derivatives (E_1P and E_2P). In the transport cycle Na^+ catalyses the phosphorylation of the enzyme by ATP and is then transported as a result of the transition, $E_1P \rightarrow E_2P$, while K^+ catalyses the dephosphorylation of E_2P and is transported as a result of the transition, $E_2 \rightarrow E_1$. Furthermore, when special conditions are imposed these two "halves" of the Na,K-pump cycle have been shown to operate independently giving Na^+/Na^+ exchange and uniport Na^+-transport in the case of the E_1P/E_2P system and K^+/K^+ exchange and K^+-uniport in the case of the E_1/E_2 system. The separation of (Na,K)-transport into these two sometimes non-energy-dependent half-systems makes it possible to make useful comparisons with the mechanism of glucose transport detailed above.

Evidence for the existence of an equilibrium between E_1 (stabilized by Na^+) and E_2 (stabilised by K^+) forms of the non-phosphorylated (Na,K)-ATPase comes from a number of sources. The two conformations have different reactivities with iodonapthylazide (37), intrinsic (tryptophan) fluorescence (38), circular dichroism spectra (39), susceptibilities to hydrolysis by chymotrypsin and trypsin (40), affinities for ATP and TNP-ATP (41,42) and rates of thermal denaturation (42,43). Fluorescence studies (38) are consistent with exposure of a tryptophan residue to a more hydrophilic environment during the transition $E_2 \rightarrow E_1$, and the relatively high rate of thermal denaturation of E_1 is also consistent with this conformation of the enzyme being in a relatively highly hydrated state. Jorgensen (43) showed that denaturation of the Na,K-ATPase was associated with large positive enthalpies and entropies of activation with a similar transition state formed from both E_1 and E_2 during thermal denaturation. Jorgensen (43) also showed the equilibrium between the E_1 and E_2 forms was modulated by nucleotides with ADP partially converting $E_2(K^+)$ to the heat-sensitive E_1,K^+ conformation.

The equilibrium between the E_1 and E_2 forms of the enzyme is dependent on the ligands present. In the presence of potassium ions the equilibrium lies very much in favour of the E_2 form (with potassium bound in an occluded complex) which has a higher intrinsic fluorescence and affinity for potassium than the E_1 form (38). However, the presence of ATP (which is bound to E_1 with higher affinity than to E_2) the equilibrium constants is decreased by about 500-fold so that at 21°C the equilibrium constants between the E_1 and E_2 conformations (with potassium and ATP complexed) is close to unity (45).

An equilibrium between two phosphorylated forms of the enzyme (E_1P and E_2P), is also well established. When the enzyme contains bound Na^+ the carboxyl group of an aspartyl residue can be phosphorylated by ATP in the presence of Mg^{2+} (46) in a readily reversible reaction with a small change in free energy (47). However, under normal conditions, E_1P rapidly isomerizes to E_2P with the equilibrium very much in favour of the latter phosphoenzyme. The conformational change does not involve any change in covalent bonding, yet the acyl phosphate bond in E_2P undergoes a substantial change in properties since it can be formed either from E_1P or by reaction of inorganic phosphate with the E_2 conformation of the enzyme (48). A similar phenomenon is found with the Ca-ATPase (49) and de Meis (50) has suggested that the "low energy" character of the enzyme's acyl phosphate bond is due to exclusion of water from its immediate environment, and this is consistent with the fact that dimethyl sulfoxide promotes phosphorylation of both the Ca-ATPase and the Na,K-ATPase by inorganic phosphate (51). During operation of the Na,K-pump E_2P is rendered sensitive to hydrolysis by the binding of potassium, which also catalyses rapid exchange of ^{18}O between water and the enzyme-bound phosphate (52).

K^+-transport by the Na, K-pump

As indicated above, by analogy with the glucose carrier it is possible to represent the "K-side" of the Na, K-pump as shown in Fig. 7, where the two carrier conformations (E_1 and E_2) exist in equilibrium with each other with and without bound K^+. The intermediate, $E_2(K^+)$, also takes part in conventional representations of the overall mechanism of active (Na^+,K^+)-transport and in fact the transition, $E_2(K^+) \rightarrow E_1,K^+$ has been shown to be rate-limiting for Na,K-ATPase activity, at least at low concentrations of ATP (53). The rate limiting nature of this transition is also emphasised in the presence of DMSO and deutrium oxide which appear to stabilise $E_2(K^+)$ at the expense of E_1 and E_1,K^+ (50).

If the "K-side" of the Na,K-pump operates in this way then the pump should catalyse K^+-K^+-exchange and uniport movements of K^+ in ways

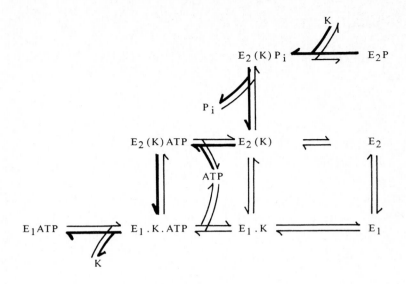

Fig. 7

Transitions of the Na,K-pump (E_1,E_2) involved in potassium transport. The main pathway for active transport of K is indicated by the bold arrows. E_2(K) is an occluded complex of K from which K is released at a signficant rate only after conversion to E_1.K.ATP or phosphorylation by inorganic phosphate to give E_2P.K.

analogous to the operation of the glucose carrier, provided that conditions are such that the "Na-side" of the Na,K-pump is not operating. In fact ouabain-sensitive K^+-K^+-exchange and K^+-for-nothing (zero-trans K^+) transport have been demonstrated in human erythrocyte ghost containing no Na^+ (54). These fluxes are very slow (relative to glucose transport allowing for the relative numbers of glucose carriers and Na,K-pumps present per cell) and are normally measured in the presence of (cytoplasmic) ATP or ADP, which accelerates conversion of E_2(K^+)) to E_1,K^+, and inorganic phosphate, which binds to E_2 (55,56). However, even slower K^+-fluxes through the Na,K-pump occur in red cells in the absence of ATP and phosphate (56) and similar Rb^+ fluxes have oberved in proteoliposomes containing reconstituted Na,K-ATPase (55,57).

Why are K^+-fluxes through the "K-side" of the Na,K-pump so slow relative to glucose transport through its carrier? The answer to this question probably lies in the nature of the intermediate transition states of the two transport systems. The transition state of the glucose carrier is probably very unstable under normal conditions, accounting for only a small proportion of the total number of glucose carrier-complexes, and the rates of conversion of the transition state to the inward- and outward-facing glucose-carrier conformations are not rate-limiting for transport. In constrast the transition state of the "K-side" of the Na,K-pump, the "occluded" K^+ (or Rb^+) complex, is stable enough to remain intact during rapid passage through a column of cation-exchange resin (27) and probably makes up a significant proportion of the total number of K^+-pump complexes, particularly in the absence of ATP.

Binding of ATP (or ADP) is also necessary to accelerate conversion of the occluded (K^+) complex to $E_1(K^+)$ and hence for dissociation of the bound K^+ (27). Clearly in such a system there could be little transport in the absence of ATP since in that case most of the carrier would be in the form of the occluded K^+-complex. In a similar way binding of inorganic phosphate appears to be necessary for conversion of the occluded form of $E_2(K^+)$ to a form (E_1, K^+) from which K^+ can dissociate rapidly (58). In this context it is notable that both ATP and inorganic phosphate are hydrophilic molecules the binding of which seems to tip the balance of the E_1-E_2 system towards a relatively hydrophilic forms, whereas reduction of the water concentration (by substitution with DMSO or deuterim oxide) favours the hydrophobic E_2-state (53) with occluded K^+.

The mechanism by which ATP or ADP causes the occluded K^+pump complex to adopt the E_1, K^+ conformation with resultant dissociation of K^+ and completion of the transport cycle is clearly of considerable interest. The binding site for K^+ must presumably be in the hydrophobic (α-helical) membrane-bound domain of the Na,K-pump, while the site for ATP is apparently in the relatively hydrophilic and presumably globular part of the protein (34,35). It seems reasonable to propose that the binding of ATP results in a change in conformation of the globular domain, and that as a consequence of this the E_1, K^+ form of the membrane-bound domain is stabilised relative to $E_2(K^+)$ and the occluded K^+ complex. In a similar way it would appear likely that stimulation of K^+ transport by inorganic phosphate arises because phosphate bindng leads to conversion of $E_2(K^+)$ to E_2PK^+ with concomittent de-occlusion of K^+ as a consequence of a change in the membrane domain of the enzyme. Unfortunately studies of the temperature dependences of the equilibrium between the E1 and E2 conformations, and of K^+-fluxes and other processes related to the "K-side" of the Na, K-pump are not sufficiently detailed to make possible a full description of the thermodynamics of this system.

326

Na⁺-transport by the Na, K-pump

The "Na-side" of the Na,K-pump can be represented by the diagram in Fig. 8 in which Na^+-transport is associated with the conformational change between E_1P and E_2P. If this is a correct representation, the system should catalyse Na^+/Na^+ exchange transport in the absence of potassium, and this process is in fact observed in both red blood cells and squid giant axons containing low ratios of ATP:ADP (59, 60). The significance of this

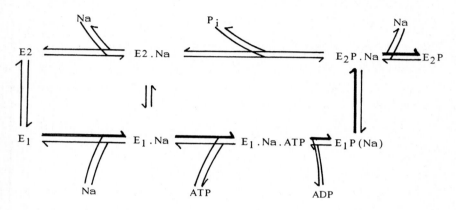

Fig. 8

Transitions of the Na,K—ATPase involved in sodium transport. The main pathway for active transport is indicated by bold arrows. $E_1P(Na)$ is an occluded complex of Na and a second occluded complex ($E^*P(Na)$) may lie between $E_1P(Na)$ and $E_2P.Na$. Release of Na from the occluded state requires either reaction with ADP (to give $E_1.Na.ATP$) or a conformational change to $E_2P.Na$.

ADP-requirement could well be that $E_1P(Na)$ is an occluded complex analogous to $E_2(K)$ and that reaction with ADP (to give ATP) is necessary in order to accelerate release of the occluded Na^+ catalyzing Na^+/Na^+ exchange. The equilibrium between the sodium complexes of E_1P and E_2P has been investigated in some detail and both kinetic studies (61) and the classical "cold chase" experiments of Post et al. (33) indicate that under normal conditions (mM concentrations of Mg^{2+}) the equilibrium lies very much in favour of the Na^+ complex of E_2P. However the E_1P form is much more important at low Mg^{2+} concentrations and conversion of E_1P to E_2P can be blocked by N-ethylmaleimide (62), which also blocks the overall ATPase activity of the Na,K-pump. More recent studies (63-67) indicate the existence of additional form(s) of phosphoenzyme, particularly at high concentrations of sodium, since the sum of the ADP-sensitive (E_1P) and the postassium-sensitive forms of the phosphoenzyme (E_2P) exceeds the total amount of phosphenzyme. Studies of the kinetics of changes in light scattering and fluorescence (68) are also indicative of interconversions between multiple forms of phosphoenzyme and as a consequence of these studies an additional species, E^*P (Na$^+$) has been proposed. It is possible that this intermediate is also an occluded complex which plays a role in Na-transport analogous to that of E_2(K) in potassium-transport via the Na,K-pump.

By analogy with the properties of the glucose carrier, the "Na-side" of the Na,K-pump should also catalyse "zero-trans" Na$^+$-transport either by repeatedly using the phosphoenzyme cycle without hydrolysis of ATP, or by hydrolysis of E_2P and return of Na$^+$ via the E_2Na$^+$ and E_1Na$^+$ intermediates. Only the latter type of ATP-hydrolysis-dependent uniport Na$^+$-transport has been reported (69,70) and this is probably because the rate of spontaneous hydrolysis of E_2P (in the absence of K$^+$) is faster than the rate of conversion of (uncomplexed) E_2P to E_1P. It is possible that ATP-hydrolysis-independent uniport transport of Na$^+$ does in fact occur, but at a rate that is very slow relative to the hydrolytic pathway.

The equilibrium between E_1Na$^+$ and E_2Na$^+$ appears to lie very much in favour of the former. Evidence for this includes the fact that Na$^+$ stabilises a conformation of the enzyme that is readily distinguished from the K$^+$ form (E_2K) by both its intrinsic (tryptophan) fluorescence (38) and the peptide fragments released from it during digestion with trypsin or chymotrypsin (40). The "early burst" of hydrolysis (71,53) observed after adding ATP to the Na$^+$-pretreated Na,K-pump, but not to the K$^+$-pretreated pump is also consistent with the idea of a relatively stable E_1Na and the concept of a slow conversion of E_2(K$^+$) to E_1Na$^+$ unless there is a sufficient concentration of ATP present to bind to E_2(K$^+$) and accelerate dissociation of K$^+$. In contrast to the non-phosphorylated enzyme, the equilibrium between E_1P and E_2P lies very much in favour of the latter as discussed above. However treatment with chymotrypsin or N-ethylmaleimide (72) modifies the equilibrium so that an occluded complex of E_1P(Na$^+$) predominates (73). In the native enzyme the forces which lead to stabilisation of E_2P,Na relative E_1P(Na$^+$) seem likely to act in ways analogous to those suggested above for the effect of ATP on the transition E_2(K$^+$) to E_1,K$^+$. Addition of the hydrophilic and charged phosphate group to the globular domain of E_1Na$^+$ may well lead to a rearrangement of intra-molecular bonds in this domain which in turn cause reorientation of the membrane domain in such a way that the Na-binding sites are transferred from the inward-facing (occluded) to outward facing (non-occluded) state. Furthermore oligomycin, which inhibits transport of Na$^+$ and K$^+$ by pump and promotes an increase in the affinity of Na$^+$ binding, appears to do so by stablising the (occluded) $E_1P(Na^+)$ conformation (24).

Overall, the picture of Na,K-transport that emerges from these considerations is one of alternative pump conformations whose relative stabilities depend on whether ATP is bound as the whole molecule or has transfered its -phosphate to the enzyme, and of a system in which non-transporting events occurring in the globular domain modulate the transporting events in the α-helical membrane domain. Clearly for efficient pumping of Na$^+$ and K$^+$ such a system depends on "leakage" pathways, such as hydrolysis of the phospho-enzyme in the absence of K$^+$ or the transition E_2P to E_1P in the absence of Na$^+$, occuring at negligible rates relative to the main cycle. In this relation it is interesting to speculate that the β-subunit of the Na,K-pump could play a part by interacting with the α-subunit in such a way that these unwanted transitions are retarded.

7.2.4. Co-transport systems

In many instances the facilitated diffusion of metabolites is found to be linked to the transport of another species, normally sodium ions (for example in the carrier systems for glucose and some amino acids in intestinal and renal brush border membranes) or hydrogen ions in (mitochondrial and bacterial carriers systems). See reviews by Wilson (75), Gunn (76) and La Noue & Schoolwerth (77). Such systems permit accumulation of metabolites inside the cell or mitochondrion since metabolite transport is effectively linked to the electrochemical gradient of the co-transported ion and this gradient is maintained by a second independent mechanism (normally the Na,K-pump in the case of the sodium gradient or electron transport in the case of the hydrogen ion gradient across the mitochondrial or bacterial membrane). In the case of glucose-sodium co-transport this electrochemical gradient (μ_{Na}) is given by the expression

$$\mu_{Na} = \mu^0_{Na} + RT\ln([Na_{in}]/[Na_{out}]) + zFE_m$$

where E_m is the membrane potential and $[Na_{in}]$ and $[Na_{out}]$ are the concentrations of sodium inside and outside the cell. The membrane potential influences transport because there is a net charge on the carried species (glucose plus Na$^+$). In electroneutral systems where there is no net charge (eg contransport of the pyruvate ion with protons across the mitchondrial membrane) there is no electrical term the pyruvate gradient is determined only by the concentraion gradient for protons.

Co-transport systems can be represented formally by an extension of the

4-state carrier model as shown below (Fig 9) in which the carrier can undergo conformational transitions (comparable to those described for the simple glucose carrier) (i) in the free state, (ii) with either sodium (complex not shown) or the metabolite bound, and (iii) with both substrate and metabolite bound. However, it is clear that efficient operation of a system of this kind depends upon the rates of reorientation of the monocomplexes being very much slower than the rates of reorientation of the free carrier and the carrier-metabolite complex.

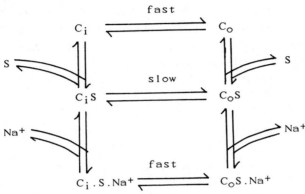

Fig. 9

A scheme for co-transport of a substrate (S) with Na^+. The transitions between the two states of monocomplexes of the carrier with substrate or Na^+ (not shown) are very slow so that transport almost always involves simultaneous movement of both substrate and Na^+.

At the molecular level there are at least two possible ways in which cotransport systems could operate with the two transported species bound either at the same or at separate binding sites. In the case of electroneutral proton-cotransport systems like the mitochondrial pyruvate carrier it seems more likely that there is only one binding site which accomodates (protonated) pyruvic acid. This would be more hydrophobic than the unprotonated pyruvate and hence would presumably be able to reach the transition state much more readily than then ionic pyruvate. It can also be envisaged that the substrate binding site would lack an anionic residue and hence be unable to bind a proton in the absence of pyruvate.

In contrast it seems unlikely that binding of both glucose and sodium occur at a single binding site. In this case it is more likley that the carrier molecule contains two separate binding sites (domains), but that attainment of transition state with only one binding site occupied is improbable. The mechanism by which the presence of both cotransported species facilitates transport stage and hence, presumbly, attainment of an intermediate transition state with both sodium and glucose occluded is unknown, but it is perhaps worth suggesting a possible, if entirely speculative scheme. It seems likely that the Na$^+$ binding site involves a negatively charged amino acid side chain (presumably a carboxyl or an aspartyl or glutamyl residue) in the hydrophobic domain of the carrier. However, in the absence of bound Na$^+$, this carboxyl might take part in an interaction with (directly or indirectly) another part of the carrier such as positively charged lysine side chain. If a movement in the position of this lysine residue were also necessary for the transport-related conformational change in the part of the molecule involved in glucose transport this interaction between the carboxyl and lysine residue would provide an explanation for the difficulty in reaching either the transition state for Na$^+$ or the transition state for glucose without having both ligands present. The results of carboxyl/lysine interactions would be that the carrier transition state would rarely be reached without Na$^+$ and glucose bound simultaneously.

7.3 MUSCLE CONTRACTION

Superficially the mechanism of muscle contraction has little in common with either the mode of operation of a facilitated diffusion system like the glucose carrier or the ATP-driven transport of sodium and potassium by the Na,K-pump. However it turns out that at the molecular level significant similarities do exist for these three systems. Detailed studies of the mechanics of muscle contraction have shown that the force generated during contraction of skeletal muscle is clearly associated with a cycle of events including (i) attachment of myosin head groups (which protrude from the "thick filaments") to actin monomers within the "thin filaments, (ii) reorientation of the myosin head-groups in a process which generates tension if the "thin filament" of actin is not free to move and (iii) detachment of the myosin head groups from the "thin filaments" so that the cycle can be repeated. Interpretation of the molecular mechanism of muscle contraction is therefore dependent on an understanding of the forces which promote the repeated operation of this cycle of contractile events.

The energy source for muscle contraction is known to be the hydrolysis of ATP by myosin, which consists of two main parts, an α-helical, supercoiled rod-shaped region and a pair of globular "heads", which can be detatched from the α-helical rods by treatment with papain to given so-called S1-myosin which contains the active centre of ATPase activity. The work of Lymn & Taylor revealed some remarkable properties of myosin ATPase. Quenched-flow experiments in which myosin was mixed with γ-(^{32}P)ATP a few milliseconds before the reaction was stopped with perchloric acid, revealed that there was an "early burst" of release of inorganic phosphate followed by a much slower (steady-state) hydrolysis of ATP. Furthermore when similar experiments were carried out in the presence of actin the steady-state hydrolysis of ATP was greatly accelerated to a rate comparable to that observed near the start of the "early burst" of hydrolysis in the absence of actin. In additional studies in which myosin (or actin plus myosin) was mixed with (γ-^{32}P,^{14}C)ATP, then separated from substrate and dissociated products by rapid filtration through a column of Sephadex, Lymn & Taylor demonstrated (83) that the products of ATP hydrolysis (^{14}C)ADP and (γ-^{32}P)phosphate) dissociated from myosin very slowly in the absence of actin, and that dissociation was greatly accelerated in the presence of actin. Thus, in the presence of actin, myosin appeared to undergo a change in conformation, which permitted released of the bound products of ATP hydrolysis (Fig.10).

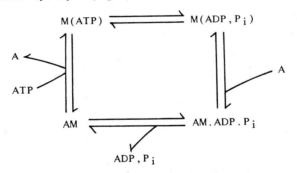

Fig. 10

The Lymn & Taylor scheme for hydrolysis of ATP by myosin (M) in the presence of actin (A). Myosin retains the products of hydrolysis in a tighly bound form until binding of actin facilitates release of ADP and P_i. Dissociation of the actomyosin complex requires binding of ATP to myosin.

Further insight into the conformational changes undergone by myosin during ATP hydrolysis was obtained by measuring intrinsic (typtophan) fluorescence (84). Binding to myosin of ATP, ADP and non- or slowly-hydrolysable ATP-analogues such as (γ-S)ATP leads to an enhancement of protein fluorescence which is suggestive of a substrate-induced change in conformation (85). Interestingly the rate of this change in fluorescence has been shown to be a hyperbolic, rather than a linear function of the concentration of nucleotide (86,87) and when (γ-^{32}P) ATP is used as substrate it is found not to be readily exhangeable (before hydrolysis) with unlabelled ATP added later. These findings are consistent with binding of ATP to myosin occuring in a two-stage process

$$M + ATP \; \rightleftharpoons \; M.ATP \; \rightleftharpoons \; M^*.ATP,$$

in which a simple Michaelis complex is formed rapidly, then converted (more slowly and as a result of a conformational change in the protein) to a tightly bound myosin-ATP complex. This is reminiscent of the rapid reversible binding of glucose to its carrier, followed by the conformational change leading to the transition state of the carrier with glucose occluded.

When ATP is hydrolysed the myosin-products complex undergoes a further enhancement of fluorescence (88, 89). Comparison of the time courses of fluorescence changes and the conversion of ATP to ADP and inorganic phosphate during hydrolysis of ATP by a molar excess of myosin enabled Bagshaw et al., (80) to calculate the equilibrium complex of the reaction:

$$Myosin.ATP \; \rightleftharpoons \; Myosin.ADP,P_i$$

to be about 10 (compared with about 10^7 for hydrolysis of ATP in aqueous solution). The comparative ease of reversal of hydrolysis of myosin.ATP was confirmed by Bagshaw et al., (91,92), who measured ^{18}O-exchange between water and ATP, and Mannherz et al., (93), who showed that ATP could be synthesised by addition of inorganic phosphate to a pre-formed complex of myosin,ADP. Furthermore Taylor (94) demonstrated that the equilibrium between myosin,ATP and myosin.ADP,P$_i$ was highly sensitive to changes in both temperature and pH, so that ATP could be synthesized from myosin.ADP,P$_i$ by a combined "jump" of temperature and pH. The reason for the low value of the equilibrium constant of ATP hydrolysis at the active site of myosin may be the occlusion of ATP and its hydrolysis products at the active site with resultant shielding from water and prevention of hydration.

Studies of the temperature-dependence of myosin conformational changes have made it possible to build up a picture of the thermodynamics of these changes. Studies of UV-difference spectra enabled Morita (95,97) to show that, at 25°C, two types of myosin-nucleotide complex can be observed, with one predominant during ATP hydrolysis or in the presence of AMPPNP and the second predominant in the presence of ADP. In contrast at 1°C the ADP-type spectrum is observed in the presence of either ADP or AMPPNP, suggesting a temperature-dependent equilibrium between two myosin conformations which have been designated the "R" (ATP-induced) and "T" (ADP-induced) states by Shriver (101). The structural significance of these changes in UV spectrum and fluorescence is probably a change in environment of tryptophan (and possibly tyrosine residues) in the myosin. These may be relatively well exposed to solvent in the "T" conformation, but buried within the hydrophobic core of the protein in the "R" conformation.

In keeping with the above UV-spectroscopic investigations $(^{31}P)NMR$ studies (102) confirm that complexes between myosin and AMPPNP or ADP both exist as equilibrium mixtures of two species and the temperature-dependence of the equilibrium constant for the "R" and"T" forms of myosin-ADP indicates an enthalpy change of 63 kJmol^{-1} and an entropy change of 231 kJmol^{-1} (both for mysoin subfragment 1). The existence of "R" and "T" forms of myosin-nucloeotide complexes implies that myosin itself should be be capable of undergoing a similar transition. This apears to be the case since, after labelling with an (^{19}F)-probe, $(^{19}F)NMR$ studies have demonstrated a strongly temperature-dependent equilibrium between two conformational states of myosin in the absence of nucleotides. Enthalpies, calculated from van't Hoff plots are 151 kJmol^{-1} for S1-myosin and 294 kJmol^{-1} for heavy menomyosin, while entropies are 441 J.K^{-1}mol^{-1} (S1 myosin) and 1008 J.K^{-1}mol^{-1} heavy meromyosin. The large difference between the thermodynamic parameters for myosin S1 and subfragment 1 are interesting since they imply that a conformatinal change in the globular (S1) region of myosin is accompanied by a change in the helical region (present in subfragment 1 but not myosin S1). This provides a possible mechanism by which a change in conformation of the globular domain of mysin (in the thick filament of a muscle) could be transmitted to actin in the thin filaments and hence give rise to a force causing either sliding of the filaments relative to one another, or, when no relative movements of the filaments takes place, to tension in the S2-part of the heavy meromysin and the muscle as a whole. This situation is reminiscent of the suggestion (made above) that globular domain of the glucose carrier can influence the orientation of the α-helices of the carrier in the membrane.

A comparison between the thermodynamics of conformatlonal changes of the glucose carrier and myosin is instructive (table 3). The changes in enthalpy and entropy associated with reorientation of the glucose carrier are similar in magniture to those occuring during the change from "R" to "T" states (or L- to H-states) of the ATP complex of myosin and so the intra-molecular changes involved in the two conformatinal changes are likely to be of similar complexity. However, whereas for the glucose carrier ligand binding has a relatively small effect on the thermodynamics of the conformational change, in the case of myosin the enthalpy and entropy changes are much greater in the absence of ligand than when ATP is bound. Thus in

Table 3

Comparison of enthalpy and entropy changes associated with conformational transitions and substrate binding to myosin (subfragment S1) (5) and the glucose carrier of the human red cell (20). Data for myosin are from Kodama (5) and the low and high-temperature forms of myosin (M_L and M_H) are equivalent to Shriver's (98) R- and T- forms.

System	ΔH° kJ mol^{-1}	ΔS° Jk^{-1}mol^{-1}	Temperature for $\Delta G^\circ = 0$ (K)
1. Myosin			
$M_L \to M_H$	+103	+373	276
$M_L(ADP) \to M_H(ADP)$	+64	+236	271
$M_L(AMPPNP) \to M_H(AMPPNP)$	+18	+60	298
$M_L + ADP \to (ADP)$	-35	-18	
$M_L + AMPPNP \to M_L(AMPPNP)$	+35	+244	
$M_R + ADP \to M_R(ADP)$	-74	-155	
$M_R + AMPPNP \to M_R(AMPPNP)$	-52	-69	
2. Glucose carrier			
$C_{in} \to C_{out}$	+46	+145	318
$C_{in}(glucose) \to C_{out}(glucose)$	+56	+185	304
$C_{in} + glucose \to C_{in}(glucose)$	-4.95	+17	
$C_{out} + glucose \to C_{out}(glucose)$	+5.51	+58	

myosin substrate binding can be regarded as controlling the equilibrium between the two conformational states of the macromolecule. This situation is also reflected in the thermodynamics of substrate binding in the two systems. Glucose binding to the two conformations of its carrier occurs with small changes in enthalpy and entropy which contrast with the much larger enthalpies and entropies of binding of ADP and the ATP analogue, AMPPNP to both the "R" and "T" conformations of myosin (98). The difference between enthalpies and entropies of binding of AMPPNP to the "R" andf "T" forms of myosin is particularly notable with the large negative values for binding to the "T" conformation signalling formation of strong (presumably ionic) bonds between ligand and protein, and the large positive values for binding to the

"R" conformation indicating a very different mode of binding, presumably as a result of a change in the binding site caused by the conformational change in the macromolecule.

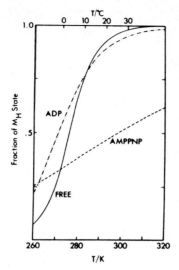

Fig. 11

     Effects of temperature and binding of ADP or AMPPNP on the equilibrium between the high energy (M_H) and low energy (M_L) states of myosin. M_L or M_H are equivalent the R- and T- forms of myosin (98).

When the temperature dependences of the distribution of myosin between the "R" and the "T" states are (equivalent to M_L and M_H) calculated (Fig. 11) it is clear that, at physiological temperature, binding of AMPPNP (or presumably ATP) converts myosin form the "T" state to a delicately balanced equilibrium between two conformations, presumably by interfering with some of the intra-molecular forces responsible for the "tight" conformation of uncomplexed myosin. In this connection it is notable that the free energy of binding of ATP to myosin matches the difference between enthalpies of conversion of conversion of myosin between its "R" and "T" conformations in the absence and presence of ATP. Futhermore, the poise of the equilibrium between "R" and "T" forms of the myosin-nucleotide complexes, is sensitive to whether ATP itself or ADP is bound so that repeated cycles of ATP-binding, hydrolysis and dissociation of products can drive the myosin repeatedly between the two conformational states.

The presence of actin has profound effects on the properties of myosin causing not only a greatly increased rate of hydrolysis of ATP but also an altered pattern of myosin hydrolysis by trypsin (102, 103) and substantially decreased the mobility (assessed by high resolution proton NMR) of amino acid side chains in the region of the molecule whose tryptic hydrolysis is modified by actin (104). In addition the fluorescence depolarisation studies of Highsmith (105) showed that binding of actin to myosin is endothermic (van't Hoff enthalpy = 39 kJ.mol^{-1}, associated with a positive entropy change (ΔS = 300 J.K^{-1}mol^{-1}) and highly dependent on ionic strength (106), with high ionic strength promoting dissociation of the actomyosin complex. These results are consistent with the idea that binding of actin to myosin is an "entropy-driven" process in which binding involves dissociation of water from (ionised) binding sites at the surface of the two proteins.

In keeping with the above actin-induced changes in myosin conformation, introduction of actin into the system with myosin and ATP also has a profound affect on the relative stabilites of the "R" and "T" states of myosin. Binding of action to the "R"-state of the myosin, ADP, P$_i$ complex favours conversion of the myosin-products complex to the "T"-state with a consequent decrease in affinity for the products and change in orientation of the helices in the S2-part of heavy meromyosin. Return of the myosin to the "R" state is effected by the binding of ATP, which also causes dissociation of the actomyosin complex so that the cycle of ATP hydrolysis and cross-bridge formation can be repeated. Thus the contractile events can be seen as a consequence of oscillation of myosin between two conformations one of which is favoured by binding of actin and the accompanying dissociation of water, and the other by ATP-binding and water dissociation from another site on the myosin. It is ligand-binding and water dissociation, rather than ATP hydrolysis, that provides the energy for these events and appropriately for this concept, Marston & Tregear (107) have shown that the non-hydrolysable ATP-analogue, AMPPNP, can induce the actomyosin complex (in the rigour state without bound nucleotides) to revert from the "T" to the "R" state. Furthermore, in insect flight muscle, this AMPPNP-induced conformational change has been shown, by electron microscopy and X-ray diffraction studies, to be associated with movement of cross-bridges between the thick and thin filaments and the AMPPNP-induced "R"-state can be reversed back to the rigour "T" state by washing (107). Thus the micro-contractile cycle can be repeated by alternate addition and removal of the non-hydrolysable nucleotide analogue.

7.4 SYNTHESIS OF ATP IN MIOCHONDRIA

In the mitochondrion it is now generally accepted that the proton gradient across the inner memrane, generated by electron transport, provides the energy for the synthesis of ATP from ADP and inorganic phosphate (108), even though the question of whether locally generated protons or "bulk phase" protons are involved in the coupling process remains open.   Assuming that "bulk phase" forces ar responsible, the free energy available for ATP synthesis in this system can be calculated from the difference between the electrochemical potentials in the mitochondrial matrix space (phase 1) and the surrounding medium (phase 2) as follows.

$$\mu H^+(1) = \mu^O H^+(1) + RT\ln[H^+](1) + zF\varphi_1$$
$$\mu H^+(2) = \mu^O H^+(2) + RT\ln[H^+](2) + zF\varphi 2$$

where $\varphi 1$ and $\varphi 2$ are the Galvani potentials in the two phases. Hence

$$\mu H^+(1) - \mu H^+(2) = RT1([H+](1)/[H+](2)) + zF(\varphi 1 - \varphi 2)$$

or $\Delta \mu = -2.303 RT\Delta pH + zFEm$

where $\Delta pH = pH(1) - pH(2)$, $Em = \varphi 1 - \varphi 2$, and $z = +1$

Mitchell's expression for the proton motive force (Δp) is obtained by dividing this expression by F:

$\Delta p = 2.303(RT/F)\Delta pH + Em.$

This proton motive force must supply the energy for ATP synthesis under the conditions prevailing in the mitochondrion.   Here typical substrate concentrations are 5 μM ATP, 0.1 mM ADP and 5 mM inorganic phosphate so that, assuming that the activity of water is unity, the free energy required for ATP synthesis is given by

$$\Delta G = \Delta G^O + RT\ln\frac{[ATP][H_2O]}{[ADP][P_i]}$$

or about 58 kJmol^{-1} assuming that the standard free energy for ATP synthesis is 31 kJmol^{-1}.   This is equivalent to a proton motive force of about 240mV (if 2 protons are coupled to synthesis of each ATP molecule) or about 180mV (if 3 protons are coupled to the synthesis of ATP as now seems more likely (109)).   In fact measured values of pH differences and membrane potentials across the inner membranes of respiring mitochondria indicate a proton motive force of this magnitude (110) in keeping with Mitchell's chemiosmotic hypothesis for ATP synthesis.

The site of mitochondrial ATP synthesis is the so-called F_1-ATPase located in the "Fernandez-Moran" particles on the inner surface of the mitochondrial cristal membrane (111).   This particle is a complex assembly of subunits (112), and its is linked to the membrane by additional proteins as indicated in Fig. 12.   The membrane-bound Fo complex is believed to contain a proton-conducting channel (113) which provides a pathway through which the electrochemical gradient can reach and influence the phosphate bond-

338

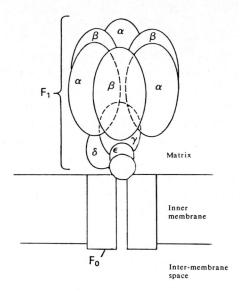

Fig. 12

The mitochondrial F_0/F_1 ATPase.   Subunits of F_0 are not shown.

synthesizing active centres in the F1 particle.

The mechanism by which ATP is synthesised has not been elucidated in detail but it is notable that the F1 particles have ATPase activity and bind ATP with very high affinity ($Ka = 10^{12}$ M^{-1}) (114).   This is similar to the association constant for binding of ATP to myosin, and the F1-ATPase also appears to be more closely related to myosin than the (Na,K)-ATPase in that there is no evidence for the existence of a covalent enzyme intermediate during ATP hydrolysis by the F1-ATPase (115).   In keeping with this analogy Penefsky (116) has shown, by acid quench experiments, that the equilibrium constant for the reaction

$$ADP + P_i = ATP + H_2O$$

at the active site of the F1-ATPase is close to unity, and it has been proposed (117) that the critical step in the operation of the F1-ATPase as an ATP synthesizing system is the release of the tightly bound ATP from the active site.   This idea is supported by the fact that, like myosin, the F1-ATPase catalyses rapid exchange of ^{18}O between water and inorganic phosphate (118,119) and by the finding (114) that in sub-mitochondrial particles the association constant for binding of ATP to F1 is decreased by the respiratory

substrate, succinate, in an effect which is prevented by uncoupling agents
(inducers of proton-permeability) and which is therefore presumably dependent
on the generation of proton motive force.   The concept that the F1-ATPase
can exist in two conformations (with high   and low affinities for ATP) is
supported by the findings that the sensitivity of F1 to trypsin varies according
whether nucleotides (and inorganic phosphate) are present (120), and that ATP
can induce a change in fluorescence of the complex formed between aurovertin
and F1 (121).   In addition circular dichrosim measurements have shown that
ATP can induce a conformational change in isolated α- and β- subunits of the
F1-ATPase (122), while (123) Hatefi has demonstrated that uncoupling agents
increase the Km for ADP and inorganic phosphate as well as decreasing the
Vmax for ATP synthesis by mitochondria.

Perhaps the simplest way of viewing this process is in terms of the
following cycle (Fig. 13) in which the F1-ATPase complex can exists in one of

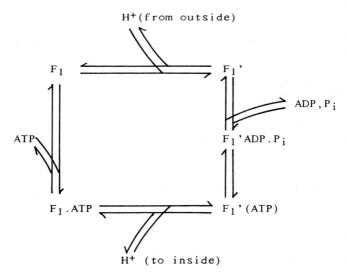

Fig. 13

A scheme for synthesis of ATP via the mitrochondrial F_1-ATPase.
Protons are envisaged as entering the system for the extramitocondrial phase
and leaving with the mitochondrial matrix.   Release of tightly-bound ATP
from F_1'(ATP) requires loss of proton(s).

two conformations, F1 and (a putatively protonated form) F1'.   By way of a
hypothesis one may suppose that the equilibrium between F1 and F1' lies in
the favour of (the protonated) F1', especially in an energised system.   However
the binding of ADP and inorganic phosphate could be accompanied by a
conformational change (possibly a rearrangment of subunits) which leads to

both dissociation of protons into the mitochondrial matrix compartment and production of ATP at the active site. Completion of this cycle (and hence gross ATP synthesis) would then require a change in conformation from Fl(ATP) to Fl'(ATP) which would be dependent on the arrival of a proton from the outside compartment (via Fo). Critical features of this hypothetical mechanism are (i) a high specificity for binding of ADP plus inorganic phosphate (rather than ATP) at the active site of the Fl' conformation and (ii) restricted geometry of the system such that the cycle is linked to the trans-membrane energy gradient by obligatory release of protons at the matrix side and uptake of protons from the outside of the mitochondrion. In the absence of a proton gradient (in uncoupled or nonenergised mitochondria) ATP hydrolysis by a system of this type could be favoured by the presence of an increased concentration of Fl' (caused by the increased elelctrochemical potential for protons in the matrix compartment) and the putative reaction

$$Fl + H^+ = Fl'$$

together with a shift in the equilbrium

$$Fl'(ATP) = Fl(ATP) + H^+$$

to the right as a result of the decrease in external protons. The electrical potential across the mitochondrial inner membrane could have effects similar to proton concentration if the Fl and Fl' conformations involve appropriate transmembrane dipoles, and in keeping with this electric field-induced synthesis of ATP has been demonstrated in both mitochondria (125) and submitochondria particles (125).

An interesting feature of the Fl-ATPase (126) is that when ATPase activity is measured with an excess of Fl over ATP (this results in occupation of only one ATP binding site per Fl since the afffinities of the second and third sites are much lower than for the first) the rate of hydrolysis is extremely slow. However then ATP is added in excess of the Fl concentration ATP hydrolysis is accelerated by up to 10^6 fold. The significance of this could be that the rate of one or both the transitions, Fl'(ADP,Pi) → Fl(ADP,Pi) or Fl(ATP) → Fl'(ATP) is limiting, but facilitated by the occurrence of a similar (but opposite) transition in an adjacent active site. This can be regarded as analogous to contransport systems (such as Na+/glucose cotransport in mammalian enterocytes) in which the transport-effecting conformational change of the carrier requires the presence of both bound species, presumably at separate binding sites.

7.5 CONCLUSION

The theme common to all the systems discussed above is that in each case there is a protein, or protein complex, which can exist in two conformational states comparable to the R- and T- states of haemoglobin. In the non-energy transducing glucose carrier these two conformations are (at physiologcal temperature) in a state of more or less evenly balanced equilibrium which facilitates transport of glucose in either direcion across the membrane. In contrast, in the energy transducing systems the equilibrium between enzyme conformations is far from evenly poised, and in order to allow the process to proceed (be it active Na+/K+-transport, muscle contraction or mitochondrial ATP synthesis) energy must be put into the system. In the cases of active cation transport and muscle contraction this energy is input in the form of ATP-binding, which has the effect of re-aligning the equilibrium between the conformational states of the protein, while in the case of the F1-ATPase it appears that it is the proton gradient that alters the affinity of the enzyme for ATP. Other points of similarity are that in all the systems intermediates with water excluded from the binding site of the enzyme or carrier appear to be vitally important, and that (as in the case of haemoglobin) events in one part of the enzyme (typically binding of ATP to the globular domain of the Na,K-ATPase) have long range effects in another part of the system (eg on the occlusion of potassium ions in the membrane (presumably α-helical) domain of the Na,K-ATPase). Overall it seems clear that energy transduction is achieved by linkage of energetic changes in protein conformation to events taking place in the surrounding bulk solvent, and that the solvent itself (water) plays an important part in determining the stabilities of the various protein conformations.

References

1. Jencks, W.P. (1980) Adv. Enzymol. Relat. Areas Mol. Biol. 51, 75-106.
2. Hill, T.L. (1977) "Free Energy Transduction in Biology." Academic Press, New York.
3. Knowles, J.R. (1980) Ann. Rev. Biochem. 49, 877-919.
4. Tanford, C. (1983) Ann. Rev. Biochem. 52, 279-409.
5. Kodama, T. (1985) Physiol. Rev. 65, 467-551.
6. Lowe, A.G. & Walmsley, A.R> (1986) Biochim. biophys. Acta 857, 146-
7. Mueckler, M., Caruso, C., Baldwin, S.A., Panico, M., Blench, I., Morris, H.R. Allard, W.J., Lienhard, G.E. & Lodish, H.F. (1985) Science 229, 941-945.
8. Lacko, L., Wittle, B & Kromphardt, H. (1972) Eur. J. Biochem. 25, 447-454
9. Haldane, J.B.S. (1930) "Enzymes". Longmans, U.K.

342

10. Wang, J-F, Falke, J.J. & Chan, S.I. (1986) Proc. Natl. Acad. Sci, U.S. 83, 3277-3281.

11. Miller, D.M. (1971) Biophys. J. 11, 915-923.

12. Brahm, J. (1983) J. Physiol. 339, 339-354.

13. Wheeler, T.J. (1986) Biochim. biophys. Acta,. 362, 387-398.

14. Widdas, W.F. (1980) Curr. Topics Bioenerget. 14, 165-.

15. Krupka, R.M. & Deves, R. (1981) J. Biol. Chem. 256, 5410-5416.

16. Gorga, F.R. & Lienhard, G.E. (1981) Biochemistry, 20, 5108-5113.

17. Hankin, B.L., Lieb, W.R. & Stein, W.D. (1972) Biochim. Biophys. Acta 288, 114-126.

18. Karlish, S.J.D., Leib, W.R. Ram, D. & Stein, W.D. (1972) Biochim Biophys Acta 225, 126-132.

19. Baker, G.F. & Naftalin, R.J. (1979) Biochem. Biophys Acrta 550, 474-484.

20. Walmsley, A.R. & Lowe, A.G. (1987) Biochim. Biophys. Acta, 901, 229-238.

21. Levine, M., Levine, S.& Jones, M.N. (1971) Biochim. Biophys. Acta 225, 291-300.

22. Shiio, H. (1958) J. Am. Chem. Soc. 80, 70-73.

23. Kahlenberg, A., Urman, B & Dolansky, D. (1971) Biochemistry 10, 3154-31632.

24. Kahlenberg, A. & Dolansky, D. (1972) Can J. Biochem. 50, 638-643.

25. Barnett, J.E.G., Holman, G.D. & Munday, K.A. (1973) Biochem. J. 131, 211-221.

26. Barnett, J.E.G., Holman, G.D., Chalkey, R.A. & Munday, K.A. (1975) Biochem. J. 145, 417-429.

27. Glynn, I.M. & Richards, D.E. (1982) Physiol. 330, 17-43.

28. Tefft, R.E., Carruthers, A. & Melchior, D.C. (1986) Biochemistry 25, 3709-3718.

29. Wheeler, T.J. & Hinkle, P.C. (1981) J. Biol. Chem. 256, 8907-8914.

30. Fersht, A. (1977) "Enzyme Structure and Mechanism." W.H. Freeman & Co.

31. Sen. A.K. & Post, R.L. (1964) J. Biol. Chem. 239, 345-352.

32. Albers, R.W. (1967) Am. Rev. Biochem. 36, 727-756.

33. Post, R.L., Kume, T., Tobin, T. Orcutt, B. & Sen A.K. (1969) J. Gen. Physiol. 54, 306-3269.

34. Shull, G.E., Schwarz, A. & Lingrel, J.B. (1985) Nature 316, 691-695.

35. Noguchi, S., Noda, M., Takahaski, H., Ohta, T., Hirose, T., Inayama, S., & Hayashida, H. (1985) Nature, 316, 733-736.

36. Shull, G.E., Lane, L.K. & Lingrel, J,B. (1986) Nature, 321, 429-431.

37. Karlish, S.J.D., Jorgensen, P.L. & Gitter, C. (1977) Nature 269, 715-717.

343

38. Karlish, S.J.D. & Yates, D.W. (1978) Biochim. Biophys. Acta. 527, 115-130.

39. Gresalfi, T.J. & Wallace, B.A. (1985) J. Biol. Chem. 259, 2622-2628.

40. Jorgensen, P.L. (1975) Biochim. biophys. Acta, 401, 399-415.

41. Karlish, S.J.D. (1980) J. Bioenerget, Biomembr. 12, 111-136.

42. Moczydlowski, E.G. & Fortes, P.A.G. (1981) J. Biol. Chem. 256, 2346-2356.

43. Jorgensen, P.L. (1985) "The Sodium Pump" (ed. Glynn, I. & Ellroy, C.) pp. 83-96.

44. Fischer, T.H. (1983) Biochem. J. 211, 771-774.

44. Beauge, L.A. & Glynn, I.M. (1980) J. Physiol. 299, 367-383.

46. Albers, R.W. Fahn, S. & Koval, G.J. (1963) Proc. Natl. Acad. Sci US 50, 474-481.

47. Post, R.L. & Kume, S. (1973) J. Biol. Chem. 248, 6993-7000.

48. Albers, R.W., Koval, G.J. & Siegel, G.J. (1968) Mol. Pharmacol. 4, 324-326.

49. De Meis, L., Martins, O.B. & Aloes, E.W. (1980) Biochemistry 19, 4252-4261.

50. De Meis, L. (1981) "The Sarcoplasmic Reticulum," New York, Wiley.

51. De Meis, L., Otero, A.S. Martins, O.B., Alves, E., Inesi, G. & Nakamoto, R. (1982) J. Biol. Chem. 257, 4993-4998.

52. Dahms, A.S. & Boyer, P.D. (1973) J. Biol. Chem. 248, 3155-3162.

53. Lowe, A.G. & Smart, J. (1977) Biochim. Biophys. Acta 481, 695-705.

54. Simons, T.J.B. (1974) J. Physiol. 237, 123-155.

55. Kaplan, J.H. & Kenney, L.J. (1982) Ann. New York Acad. Sci. 402, 292-295.

56. Kenney, L.J. & Kaplan, J.H. (1985) "The Soidum Pump" (ed. Glynn, I. & Ellroy, C.) pp. 535-539.

57. Karlish, S.J.D. & Stein, W.D. (1982) J. Physiol. 328,295-316.

58. Kaplan, J.H., Kenney, L.J. & Webb, M.R. (1985) "The Sodium Pump" (ed. Glynn, I. & Ellroy, C.) pp. 415-421.

59. Garrahan, P.J. & Glynn, I.M. (1962) J. Physiol. 192, 189-216.

60. De Weer, P. (1970) J. Gen, Physiol. 56, 583-620.

61. Hobbs, A.S. Alberts, R.W. & Froehlich, J.P. (1985) "The Sodium Pump" (ed. Glynn, I. & Ellroy, C.) pp. 355-361.

62. Fahn, S., Hurley, M.R., Koval, G.J., Goebel, R. & Berman, M. (1976) J. Biol. Chem. 241, 1890-1895.

63. Hara, Y. & Nakao, M. (1981) J. Biochem. 90, 923-931.

64. Norby, J.G., Klodos, I. & Christiansen, N.O. (1983) J. Gen. Physiol. 82, 725-759.

344

65. Yoda, A. & Yoda, S. (1982) Mol. Pharmacol. <u>22</u>, 693-699.

66. Yoda, A. & Yoda, S. (1986) J. Biol. Chem. <u>261</u>, 1147-1152.

67. Klodos, I. & Norby, J.G. (1987) Biochim. Biophys. Acta <u>897</u>, 302-314.

68. Taniguchi, K. Suzuli, K., Sasaki, T., Shimokobe, H. & Iida, S. (1986) J. Biol. Chem. <u>261</u>, 3272-3281.

69. Garrahan, P.J. & Glynn, I.M. (1967) J. Physiol. <u>192</u>, 175-188.

70. Glynn, I.M. & Karlish, S.J.D. (1976) J. Physiol. <u>256</u>, 465-496.

71. Froehlich, J.P., Albers, R.W., Koval, G.J., Goebel, R. & Berman, M. (1976) J. Biol. Chem. <u>251</u>, 2186-2188.

72. Jorgensen, P.L., Skriver, E., Herbert, H. & Maunsbach, A.B. (1982) Ann. N.Y. Acad. Sci. <u>402</u>, 2207-224.

73. Glynn, I.M., Hara, Y. & Richards, D.E. (1984) J. Physiol. <u>351</u>, 531-547.

74. Skou, J.C. & Esmann, M. (1983) Biochim Biophys Acta, <u>727</u>, 101-107.

75. Wilson, D.B. (1978) Ann. Rev. Biochem. <u>47</u>, 933-965.

76. Glynn, R.B. (1979) Ann. Rev. Physiol. <u>42</u>, 249-259.

77. LaNoue, K.F. & Schoolwerth, A.C. (1979). Ann. Rev. Biochem. <u>48</u>, 871-922.

78. Huxley, H.E., Simmons, R.M., Faruqi, A.R., Kress, M. Bordas, J. & Koch, M.H.J. (1981) Proc. Natl. Acad. Sci. USA, <u>78</u>, 2297-2301.

79. Lowey, S., Slayter, H.S., Weeds, A.G. & Baker, H. (1969) J. Mol. Biol. <u>42</u>, 1-29.

81. Starr, R. & Offer, G. (1973) J. Mol. <u>81</u>, 17-31.

82. Lymn, R.W. & Taylor, E.W. (1970) Biochemistry, <u>9</u>, 2975-2983.

83. Lymn, R.W. & Taylor, E.W. (1971) Biochemistry, <u>10</u>, 4617-4624.

84. Werber, M.M., Szent-Gyorgyi, A.G. & Fasman, G.D. (1972), Biochemsitry <u>11</u>, 2872-2883.

85. Bagshaw, C.R. & Trentham, D.R. (1974) Biochem. J. <u>141</u>, 331-349.

86. Bagshaw, C.R., Eccleston, J.F., Eckestein, F., Goody, R.S., Gutfreund, H. & Trentham, D.R. (1974) Biochem. J. <u>141</u>, 351-364.

87. Chock, S.P. (1981) J. Biol. Chem. <u>256</u>, 10954-10966.

88. Bagshaw, C.R. & Trentham, D.R. (1974) Biochem. J. <u>141</u>, 131-142.

89. Bagshaw, C.R, Trentham, D.R., Wolcott, R.G. & Boyer, P.D., (1975) Proc. Natl. Acad. Sci. USA <u>72</u>, 2592-2596.

90. Bagshaw, C.R., Eccleston, J.F., Eckstein, F., Goody, R.S., Gutfreund, H. & Trentham, D.R. (1974) Biochem. J. <u>141</u> 351-364.

91. Bagshaw, C.R.& Tretham, D.R. (1973) Biochem. J. <u>133</u>, 323-328.

92. Trentham, D.R., Eccleston, J.F. & Bagshaw, C.R. (1976) Quart J. Biophys, <u>9</u>, 217-281.

93. Mannherz, H.G., Schenck, H., Goody, R.S. (1974) Eur. J. Biochem. <u>48</u>, 287-295.

94. Taylor, E.W. (1977). Biochemistry <u>16</u>, 732-

95.   Morita, J. (1967) J. Biol. Chem. 242, 4501-4506.

96.   Morita, J. (1977) J. Biochem. 81, 313-320.

97.   Morita, J. & Ishigami, F. (1977) Biochem. 81, 305-312.

98.   Shriver, J. (1984) Trends Biochem. Sci. 322-328.

99.   Shriver, J.W. & Sykes, B.D. (1982) Biochemistry, 21, 3022-3028.

100   Shriver, J.W. & Sykes, B.D. (1982) Biochemistry, 20, 2004-2012.

101   Shriver, J.W. & Sykes, B.D. (1981) Biochemistry, 20, 6357-6362.

102   Mornet, D., Pantel, P., Audemard, E. & Kassar, R. (1979) Eur. J. Biochem, 100, 421-431.

103   Yamamoto, K. & Sekine, T. (1979) J. Biochem. 86, 1855-1862.

104   Highsmith, S., Akasaka, K., Konrad, M., Goody, R., Holmes, K., Wade-Jardetzky, N. & Jardetzky, O. (1979) Biochemistry 18, 4238-4244.

105   Highsmith, S. (1977) Biochim. Biophys. Acta 180, 404-408.

106   Highsmith, S. (1976) J. Biol. Chem. 251, 6170-6172.

107   Marston, S., Tregear, R.T., Rodger, C.D. & Clarke, M. (1979) J. Mol. Biol. 128, 111-126.

108   Mitchell, P. (1961) Nature, 191, 144-148.

109   Ferguson, S.J. & Sorgato, M.C. (1982) Ann. Rev. Biochem. 51, 185-217.

110   Nicholls, D.G. (1974) Eur. J. Biochem. 50, 305-315.

111   Kagawa, Y. & Racker, E. (1966) J. Biol-Chem. 241, 2475-2482.

112   Senior, A.E. (1983) Biochim. Biophys. Acta 726, 81-95.

113   Sebald, W. & Hoppe, (1979) Curr. Topics Bioenerget, 12, 1-64.

114   Penefsky, H.S. (1985) J. Biol. Chem. 259, 13735-13741.

115   Webb, M.R., Grubmeyer, C., Penefsky, H.S. & Trentham, D.R. (1980) J. Biol. Chem. 255, 11637-11639.

116   Penefsky, H.S. (1986) J. Biol. Chem. 260, 13728-13734.

117   Boyer, P.D., Cross, R.L. & Momsen, W. (1973) Proc. Natl. Acad. Sci. USA 70, 2837-2839.

118   Mitchell, R.A., Hill,R.D. & Boyer, P.D. (1967) J. Biol. Chem. 242, 1793-1801.

119   Hinkle, P.C., Penefsky, H.S. & Racker, E. (1967) J. Biol. Chem. 242, 1788-1792.

120   Di Pietro, A., Godinot, C. & Gautheron, D.C. (1982) Biochemsitry, 22, 785-792.

121   Chang, T.M. & Guillroy, R.J. (1981) J. Biol. Chem. 256, 8318-8323.

122   Ohta, S., Tsuboi, M., Ohshima, T. Yoshida, M., & Kagawa, Y. (1980) J. Biochem. 1609-1617.

123   Hatefi, Y., Yagi, T., Phelps, D.C., Wong, S-Y., Vik, S.B. & Galante, Y.M. (1982) Proc. Natl. Acad. Sci. USA 79, 1756-1760.

124   Hamamoto, T., Ohno, K. & Kagawa, Y.J. Biochem. 91, 1759-1766.

125 Teissie, J., Knox, B.E., Tsong, T.Y., & Wehrle, J. (1981) Proc. Natl. Acad. Sci. USA 78, 7473-7477.

126 Grubmeyer, C., Cross, R.L. & Penefsky, H.S. (1982) J. Biol. Chem. 257, 12092-12100.

CHAPTER 8

THERMODYNAMICS AND METABOLISM

B. CRABTREE & B.A. NICHOLSON

8.1   INTRODUCTION

In this chapter we intend to explain how thermodynamic functions such as enthalpy, entropy and Gibbs free energy may be applied to metabolic systems. We have decided to take a somewhat broad approach, emphasizing basic principles, since we feel that there are already sufficient texts dealing with the finer points of the subject, especially the many detailed algebraic manipulations that so often represent a student's only knowledge of thermodynamics. We have also adopted a kinetic framework for the discussions because this gives a much better appreciation of the dynamics of these systems. However, we shall not discuss the control of metabolic fluxes, since this subject is best treated by a purely kinetic approach (e.g. Kacser & Burns, 1973; Crabtree & Newsholme, 1985). Nor will we discuss the application of so-called 'irreversible thermodynamics' (e.g. Stucki, 1983) since, in our opinion, this somewhat abstract topic has so far proved to be of somewhat limited value for understanding metabolism, compared with the application of reaction kinetics based on the law of mass action. Finally, for reasons of space we are unable to include details of how changes in thermodynamic functions are measured experimentally or calculated from fundamental values: these may be found in most standard texts, for example Glasstone & Lewis (1963), Klotz (1967).

8.2   METABOLIC SYSTEMS AND STATES

A metabolic system can be represented by the conversion of a substance X into Y, i.e.

X ⟶ Y.

X and Y may be different chemical substances, in which case the system represents a chemical transformation of X, usually catalysed by an enzyme. Alternatively, X and Y may be the same substance, in which case the system represents a translocation of X from one site (or compartment) to another: this may involve a protein carrier-system, for example, the transport of glucose across the cell membrane of muscle or adipose tissue. Moreover, the conversion of X into Y may represent several consecutive chemical or enzyme-catalysed reactions; for example, it could represent the conversion of glucose ($\equiv$X) into lactate ($\equiv$Y) by the sequence of reactions known as glycolysis.

The state of a metabolic system such as X ──→Y is determined by factors such as temperature, pressure and the concentrations of X and Y. However, for most metabolic systems the temperature and pressure may be assumed to be constant, so that the only significant factor influencing the state of such a system is the concentrations of the intermediates.

There is no reaction supplying X to the above system, nor is there any reaction that removes Y, so that no material is exchanged with the surroundings. This type of system is therefore known as a closed system and would proceed until a state of equilibrium is attained. At equilibrium there is still a conversion of X into Y (v_f), but this is counterbalanced by an equal and opposite conversion of Y into X (v_r), so that the net flux of X to Y is zero, i.e.

$$X \underset{v_r}{\overset{v_f}{\rightleftharpoons}} Y,$$

where $v_f = v_r$. Thus if a quantity of radioactively-labelled X (small enough to result in no significant disturbance of the equilibrium) was added to the system, there would be an initial flux of radioactive X to Y equal to v_f. This equilibrium state is an example of a kinetically-reversible reaction, i.e. one for which the rate of the reverse process, v_r, is not negligible compared with the net flux, v_f-v_r:equilibrium occurs when the net flux is zero. For a closed system, equilibrium is the only state in which the concentrations of intermediates such as X and Y do not change with time: if a closed system is not at equilibrium [X] and [Y] change continuously until equilibrium is reached.

Most metabolic systems and all metabolic pathways do not represent closed systems. Instead they represent systems in which there is a continuous supply of X and removal of Y,

$$\longrightarrow X \rightleftharpoons Y \longrightarrow$$

This continuous flux of X into Y, and Y out of the system, offsets the tendency of X and Y to equilibrate. Such a system is termed an open system because it continuously exchanges material with its surroundings. Unlike a closed system, for which equilibrium is the only stable state (i.e. when the concentrations of the intermediates do not vary with time), an open system can exist in a stable state which is not at equilibrium: this is termed a steady state and provides a very useful approximation to metabolic systems. Thus, if the rate of supply of X to the above system is constant and if the rate of the reaction removing Y increases as [Y] increases, [Y] reaches a constant value at which the rate of removal of Y is equal to the rate of its formation from X. If the rate of conversion of X into Y increases as [X] increases, [X] also reaches a constant value at which the rate of conversion of X into Y is equal to the rate at which X is supplied to the system. Consequently, as long as there is a constant supply of X to the system, the concentrations of both

X and Y attain constant values and provide a net flux of X to Y that is equal to the rate at which X is supplied from the surroundings: the system is then in a steady state. Since steady-state systems are referred to extensively in this chapter, it is necessary to discuss briefly two other important factors that are required for the production and maintenance of this state (for further details see Crabtree & Newsholme, 1985).

First, the constant rate of supply of X must ultimately result from a reaction whose rate is not significantly affected by changes in the concentration of its substrate. Thus in the system,

$$S \longrightarrow X \rightleftharpoons Y \longrightarrow$$

the first reaction of the sequence S→X, must be insensitive to changes in [S]: if this were not so, the rate of this reaction, and hence the rate of supply of X, would decrease continuously as [S] decreased and a constant flux of S to X would be impossible. The condition of insensitivity to a substrate (i.e. zero order kinetics with respect to S) is easily achieved by enzymic catalysis, since all that is required is for [S] to be maintained at a value sufficient to saturate (i.e. occupy all the binding sites on) the enzyme catalysing the conversion of S into X. Furthermore, if the reaction(s) supplying X is saturated with S, it must be unidirectional or kinetically irreversible, i.e. the rate of the reverse process, X→S, must be insignificant compared with the net flux of S to X. This is because saturation of the enzyme implies that all the enzyme molecules are involved in binding S and subsequently converting it into X: therefore no enzyme molecules are available to catalyse the reverse process. Such reactions have been termed flux-generating, and a metabolic example is the conversion of glycogen into glucose 1-phosphate in tissues such as muscle. In resting muscle the enzyme catalysing this reaction, phosphorylase, is saturated with its substrate, glycogen, and is thus able to generate a steady-state flux of glycogen to glucose 1-phosphate and hence to glycolytic end-products such as lactate. In contracting muscle, this flux serves to regenerate ATP, especially under anaerobic or hypoxic conditions. However, since the net flux of glycogen breakdown is large relative to the intitial concentration of glycogen in contracting muscle, phosphorylase can only generate a steady-state flux for a short period of time: eventually the concentration of glycogen falls below that required for saturation and the steady-state conditions break down. This is not serious for many muscles because, by the time the glycogen concentration has fallen to non-saturating values, other fuels such as glucose and fatty acids have been mobilised and replace glycogen: however, those muscles (e.g. vertebrate white muscle) that utilise glycogen as the major fuel for contraction can only contract for a relatively short time (see Crabtree & Newsholme, 1975). In general, other systems are needed to maintain the concentrations of intermediates such as S in the hypothetical system at a value sufficient to saturate (or almost saturate) the enzyme catalysing the relevant flux-generating step.

Second, to provide a constant removal of the end product(s), such as Y, the terminal reaction of the system must also be kinetically irreversible; otherwise the continuous accumulation of intermediates beyond Y would influence [Y] via the reverse process of this reaction, and a constant [Y] and hence a steady state between X and Y would be impossible. Consequently, a metabolic steady-state system is protected at each 'end' by kinetically-irreversible reactions, the first of which is saturated with its substrate: these reactions effectively buffer the system and thereby enable a steady-state flux to exist despite fluctuations in concentrations of intermediates preceding and beyond the system.

Finally, although the steady state provides a useful approximation to most metabolic systems, there are important examples of non-steady-state systems. For example, the concentrations of some glycolytic intermediates may be made to oscillate under certain conditions (see Hess & Boiteux, 1971). Also, when the magnitude of a steady-state flux is changed by a metabolic regulator, there is a transient period during which the concentrations of the intermediates change in order to re-establish steady-state conditions: thus if the rate of supply of X to the hypothetical system is increased to a new constant value, [X] and [Y] both increase until the rates of X→Y and the terminal reaction both become equal to the new flux (for a discussion of some important non-steady states, see Crabtree & Newsholme, 1985).

8.3   THERMODYNAMIC LAWS AND FUNCTIONS

8.3.1   The first law of thermodynamics: internal energy

In the following system,

$$\rightarrow X \rightleftharpoons Y \rightarrow$$

heat may be produced (or absorbed from the surroundings) and/or work may be done (as in muscle) during the conversion or translocation of X to Y. Let q represent the heat absorbed by the system and let w_t represent the total work done by the system when one mole of X is converted into Y. If heat is produced, q is negative; if work is done on the system, w_t is negative. Then the total increase in the energy of the system per mole of X converted into Y is equal to $q - w_t$. By the first law of thermodynamics (often termed the law of conservation of energy) this change in energy must be accompanied by an equal and opposite change in energy elsewhere in the system, which implies that there is a difference in energy between one mole of X and one mole of Y. This energy is known as internal energy, U, so that by the first law of thermodynamics, the change in internal energy, ΔU, when one mole of X is converted into Y is given by the equation

$\Delta U = q - w_t$.

Since ΔU is the difference in energy between X and Y, its value is not affected by the nature of the pathway between X and Y; a function with this property is termed a <u>state function</u>. In contrast, neither the heat absorbed, q, nor the work done, w_t, are state functions since they depend on the path between X and Y. For example, in muscle the hydrolysis of ATP to ADP by the myofibrillar ATPase is accompanied by the performance of work, whereas no work results if this same reaction is carried out non-enzymically, i.e. by a different path; nevertheless, the change in internal energy of one mole of ATP when converted into ADP is the same in both cases. It must also be stressed that ΔU is not significantly affected by the concentrations of the initial and final intermediates, such as X and Y.

8.3.2 Enthalpy

If the system, X⟶Y, involves a change in volume, e.g. if it contains gaseous reactants, the system has to do work on the surroundings corresponding to the net change in volume. At constant pressure, P, a volume change, ΔV, requires an amount of work equal to $P\Delta V$. If the volume increases, ΔV, and hence $P\Delta V$, is positive, so that work is done <u>by</u> the system: if the volume decreases, ΔV, and hence $P\Delta V$, is negative, so that work is done <u>on</u> the system. Consequently, the total work done by a system, w_t, is the sum of the work due to the net volume change, $P\Delta V$, and the remaining, useful, work, w:

$w_t = w + P\Delta V$.

(The useful work, w, may for example represent work done during muscular contraction or during the generation of electrical currents.)

Consequently,

$\Delta U = q - w_t$
$ = q - w - P\Delta V$

so that

$\underline{\Delta U + P\Delta V = q - w}$

The function, $\Delta U + P\Delta V$, on the left hand side of this equation is known as the change in <u>enthalpy</u>, ΔH, so that

$\Delta H = q - w$.

This shows that ΔH is analogous to ΔU $(= q - w_t)$. However, since ΔH contains and thus allows for the 'non-useful' work, $P\Delta V$, and since metabolic systems operate at constant pressure, ΔH is a more useful function for describing the heat and work produced by these systems. Like ΔU, ΔH is a state function and its value is not significantly affected by changes in the concentrations of the initial and final

352

intermediates.* However, by definition, the value of ΔH is a function of the pressure, P; therefore, for most metabolic systems (with the exception of species that live deep in the ocean) the relevant values for ΔH are those at atmospheric pressure.

Enthalpy represents a conventional energy, i.e. it obeys the first law of thermodynamics; because both the internal energy and the work done during a change in volume obey this law. Consequently, the value of ΔH (or ΔU) for a complex system is equal to the sum of the ΔH (or ΔU) values of the component reactions. Thus, for the coupled system,

$$\longrightarrow X \longrightarrow Y \longrightarrow Z \longrightarrow$$

$$\Delta H_{(X \longrightarrow Z)} \quad = \quad \Delta H_{(X \longrightarrow Y)} \quad + \quad \Delta H_{(Y \longrightarrow Z)},$$

in which the ΔH values are the enthalpy changes per mole for the reactions stated in the subscripts. This relationship is frequently referred to as 'Hess's law of constant heat summation' and was actually discovered before the more general law of conservation of energy: it enables the ΔH value of a reaction to be calculated from standard ΔH values derived from heats of formation and heats of combustion. (Details of the measurement and use of standard enthalpy changes may be found in most thermodynamics or physical chemistry texts, e.g. Glasstone & Lewis, 1963; Klotz, 1967.)

However, it must be stressed that a relationship such as Hess's law would apply to any state function. By definition, the change in a state function is determined only by the initial and final states of the system. Thus, in the above system, the change in a hypothetical state function, SF, for a path involving the direct conversion of X into Z, $\Delta SF_{(X \longrightarrow Z)}$, must be the same as that for a path involving Y as intermediate. Since the total change in SF via the latter path is the sum of the changes for the successive reactions,

$$\Delta SF_{(X \longrightarrow Z)} \quad = \quad \Delta SF_{(X \longrightarrow F)} \quad + \quad \Delta SF_{(Y \longrightarrow Z)}.$$

Although this relationship bears a superficial resemblance to one resulting from the first law of thermodynamics (see above), a state function need not obey this law: thus, although internal energy and enthalpy both obey this law, two other important state functions, entropy and Gibbs free energy, do not (sections 3.5 and 3.6). Consequently, the summation of component state functions does not necessarily reflect the operation of the first law of thermodynamics.

---

*In principle ΔH for a reaction is independent of concentrations of reactants and products, in practice if there are large deviations from ideality ΔH at finite concentrations can deviate from ΔH at infinite dilution.

Finally, it must be emphasised that, unless otherwise stated, both ΔU and ΔH refer to changes when one mole of substrate is converted into product, or when one mole of a substance is transferred from one compartment to another. Therefore, if n moles of a substrate are converted into a product, the total absolute changes in internal energy and enthalpy are $n\Delta U$ and $n\Delta H$, respectively. For a closed system (section 2) the total change of enthalpy, $n\Delta H$, depends on the initial displacement of the reaction from equilibrium. Since such a system can only proceed to equilibrium, the greater the intitial displacement from equilibrium, the greater is the amount, n, of substrate converted into product as equilibrium is approached and therefore the greater is the total change of enthalpy, $n\Delta H$. However, this relationship between total change of enthalpy and initial displacement from equilibrium only applies to a closed system: it cannot be applied to reactions in metabolic pathways, i.e. steady-state systems. A reaction in a steady-state system is never allowed to come to equilibrium (section 2) and therefore it is meaningless to refer to a total change of enthalpy, except during a given time interval: moreover, this total change is _independent_ of displacement from equilibrium. For example, if the steady-state flux through the system

$$\longrightarrow X \rightleftharpoons Y \longrightarrow$$

is J mole/sec, the rate of change of enthalpy during the conversion of X into Y is equal to $J\Delta H$ units/sec. Since neither J nor ΔH are affected by the displacement of reaction $X \longrightarrow Y$ from equilibrium, the rate of change of enthalpy and hence the total change during a given time interval, t, ($= J.t.\Delta H$) is the same whether this reaction is close to, or far-displaced from, equilibrium.

8.3.3 Metabolic heat production

If no work is done by a system, i.e. if w is zero, ΔH is equal to q: consequently, under these conditions the heat absorbed or produced by a system, at constant pressure, may be equated with ΔH for that system. If ΔH is positive, so that q is positive, heat is absorbed and the reaction or system is said to be _endothermic_. If ΔH is negative, so that q is negative, heat is produced and the reaction or system is said to be _exothermic_: this is the case for most metabolic reactions. Values of ΔH (at atmospheric pressure) for some metabolic reactions are given in Table 1.

TABLE 1

Values of ΔH for some metabolic reactions

Reaction	ΔH(kJ/mole of organic substrate)
palmitate + $23O_2 \longrightarrow 16CO_2 + 16H_2O$	-10040
glucose + $6O_2 \longrightarrow 6CO_2 + 6H_2O$	$- 2820$
2glutamate + $9O_2 \longrightarrow 9CO_2 + 7H_2O +$ urea	$- 1934$
2alanine + $6O_2 \longrightarrow 5CO_2 + 5H_2O +$ urea	$- 1307$
pyruvate + NADH $\longrightarrow$ lactate + NAD	$- 46$ (at pH 7.3)
ATP $\longrightarrow$ ADP + P_i	$- 20$

Calculated from data in Huennekens & Whiteley (1960), Blaxter (1962) and Dawes (1972).

Values of ΔH for the complete oxidation of carbohydrates and fats and for the oxidation of amino acids as far as urea are of special importance since they are used to calculate the <u>calorific or energy value</u> of a food. This is the amount of heat <u>released</u>, and is therefore equal to $-\Delta H$, when one gramme of the food is oxidized to its metabolic end products. (It should be noted that, in this case, ΔH is expressed on a gramme instead of a mole basis). By the first law of thermodynamics, the energy value of a food can be calculated by adding together the heat produced by its component metabolizable fuels, carbohydrate, fat and protein. Thus, if one gramme of food contains c, f and p grammes of metabolizable carbohydrate, fat and protein, respectively, its energy value, $-\Delta H$, is given by the equation,

$$-\Delta H = c(-\Delta H_{carbohydrate \rightarrow CO2}) + f(-\Delta H_{fat \rightarrow CO2}) + p(-\Delta H_{protein \rightarrow urea}),$$

in which the values of $-\Delta H_{carbohydrate \rightarrow CO2}$, $-\Delta H_{fat \rightarrow CO2}$ and $-\Delta H_{protein \rightarrow urea}$ are approximately 17.6, 40 and 18 kJ/g, respectively. For example, if one gramme of a food contains 0.27 g protein, 0.28 g fat and 0.38 g carbohydrate, its energy value is equal to (0.38 x 17.6 + 0.28 x 40 + 0.27 x 18) i.e. approximately 22.7 kJ/g.

Since ΔH obeys the first law of thermodynamics and is not significantly influenced by the concentrations of the intermediates, the change in enthalpy of a metabolizable food during its oxidation in an organism is the same as that in a calorimeter. Thus, if an animal requires M units of energy (i.e. dissipates M units of heat plus work) in one day and if a metabolizable food releases $-\Delta H$ units of energy per gramme during its oxidation to the appropriate metabolic end products (e.g. CO_2, urea), the energy requirements of the animal can be met by supplying $M/(-\Delta H)$ grammes of that food per day. Such calculations are the basis of the familiar 'calorie-controlled diets' (for further details see Blaxter, 1962; McDonald et al, 1973; Newsholme & Leech, 1983).

Since $-\Delta H$ represents the heat produced by a metabolic system or reaction in vivo (when no work is done) a knowledge of its value can help to assess the importance of mechanisms proposed for heat production in animal tissues. For example, it has been shown that bumblebees, which do not normally maintain a constant body temperature, can nevertheless maintain the temperature of their flight muscles above 30°C whilst collecting food from flowers (Heinrich, 1972): this enables them to take off immediately without having to spend time in raising their muscle temperature to a value commensurate with flight. Newsholme et al (1972) proposed that one mechanism for producing heat under these conditions was the hydrolysis of ATP to ADP as a result of a substrate cycle between the glycolytic intermediates, fructose 6-phosphate and fructose 1,6-bisphosphate: this cycle is produced by the simultaneous activities of the enzymes phosphofructokinase and fructose bisphosphate (see also Newsholme & Crabtree, 1973):

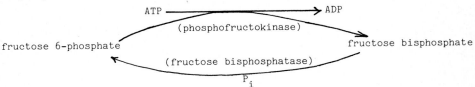

The net effect of this cycle is to hydrolyse one molecule of ATP to ADP and P_i per revolution. Since ΔH for ATP hydrolysis to ADP is approximately -20 kJ/mole (Table 1) and since no work is done, each mole of ATP hydrolysed by this substrate cycle releases 20 kJ of heat to the surroundings. However, ADP and P_i do not accumulate, but are converted into ATP as a result of the metabolism of fuels such as glucose and fatty acids. Consequently, the total heat released to the surroundings as a result of the cycle is the sum of the 20 kJ due to ATP hydrolysis by the cycle and the heat released (per mole of ATP regenerated) by the reaction,

fuel + n.ADP + $n.P_i \longrightarrow$ end products + n.ATP.

In practice it is simpler to evaluate this sum directly from the value of ΔH for the reaction, fuel $\longrightarrow$ end products, and the number of moles of ATP regenerated per mole of fuel consumed, n. Thus bumblebee flight muscles utilise only carbohydrate as a fuel and this (in the form of glucose) is oxidised completely with the formation of 38 moles of ATP per mole of glucose oxidised (see Crabtree & Newsholme, 1975). Therefore, by the first law of thermodynamics, the heat released during the oxidation of one mole of glucose to CO_2 (2820 kJ: Table 1) must be the sum of the heat released by the reactions:

$$glucose + 6O_2 + 38ADP + 38P_i \longrightarrow 6CO_2 + 6H_2O + 38ATP$$

and

$$38ATP \longrightarrow 38ADP + 38P_i.$$

Consequently, when glucose is the sole fuel and is oxidised completely, 2820 kJ of heat are released during the synthesis and subsequent hydrolysis of 38 moles of ATP: thus approximately 74 kJ are released when one mole of ATP is synthesised and subsequently hydrolysed. (This is only correct when the hydrolysis of ATP does not result in mechanical work: if work is done, the heat released is less than 74 kJ per mole of ATP). Since 74 kJ of heat are released per mole of ATP hydrolysed by the substrate cycle and subsequently resynthesised, the contribution of this cycle to heat production can be calculated from the maximum rate of cycling. Assuming that the maximum rate of cycling is equal to the maximum catalytic activity of fructose bisphosphatase, which in bumblebee flight muscles is approximately 10^{-4} moles/min per g fresh wt at 35°C (see Newsholme et al, 1972), the maximum rate of ATP hydrolysis by the cycle is 10^{-4} moles/min per g; i.e. 7.6 J/min per g. However, this value is only approximately 10% of that required to maintain the muscle temperature at a value suitable for flight during the collection of food (approximately 92 J/min per g: Heinrich, 1972). Therefore additional mechanisms are required for heat production under these conditions (see Newsholme & Crabtree, 1976).

One possible mechanism is the uncoupling of the process of oxidative phosphorylation, so that reducing equivalents pass from respiratory substrates such as NADH and succinate to O_2 without any formation of ATP (for details of oxidative phosphorylation and its uncoupling see Nicholls, 1982; Stryer, 1981). If oxidative phosphorylation is completely uncoupled, ATP can only be synthesised at the glycolytic reactions catalysed by phosphoglycerate kinase and pyruvate kinase and at the tricarboxylic-acid-cycle reaction catalysed by succinyl-CoA synthetase, and the yield of ATP is only 4 moles per mole of glucose oxidised. Consequently, the heat released per mole of ATP hydrolysed and resynthesised is 2820/4, i.e approximately 700 kJ, compared with 74 kJ when oxidative phosphorylation is fully coupled. Therefore, when oxidative phosphorylation is uncoupled, the heat resulting from the substrate cycle can be as much as 72 J/min per g, which is quite close to the rate required in vivo. Although far from being conclusive, this suggests that a simultaneous uncoupling of oxidative phosphorylation and increased substrate cycling between fructose 6-phosphate and fructose bisphosphate may be sufficient to maintain the muscle temperature above 30°C without the need for any other mechanism. Indeed, in mammalian brown adipose tissue uncoupling of oxidative phosphorylation is now established as a principal mechanism for producing heat (see Nicholls, 1982; Cannon & Nedergaard, 1985).

8.3.4 The second law of thermodynamics: entropy

As shown in the previous section the function required for calculating metabolic heat production at constant pressure is the change in enthalpy, ΔH. However, neither ΔH nor ΔU can be used to predict from which direction an isolated reaction will approach equilibrium. For example the value of ΔH or ΔU for the reaction, $X \rightleftharpoons Y$, gives no indication whether, for given initial concentrations of X and Y, the reaction approaches equilibrium from the left (i.e. in the direction ($X \rightarrow Y$), from the right (i.e. in the direction $Y \longrightarrow X$) or is actually at equilibrium. At one time it was thought that a reaction would only proceed in a direction for which ΔH was negative, i.e. in a direction that resulted in the production of heat. This was soon found to be incorrect by the discovery of spontaneous reactions that were accompanied by the absorption of heat: for example, approximately 32 kJ of heat are absorbed when one mole of KNO_3 dissolves spontaneously in water. In an open system, and hence in metabolic systems, the situation is somewhat more complex because the direction in which a reaction proceeds is determined by the net flux of material through the pathway, i.e. by the activity of the flux-generating reaction (section 2). Nevertheless, the 'spontaneity' of a metabolic reaction if it were suddenly removed from the pathway is important since it is a factor determining the steady-state concentrations of its substrates (section 5.2).

The problem of deciding in which direction a reaction will approach equilibrium in an isolated system is solved by the application of the second law of thermodynamics, which is most rigorously stated in terms of the state function, entropy, S. According to the second law, the criterion for spontaneity is that the entropy change of the system plus the entropy change of the surroundings must increase, i.e.

$$\Delta S_{system} + \Delta S_{surroundings} > 0.$$

8.3.5 Classical definition of entropy: thermodynamic reversibility

The above condition for spontaneity is that ΔS for the system plus the surroundings must increase: ΔS for the system alone may actually decrease, provided that this is offset by a greater increase in ΔS for the surroundings (see p 104 of Glasstone & Lewis, 1963). However, if the temperature is constant (a condition that may be applied to most metabolic systems), the need to evaluate ΔS for both system and surroundings may be avoided by using a function for the system alone: this function is the Gibbs free energy (or Gibbs function), and its derivation requires the classical definition of entropy and the concept of thermodynamic reversibility.

A system is said to be thermodynamically reversible if it proceeds infinitely slowly, so that an equilibrium between the system and the surroundings is never disturbed. The only metabolic reactions or systems that satisfy this condition are those that are actually <u>at</u> equilibrium: a reaction at equilibrium transmits no <u>net</u> flux in either direction, so that it may be considered to proceed 'infinitely slowly'. Consequently, any reaction or system that is not at equilibrium, i.e. one that transmits a net flux however small, is <u>thermodynamically irreversible</u>. It is very important not to confuse thermodynamic reversibility with kinetic reversibility. Only a reaction actually at equilibrium is both kinetically and thermodynamically reversible: a kinetically reversible reaction in a metabolic pathway, e.g.

$$\longrightarrow X \rightleftharpoons Y \longrightarrow \quad ,$$

is thermodynamically irreversible.

The change in entropy of a system was originally defined as the heat absorbed from the surroundings divided by the temperature, <u>when the system proceeds in a thermodynamically-reversible manner</u>; thus,

$$\Delta S = q_{rev}/T.$$

(Although entropy is more difficult to visualise when defined in this way than when defined in terms of "disorder", the two definitions are actually equivalent: see Spanner, 1964). <u>At equilibrium</u> a reaction is thermodynamically reversible, so that in this case q_{rev} is equal to ΔH and,

$$\Delta S_{eq} = \Delta H/T, \text{ or}$$

$$\Delta H - T\Delta S_{eq} = 0.$$

For a reaction or system that is not at equilibrium, for example reaction $X \longrightarrow Y$ in the system

$$\longrightarrow X \rightleftharpoons Y \longrightarrow$$

q_{rev}, and hence $T\Delta S$, is the heat that would have been absorbed if the reaction had been thermodynamically reversible <u>at the given (i.e. non-equilibrium) concentrations of X and Y</u>; in other words, if the reaction were at equilibrium at these concentrations of X and Y. This is physically impossible, so that the thermodynamically-reversible path between X and Y under these conditions is a purely imaginary reaction. Consequently, for a reaction not at equilibrium, q_{rev}, and hence $T\Delta S$, represents an <u>imaginary</u> heat and cannot be equated with the actual heat absorbed, ΔH. Nevertheless, the thermodynamically-reversible path between X and Y (even though it is imaginary) can be used to deduce some important properties of entropy. Thus, since the value of the 'imaginary equilibrium constant', [Y]/[X], and hence the nature of the thermodynamically-reversible path between X and Y, is determined uniquely by [X] and [Y], the value of q_{rev} and hence $T\Delta S$ is determined uniquely by these concentrations. Consequently, ΔS is influenced only by the initial and final concentrations, i.e. state of the system, and is thus a state function. However, entropy differs from the state functions, internal energy and enthalpy, in two important respects. First, ΔS depends directly on the concentrations of the substrates and products of a reaction. In the above system a change in the ratio, [Y]/[X], changes the nature of the thermodynamically-reversible path between X and Y; and thus changes the value of q_{rev} and hence ΔS. As stated earlier (section 3.1 and 3.2), neither ΔU nor ΔH are significantly affected by such changes in concentration. Second, and also in contrast with internal energy and enthalpy, the function $T\Delta S$ does not obey the first law of thermodynamics, in spite of having the same units as heat and energy. This is because, for a thermodynamically-irreversible process, q_{rev}, and thus $T\Delta S$, represents an imaginary heat (see above): thus $T\Delta S$ is outside the scope of the first law, which applies only to actual (i.e. real) heat and energy. The non-conformity of $T\Delta S$ to the first law may easily be overlooked, with erroneous or potentially misleading results (see section 5).

8.3.6 Gibbs free energy and the direction of spontaneous processes

The quantitative definition of ΔS given in the previous section enables the derivation of a very useful function, the Gibbs free energy. The condition for spontaneity given in section 3.4 can be multiplied throughout by the temperature, T, to give the equivalent inequality,

$$T\Delta S_{system} + T\Delta S_{surroundings} > 0.$$

Since the heat absorbed by the system is equal to ΔH, the surroundings could be restored to their original state by supplying them with an amount of heat equal to ΔH by a thermodynamically-reversible process. By definition, the change of entropy resulting from this hypothetical (imaginary) process would be $\Delta H/T$ units and, since entropy is a state function, this change of entropy must be equal and opposite to that which occurred during the operation of the system: therefore,

$$\Delta S_{surroundings} = -\Delta H/T.$$

Consequently the condition for spontaneity becomes,

$$T\Delta S_{system} -\Delta H > 0,$$

or,

$$\Delta H - T\Delta S < 0.$$

(The subscript may be dropped since henceforth ΔS refers only to the system). It has already been shown that, at equilibrium, $\Delta H - T\Delta S = 0$ (section 3.5), so that the condition for either spontaneity or equilibrium is given by the inequality.

$\Delta H - T\Delta S \leqslant 0.$

The function, (H - TS), is known as the Gibbs free energy (or Gibbs function), G, so that <u>at constant temperature and pressure</u>, ($\Delta H - T\Delta S$) is equal to ΔG. (As for the other thermodynamic functions, ΔG usually refers to the change per mole.) Consequently, the condition for spontaneity or equilibrium can be written as,

$\Delta G \leqslant 0,$

so that the sign of ΔG can be used to predict the direction of a spontaneous process without having to evaluate ΔS for both system and surroundings. For a reaction to proceed in a given direction ΔG must be negative: if the reaction is at equilibrium ΔG is zero. As will be shown in a later section, the value of ΔG for a chemical reaction is a simple function of its equilibrium constant and the concentrations of its substrates and products.

Since Gibbs free energy is a combination of enthalpy (which derives from the first law) and entropy (which derives from the second law), it expresses both laws of thermodynamics in terms of a single function. Moreover, since enthalpy and entropy are both state functions, Gibbs free energy is also a state function. This means that the value of ΔG for a complex system can be obtained as the sum of the values of ΔG for each component reaction (section 3.2). However, it must be stressed that <u>Gibbs free energy does not obey the first law of thermodynamics</u>, because one of its components, $T\Delta S$, does not obey this law (section 3.5). Consequently, the summation of component ΔG values for a complex system must not be confused with the analogous summation of energy terms resulting from the first law (e.g. Hess's law). Indeed, a major pitfall in many discussions of metabolic thermodynamics has been the assumption (implicit or explicit) that Gibbs free energy is an energy in the conventional sense, leading to erroneous statements such as 'if no work is done ΔG is dissipated as heat' (see section 5.4). Such a conclusion only applies to the enthalpy component of ΔG; the entropy component, $T\Delta S$, is subject to no such conservation.

Finally, since ΔS is directly affected by the concentrations of substrates and products (section 3.5), this will also be the case for ΔG: therefore, unlike ΔU and ΔH, the value of ΔG for a reaction <u>in vivo</u> cannot be evaluated unless the cellular concentrations of its substrates and products are known.

8.3.7 Gibbs free energy and useful work

If a reaction or system does 'useful' work, w, (i.e. work other than that due to changes in volume), the heat absorbed from the surroundings is equal to $\Delta H + w$ (because $\Delta H = q - w$: section 3.2). The condition for spontaneity (sections 3.4 and 3.6) then becomes,

$T\Delta S - (\Delta H + w) > 0,$

or,

$(\Delta H - T\Delta S) + w < 0.$

Therefore, since $(\Delta H - T\Delta S) = \Delta G$

$\Delta G + w < 0,$

so that,

$w < -\Delta G.$

If work is done by the system, w is positive and the inequality, $w < -\Delta G$, can only be satisfied if $\Delta G < 0$: this is the same condition for spontaneity as that when w is zero (section 3.6). Moreover, since ΔG is a state function, so that its value is determined only by the initial and final states of the system, whereas w may vary with the path taken (sections 3.1 and 3.6), the inequality also shows that the amount of useful work done by any system is always less than $-\Delta G$. Thus the magnitude of $-\Delta G$ is a measure of the maximum useful work that can be obtained from the system. However, this equality between maximum useful work and the magnitude of $-\Delta G$ does not mean that ΔG is equivalent to work: as stated in section 3.6 work is a conventional (real) energy, whereas free energy is not. If this seems paradoxical, especially as the units of ΔG are the same as those of conventional energy, it should be remembered that the $T\Delta S$ component of ΔG represents an imaginary heat (section 3.5): thus $T\Delta S$ and hence ΔG does not obey the first law (which applies only to real energy) despite having the same units as conventional energy.

It must also be stressed that work is not done at the expense of ΔG, i.e. the value of ΔG does not change as work is done: since ΔG is a state function, its value is the same (for fixed initial and final states) whether work is done or not. The maximum work would be derived from the change in enthalpy (i.e. from the energy corresponding to ΔH) and, if $-\Delta G$ is greater than $-\Delta H$, from heat absorbed from the surroundings. Since metabolic systems are unlikely to use energy other than chemical (internal) energy for the production of work, the maximum work done by these systems is unlikely to be greater than $-\Delta H$.

Finally, if work is done on the system, w is negative and the inequality, $w < -\Delta G$, can be satisfied by a positive value of ΔG: for example, if ΔG is + 10 units and w is -11 units, the inequality is satisfied and the system proceeds despite a positive value for ΔG. However, it should be noted that w represents work, i.e. conventional energy, and, except for photosynthesis when w represents radiant energy, this means of offsetting a positive ΔG does not occur in metabolism. The mechanisms adopted for offsetting positive values of ΔG in metabolism (discussed in section 5.2) result from the state function nature of ΔG and not from the input of conventional energy.

8.4   THERMODYNAMIC FUNCTIONS AND METABOLIC SYSTEMS

8.4.1 Introduction

It has sometimes been doubted whether thermodynamic functions such as ΔH, ΔS and ΔG can be applied to systems that are not thermodynamically reversible, and thus whether they can be used to describe metabolic systems (see Newsholme & Start, 1973). Fortunately, such doubts are groundless because these functions depend only on the initial and final states of the system and not on the path taken: therefore they apply whether a system is thermodynamically reversible or not. Indeed, at only one step in the development of these functions was the concept of thermodynamic reversibility invoked: and then only to calculate the value of ΔS for the surroundings (section 3.6). None of the functions for the system ever involved the assumption of thermodynamic reversibility.

Such doubts have arisen from a misunderstanding of the approach known as 'irreversible (or non-equilibrium) thermodynamics' (for details of this approach see Katchalsky & Curran, 1965; Spanner, 1964). Irreversible thermodynamics is primarily concerned with expressing reaction rates in terms of thermodynamic rather than kinetic functions (e.g. potentials and affinities) and many of the resulting equations can only be applied to systems or reactions that are at or very close to equilibrium (see also Banks, 1969). However, in this chapter reaction rates are expressed in terms of kinetic functions derived from the law of mass action (i.e. rate constants and concentrations) and the thermodynamic functions, ΔH, ΔS and ΔG, are used only to calculate rates of heat or work production and to examine the effects of coupled reactions and membrane potentials on metabolite concentrations and equilibria. Since the kinetic equations apply whatever the displacement of a reaction from equilibrium and since the overall thermodynamic functions are independent of the reaction mechanism, the results in this chapter apply to systems far-displaced from equilibrium as well as to those at or close to equilibrium.

8.4.2 Gibbs free energy change as a function of concentration

The equation for ΔG in the form, $\Delta G = \Delta H - T\Delta S$, is not very useful for calculating the free energy changes of metabolic reactions. However, it can be shown that, for a general reaction

substrates $\rightleftharpoons$ products,

the value of ΔG is given by the equation,

$$\Delta G = \Delta G^{\circ} + RT\ln \left\{ \frac{\prod [\text{products}]}{\prod [\text{substrates}]} \right\}$$

In this equation (which is derived in most standard texts, for example Glasstone & Lewis, 1963; Spanner, 1964) R is the gas constant (8.31 J mol^{-1} K^{-1}), T is the absolute temperature (which for convenience is assumed to be 300°K throughout this chapter) and ln denotes a logarithm to base e (= 2.303.log$_{10}$). ΔG° is a parameter termed the <u>standard free energy change</u> and is discussed in section 4.3. The symbol, Π, represents a sequential product: for example, if there are three substrates, A, B and C, the term Π[substrates] is equal to [A]x[B]x[C]. (Strictly speaking, thermodynamic activities and not concentrations should be used in the equation for ΔG, but this point will be ignored for the discussions in this chapter.) Therefore, for the reaction, X $\rightleftharpoons$ Y, in the system,

$$\longrightarrow X \rightleftharpoons Y \longrightarrow$$

$$\Delta G \;=\; \Delta G^\circ \;+\; RT\ln\left\{[Y]/[X]\right\}$$

If the system is in a steady state the ratio [Y]/[X], (often termed the mass-action ratio, Γ) is constant, so that ΔG is constant. In any other system (e.g. a closed system approaching equilibrium or an oscillating open system) the value of the mass-action ratio, Γ, and hence ΔG, varies with time. Although the mass-action ratio for reaction X $\rightleftharpoons$ Y is dimensionless, this is not generally the case: for the reaction, A + B $\rightleftharpoons$ C, the mass-action ratio is [C]/[A].[B], whose units are concentration^{-1}. Since ΔG denotes the free energy change per mole, this dimensional problem is usually resolved by expressing concentrations in molar units (moles/litre, i.e. M): however, there are some exceptions to this general rule and these are discussed in the next section.

8.4.3 Standard free energy change and its relation to equilibrium

The parameter, ΔG°, is termed the standard free energy change and is equal to ΔG when the logarithmic term is zero and thus when the value of Π[products]/Π[substrates] (i.e. the mass-action ratio) is unity; this occurs when all substrates and products are present at unit (or standard) concentrations. As stated in the previous section, the unit or standard concentration for most substances is 1 mole/litre (1M) but, for practical reasons, some substances are treated differently. Thus, in aqueous solutions, the standard concentration of water is taken to be its actual molar concentration (approximately 55M) instead of 1M. Similarly, the standard 'concentration' for a gas is taken to be a pressure of one atmosphere. In addition, because metabolic reactions usually take place at a pH close to 7 (i.e. a hydrogen ion concentration of 10^{-7}M), the standard concentration for H$^+$ is often taken to be 10^{-7}M instead of 1M: the values of ΔG and ΔG° relative to a pH of 7 are usually written as ΔG' and ΔG°'.

The units chosen for the standard concentrations, and hence for ΔG°, must also be used for the concentrations in the mass-action ratio, Γ, in order to apply the equation,

$$\Delta G \;=\; \Delta G^\circ + RT\ln \Gamma.$$

Thus, if ΔG° refers to water at 55M, gases at one atmosphere presure and a pH of 7, this must also be the case for all the concentrations in Γ. For example, if Γ involves H^+ and if the concentration of H^+ is $10^{-8}M$ (i.e. the pH is 8), this molar concentration can only be used in Γ if the standard concentration of H^+ (i.e. that for ΔG°) is 1M: if the standard (unit) concentration of H^+ is $10^{-7}M$, the 'concentration' required for Γ is $10^{-8}/10^{-7}$, i.e. 10^{-1}. Values of ΔG° (i.e. relative to pH 7) for some glycolytic reactions are given in Table 2.

At equilibrium the value of ΔG is zero (section 3.6) and the mass-action ratio under these conditions is termed the equilibrium constant K. (Strictly speaking, this should be written as K_c to denote that it refers to concentrations rather than thermodynamic activities, but this distinction is not important for the discussions in this chapter.) Consequently, at equilibrium

$$0 \;=\; \Delta G^\circ + RT\ln K,$$

so that

$$\Delta G^\circ \;=\; -RT\ln K.$$

This important relationship shows that K (the position of equilibrium, i.e. zero net flux across the reaction) and ΔG° (i.e. the free energy change at standard concentrations, which seemingly has little connection with equilibrium) are mutually equivalent. From the definition of ΔG in terms of ΔH and ΔS, ΔG° may be expressed as,

$$\Delta G^\circ \;=\; \Delta H^\circ - T\Delta S^\circ,$$

where the subscript denotes a change relative to a system in which the substrates and products are present at their standard concentrations. It must be stressed that in this equation both ΔG° and ΔS° refer to a system in which substrates and products are at standard concentrations and not to a system at equilibrium: as shown in section 3.5, at equilibrium ΔG is zero and ΔS is equal to $\Delta H/T$.

By determining the values of ΔH° and ΔS° it is possible to examine the effects of different physical forces or environments on the value of ΔG° and hence on the position of equilibrium. An interesting example of this type of analysis is the investigations of Hochachka and co-workers of the reasons for the selection of acetylcholine as a neural transmitter (see Hochachka, 1976). The concentration of acetylcholine in vivo is determined by a system in which acetylcholine is continuously synthesised and subsequently hydrolysed by reactions catalysed by choline acetylase and cholinesterase, respectively:

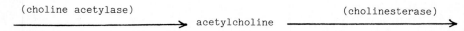

(choline acetylase)   (cholinesterase)

acetylcholine

Hochachka has shown that the binding of acetylcholine to cholinesterase, i.e. the equilibrium,

acetylcholine + cholinesterase $\rightleftharpoons$ acetylcholine-enzyme complex,

results from two physical forces: an electrostatic (coulombic) binding, which gives rise to $-\Delta H°$, and a hydrophobic binding, which gives rise to $T\Delta S°$ in the equation for the standard free energy change of the overall binding. Consequently, these two forces affect the equilibrium constant (or binding constant), i.e. the ratio, [acetylcholine-enzyme complex]/[enzyme][acetylcholine], and hence influence the affinity of cholinesterase for acetylcholine. Changes in temperature or pressure have opposite effects on the two forces: $-\Delta H°$ changes in a direction opposite to $T\Delta S°$, or $\Delta H°$ changes in the opposite direction to $-T\Delta S°$. Consequently, the net effect of temperature or pressure on the sum of $\Delta H°$ and $-T\Delta S°$, and hence on the value of $\Delta G°$, is smaller than that for a binding for which either $\Delta H°$ or $\Delta S°$ is very small or zero. For example, if $-\Delta H°$ increases from 12 to 20 units and $T\Delta S°$ decreases from 10 to 5 units, the value of $\Delta G°$ changes from $(-12-10)$, i.e. -22 units, to $(-20-5)$, i.e. -25 units: thus $\Delta G°$ decreases by only 14%, compared with a decrease of 67% or an increase of 50% if either $\Delta S°$ or $\Delta H°$, respectively, were zero. Consequently, $\Delta G°$, and hence the binding constant, is relatively insensitive to changes of temperature or pressure. This is physiologically important because large changes in the binding constant would be reflected in large changes in the affinity of cholinesterase for acetylcholine, and hence in large fluctuations of [acetylcholine] in vivo. Since the physiological state of the nervous system is a function of the concentration of acetylcholine, such large fluctuations might be dangerous if not lethal. Consequently, the relative insensitivity of this binding constant to changes in temperature or pressure, compared with those of acetylcholine-analogues that bind solely by electrostatic or hydrophobic interactions, may explain why acetylcholine has been selected as a neural transmitter.

A similar analysis of a standard free energy change in terms of its components, $\Delta H°$ and $T\Delta S°$, occurs in the analysis of reaction rates by the approach of Eyring. In this approach the rate of a reaction is assumed to be proportional to the concentration of an 'activated complex' which is at equilibrium with the substrates (or products): for further details of this approach see Glasstone & Lewis (1963). Consequently, the concentration of the activated complex, and hence the rate of the reaction at given concentrations of substrates depends on the equilibrium constant, $K^{\#}$, where,

$$K^{\#} = \frac{[\text{activated complex}]}{\Pi[\text{substrates}]}.$$

The standard free energy change equivalent to $K^{\#}$ is termed the 'standard free energy of activation', $\Delta G^{\#}$. ($\Delta G^{\#}$ must not be confused with 'activation energy', which under most circumstances may be approximated by $\Delta H^{\#}$: see Glasstone & Lewis, 1963; Dawes, 1972). As with a conventional ΔG°, $\Delta G^{\#}$ may be expressed in terms of its components $\Delta H^{\#}$ and $\Delta S^{\#}$ and, by analysing the effects of physical forces or environments on these functions, it is possible to deduce their effects on the value of $K^{\#}$ and hence on the rate of the reaction. Examples of this type of analysis for some metabolic reactions may be found in the paper by Somero & Low (1976).

The examples discussed in this section illustrate the usefulness of the standard free energy change, ΔG°, when analysing the factors that determine the position of an equilibrium, especially when the equilibrium represents the binding of a substance to an enzyme of protein. However, for reactions in metabolic pathways, the value of ΔG° is not sufficient because these reactions are not at equilibrium: the value of ΔG is required and, as shown earlier, this is determined by the actual concentrations of the substrates and products in addition to ΔG°.

8.4.4 Gibbs free energy change and displacement from equilibrium

As shown in the previous section,

$$\Delta G^{\circ} = -RT\ln K$$

so that,

$$\begin{aligned}
\Delta G &= \Delta G^{\circ} + RT\ln \Gamma \\
&= -RT\ln K + RT\ln \Gamma \\
&= -RT(\ln K - \ln \Gamma) \\
&= -RT\ln(K/\Gamma)
\end{aligned}$$

This expression for ΔG is very useful and the dimensionless ratio, K/Γ, is a simple measure of displacement from equilibrium. If a reaction is at equilibrium, K is equal to Γ, K/Γ is unity and ΔG is zero: as the reaction moves further from equilibrium the value of K/Γ increases and ΔG becomes more negative. Moreover, if K is determined under similar physical conditions (e.g. temperature, ionic strength) to Γ, any activity coefficients cancel out in the ratio, K/Γ; therefore concentrations may be used in this ratio even under conditions where thermodynamic activities should be used in the equation for ΔG in terms of ΔG° and Γ.

The ratio K/Γ may also be used to link certain thermodynamic approaches to metabolism (notably those dealing with the control of flux: see Rolleston, 1972; Newsholme & Start, 1973) to kinetics, because the value of K/Γ for a reaction,

$$\text{substrates} \underset{v_r}{\overset{v_f}{\rightleftharpoons}} \text{products}$$

is equal to the ratio of the rates of the forward and reverse processes: thus, $K/\Gamma = v_f/v_r$ (see Rolleston, 1972; Newsholme & Crabtree, 1976). This relationship has been implicit in some of the previous discussions of this chapter. Thus when $-\Delta G$ is small, so that K/Γ is close to unity, v_f must be similar to v_r: consequently, a reaction close to equilibrium is kinetically reversible. Similarly, when $-\Delta G$ is large, v_f must be very much greater than v_r: thus a reaction far displaced from equilibrium is kinetically irreversible. At equilibrium $\Delta G = 0$, so that $K/\Gamma = 1$; hence K represents the concentration ratio (Γ_o) that produces zero net flux across the reaction. This will seldom be equal to the absolute thermodynamic equilibrium constant, because K is influenced by the cellular environment, e.g. ionic strength, pH, metal ion concentrations, activity coefficients. Moreover, if the reaction is a membrane transport system the value of K may also be a function of the membrane potential (section 6).

Values of K/Γ (and hence ΔG) for metabolic reactions may be determined by measuring the concentrations of metabolic intermediates in extracts of freeze-clamped tissue (for details see Newsholme & Start, 1973): these are then used to calculate the value of Γ. Values for K, which should ideally be determined under conditions approximating, as far as possible, to those in vivo, are then used to calculate the ratio, K/Γ. Although this approach is simple in theory, in practice it may lead to large errors if the metabolites are present in more than one compartment (e.g. mitochondria and cytoplasm) and/or if a significant fraction of the tissue content of a metabolite is bound to protein (see legend to Table 2). The method will also give erroneous results if the value of K does not reflect that at the site of the reaction in vivo (see above). Another approach for calculating K/Γ is to measure the rates of the reaction processes, v_f and v_r, directly by administering radioactive substrates (for v_f) or products (for v_r): the rate of each process may then be calculated from the flux of radioactivity across the reaction and the specific radioactivity of the radioactive substrate or product. For example, Randle et al (1970) found that the flux of radioactive CO_2 into isocitrate, due to the reverse process of the reaction catalysed by isocitrate dehydrogenase, was approximately 50% of the rate of the forward process (i.e. the conversion of isocitrate into 2-oxoglutarate and CO_2) during the operation of the tricarboxylic acid cycle in a perfused rat heart. Thus, for this reaction,

$$\text{isocitrate} + \text{NAD}^+(P) + H^+ \underset{v_r}{\overset{v_f}{\rightleftharpoons}} \text{2-oxoglutarate} + CO_2 + \text{NAD(P)H},$$

$$v_r/v_f = 0.5,$$

so that,

$$v_f/v_r \quad = \quad 2$$

$$= \quad K/\Gamma .$$

Consequently, for this reaction,

$$\Delta G \quad = \quad -RT\ln(K/\Gamma)$$

$$= \quad -(8.31)(300)\ln(2)$$

$$= \quad -1.73 \text{ kJ/mole.}$$

Values of K/Γ and ΔG for many of the reactions of glycolysis in some tissues are given in Table 2. (These were calculated from values of K and total tissue contents of intermediates.) They show that most of the reactions of this pathway have relatively small values for K/Γ , and hence v_f/v_r and ΔG; the only exceptions are the reactions catalysed by hexokinase, phosphofructokinase and pyruvate kinase*. Consequently most of the reactions of this pathway are not too far displaced from equilibrium and are therefore kinetically reversible in vivo. The physiological significance of this discontinuous distribution of reversibility or ΔG along the pathway has been discussed by Buecher & Ruessmann (1964). The large negative free energy changes across the kinetically-irreversible reactions (e.g. phosphofructokinase) were likened to dams in a river and, from this analogy, these reactions were proposed to be the sites at which the flux through the pathway is controlled. Indeed, the reaction catalysed by phosphofructokinase is an important control site, at which the rate of glycolysis is regulated by several metabolites, notably ATP, AMP and citrate (see Newsholme, 1970; Newsholme & Start, 1973 for reviews).

These ideas have been extended (Newsholme & Crabtree, 1973, 1976; Crabtree & Newsholmel, 1978, 1985). Thus, in addition to serving as effective control sites for allosteric regulators, kinetically-irreversible reactions are now also known to be required for the generation and maintenance of the steady-state flux (section 2). However, since a kinetically-reversible reaction (i.e. one close to equilibrium) can be very sensitive to changes in the concentrations of its substrates, it is economical to ensure that most of the reactions of a pathway are close to equilibrium. In this way a high sensitivity of these reactions to their substrates is provided automatically (because of their reversibility) without the

---

*The reactions catalysed by aldolase and glyceraldehyde 3-phosphate dehydrogenase plus phosphoglycerate kinase are borderline cases, but are usually considered to be reversible, i.e. not too far displaced from equilibrium; especially as they are known to be able to transmit a net flux in either direction in gluconeogenic tissues such as rat liver, in contrast to the reactions catalysed by phosphorfructokinase, hexokinase and pyruvate kinase which can only be 'reversed' by different reactions (see Rolleston, 1972; Newsholme & Start, 1973).

TABLE 2

Standard free energy changes and free energy changes for some glycolytic reactions in vivo

Reaction catalysed by	$\Delta G^{o'}$ (pH=7)	K/Γ ($=v_f/v_r$)			$\Delta G'$ ($= -RT\ln$ K/Γ)		
		rat brain	rat heart	erythrocytes	rat brain	rat heart	erythrocytes
HK	-21	1.2×10^5	5.9×10^4	6.2×10^6	-29.3	-27.6	-39.2
PGI	+2.1	1.89	1.73	1(approx)	-1.6	-1.4	0(approx)
PFK	-17.3	8.1×10^3	3.5×10^4	2.4×10^4	-22.6	-26.3	-25.2
Ald	+23.1	41	11	7.1	-9.2	-6	-4.9
GAPDH + PGK	+8.06	16	94	6.9	-6.9	-11.4	-4.8
PGM	+4.75	1.5	1.25	1(approx)	-1	-0.55	0(approx)
EN	-3.36	1.02	2.64	2.18	-0.04	-2.44	-1.97
PK	-24.8	3.7×10^3	5×10^2	3.9×10^2	-19.36	-15.58	-14.95

The data in this table are compiled from Table 3.2 of Newsholme & Start (1973): the units of ΔG and ΔG^o are kJ/mole of substrate. Since Γ was determined from the total tissue content of the metabolites, some of these free energy changes may only be approximate. For example, the total contents of ATP and ADP may not reflect the cytoplasmic concentrations (which are required for glycolytic reactions) because of the presence of significant amounts of these nucleotides in the mitochondria. Moreover, some metabolites may exist as complexes with proteins or enzymes and such complexes may represent a significant fraction of the total content, especially when the latter is very small: since Γ refers to free (unbound) metabolites, the use of total tissue contents for Γ may produce significant errors in ΔG (see Rolleston, 1972; Tornheim and Loewenstein, 1976; Newsholme & Leech, 1983). HK:- hexokinase; PGI:- phosphoglucose isomerase; PFK:- phosphofructokinase; Ald:- aldolase; GAPDH + PGK:- glyceraldehyde 3-phosphate dehydrogenase plus phosphoglycerate kinase; PGM:- phosphoglycerate mutase; EN:- enolase; PK:- pyruvate kinase.

need for sophisticated mechanisms for increasing sensitivity, e.g. substrate cycling, sigmoid kinetics and interconvertible forms of the enzymes: these mechanisms are only needed at the relatively small number of irreversible reactions. The overall result of this high sensitivity to substrates is that large changes in the flux through the pathway can be accommodated with much smaller changes in the concentrations of the pathway intermediates, thereby producing a very stable metabolic system.

8.4.5 Directionality of thermodynamic and kinetic functions

Most thermodynamic and kinetic functions are defined (usually implicitly) relative to the direction of a reaction or a net flux. Thus for a reaction at equilibrium,

$X \rightleftharpoons Y$, the equilibrium constant (and thus the standard free energy change) can be defined in two ways: either as the ratio, $[Y]_{eq}/[X]_{eq}$, or as its reciprocal, $[X]_{eq}[Y]_{eq}$. Whichever of these alternatives, designated K_L and K_R, respectively, is adopted will be reflected in the sign of ΔG°. Thus,

$$\Delta G_L^\circ \quad = \quad -RTln\ K_L$$

$$= \quad -RTln\ (1/K_R)$$

$$= \quad +RTln\ K_R$$

$$= \quad -\Delta G_R^\circ.$$

Consequently, if ΔG_L° is negative, ΔG_R° is positive, and vice versa. However, for a reaction that is not at equilibrium there is only one choice for K and hence only one choice for the mass-action ratio, Γ. This is because both K and Γ are defined as $\prod$[products]/$\prod$[substrates] and, when there is a net flux, the terms 'substrates' and 'products' are defined unambiguously in the direction of this flux. For example, in the system, $\longrightarrow X \overset{E}{\rightleftharpoons} Y \longrightarrow$, X is a substrate and Y is a product of reaction E, whereas in the system, $\longrightarrow X \overset{E}{\rightleftharpoons} Y \longrightarrow$, Y is a substrate and X is a product of this reaction. When both K and Γ are defined in the direction of the net flux through a reaction (as they should always be), the value of K/Γ is always greater than unity and thus ΔG (= -RTln(K/Γ) is negative, as required for a flux in this direction. If the net flux had been erroneously considered to be in the opposite direction, the resulting values of 'K' and 'Γ' would be the reciprocals of those relative to the actual flux: thus 'K/Γ' would be less than unity and 'ΔG' would be positive. Therefore, a positive value for ΔG may simply mean that the net flux is in the direction opposite to that assumed! However, such a conclusion is only justified if there is no external work done on the system (section 3.7) or if the reaction is not kinetically coupled to some other reaction (section 5.2).

Other thermodynamic, as well as kinetic, functions also have an implicit direction. For example, the 'forward' and 'reverse' processes of a reaction are so-called because they are in the same and the opposite direction, respectively, to the net flux. Also, for a reaction, $X \rightleftharpoons Y$, any thermodynamic state function (e.g. ΔU, ΔH, ΔS, ΔG) defined in the direction $X \rightarrow Y$ is equal and opposite to that defined in the opposite direction, $Y \rightarrow X$. This is because there must be no net change in a state function when one mole of X is converted into Y and back again into X: thus, considering ΔH,

$$\Delta H_{X \rightarrow Y} \;+\; \Delta H_{Y \rightarrow X} \;=\; 0,$$

so that,

$$\Delta H_{Y \rightarrow X} \;=\; -\Delta H_{X \rightarrow Y},$$

and analogous results apply to ΔU, ΔS and ΔG. The directionality of all the thermodynamic state functions (which, as shown above, is reflected in their signs) must never be overlooked, especially when applying a relationship such as $\Delta G = \Delta H - T\Delta S$; this is only valid if all three state functions are defined relative to the same net flux.

Finally, although the choice of a direction for the equilibrium constant, K, is arbitrary for a reaction at equilibrium, once that direction has been chosen all other directional functions must conform to it. Thus if for the equilibrium,

$$X \underset{v_r}{\overset{v_f}{\rightleftharpoons}} Y,$$ K is chosen to be $[Y]_{eq}/[X]_{eq}$, i.e. in the direction $X \rightarrow Y$, the forward and reverse processes are as shown: if K had been chosen in the opposite direction, i.e. relative to $Y \rightarrow X$, the forward and reverse processes would be interchanged.

8.5 COUPLED REACTIONS AND FACTORS DETERMINING METABOLIC CONCENTRATIONS

8.5.1 Coupling in series

The reactions of a metabolic pathway, such as those of glycolysis, can be regarded as being coupled in series, i.e. in a 'head to tail' manner. In such a system, for example,

$$\rightarrow P \rightleftharpoons Q \rightleftharpoons R \rightarrow$$

the overall change of Gibbs free energy from P to R, ΔG_{net}, is related to the free energy changes across the component reactions by the equation,

$$\Delta G_{net} \;=\; \Delta G_{P \rightarrow Q} \;+\; \Delta G_{Q \rightarrow R},$$

because ΔG is a state function (section 3.6). Since ΔG_{net} is determined only by the initial and final reactants of the system, i.e. by the ratio $[R]/[P]$, for fixed values of [R] and [P] the value of ΔG_{net}, and thus the sum $(\Delta G_{P\rightarrow Q} + \Delta G_{Q\rightarrow R})$, is a constant. Therefore, in general, <u>for fixed concentrations of the initial</u> <u>substrate(s) and end product(s)</u>, the sum of the free energy changes across the component reactions of a metabolic pathway is a constant. Since all these component ΔG values must be negative (section 3.6), the constancy of their sum under these conditions implies that an increased displacement of one of the reactions from equilibrium (i.e. an increased $-\Delta G$ for the reaction) must result in a decreased $-\Delta G$ for the other reactions and must therefore bring them closer to equilibrium. For example, in the above system an increased displacement of reaction $P \rightleftharpoons Q$ from equilibrium increases $-\Delta G_{P\rightarrow Q}$: if [P] and [R] do not change, so that the sum $(\Delta G_{P\rightarrow Q} + \Delta G_{Q\rightarrow R})$, and hence $(-\Delta G_{P\rightarrow Q} - \Delta G_{Q\rightarrow R})$, is a constant, the value of $-\Delta G_{Q\rightarrow R}$ must decrease and thus reaction $Q \rightarrow R$ must become closer to equilibrium. This may explain how the discontinuous distribution of ΔG values for the reactions of glycolysis is produced <i>in vivo</i> (Table 2; section 4.4). Since the standard free energy changes of the reactions catalysed by hexokinase, phosphofructokinase and pyruvate kinase are all large and negative in the direction of the glycolytic flux (Table 2), so that their equilibria are strongly in the direction of this flux, these reactions would tend to be far displaced from equilibrium. Near-equilibrium conditions at these reactions would require the mass-action ratio, Γ, to be extremely large and, although large values for Γ may be possible for some reactions with large equilibrium constants (e.g. lactate dehydrogenase: section 5.2), they are probably not possible for these three glycolytic reactions because they would produce unacceptably-large concentration gradients of the glycolytic

---

*this summation also applies to standard free energy changes. Thus

$$\Delta G_{net} = \Delta G_{P\rightarrow Q} + \Delta G_{Q\rightarrow R}$$

$$= \Delta G^{\circ}_{P\rightarrow Q} + RT\ln [Q]/[P] + \Delta G^{\circ}_{Q\rightarrow R} + RT\ln [R]/[Q]$$

$$= (\Delta G^{\circ}_{P\rightarrow Q} + \Delta G^{\circ}_{Q\rightarrow R}) + RT\ln [R]/[P] ,$$

and since,

$$\Delta G_{net} = \Delta G^{\circ}_{net} + RT\ln [R]/[P] ,$$

a comparison of the last two equation shows that,

$$\Delta G^{\circ}_{net} = \Delta G^{\circ}_{P\rightarrow Q} + \Delta G^{\circ}_{Q\rightarrow R}.$$

Therefore the standard free energy change of a complex system is the sum of the standard free energy changes of its component reactions.

intermediates. For example, if the phosphofructokinase reaction was close to equilibrium in muscle, at the physiological concentrations of fructose 6-phosphate (approximately 3×10^{-5}M) and [ATP]/[ADP] (approximately 10), the concentration of fructose bisphosphate would be approximately 300mM, which is thirty times greater than the total concentration of inorganic and organic phosphate in the cell. However, by enhancing the natural tendency of these three glycolytic reactions to be far displaced from equilibrium in the direction of glycolysis (e.g. by providing little if any catalytic activity in excess of that required for the maximum net flux: see Crabtree & Newsholme, 1975), the other reactions of glycolysis may be brought very close to equilibrium as a result of the 'constancy' of ΔG_{net}; especially as the maximum catalytic activities of the enzymes catalysing near-equilibrium reactions are usually many times greater than the maximum flux (Krebs, 1963; Newsholme & Gevers, 1967). The overall result is a discontinuous distribution of ΔG along the pathway, with most reactions being close to equilibrium or kinetically reversible, and the metabolic advantages of this situation were outlined in section 4.4.

It must be stressed that the relationship, $\sum \Delta G$ = constant, only applies for fixed concentrations of initial and final substrates: it is not a reflection of the first law of thermodynamics, to which free energy does not conform. In contrast, the overall heat production by a pathway (if no work is done) is also the sum of the ΔH values of the component reactions, but this is a consequence of the first law and therefore does not require any restrictions on the concentrations of the initial and final substrates. Moreover, some of the component ΔH values may be positive (i.e. some reactions may be endothermic), whereas all of the component ΔG values must be negative.

A fallacy, often found in some elementary texts, is that the free energy of a pathway, such as that for glucose oxidation, must be 'released' in several consecutive steps, i.e. there must be many reactions between glucose and CO_2; because if this free energy was released in a single step there would be an unacceptably-large release of heat. Since this implies that ΔG is dissipated as heat, i.e. that Gibbs free energy obeys the first law of thermodynamics, it is clearly an incorrect explanation for the many reactions between glucose and CO_2. Furthermore, by the first law, the overall heat production of the pathway is the same whether the reaction takes place in one step or many steps (see above), so that the presence of a series of consecutive reactions will not affect the heat produced per mole of glucose oxidised. The most probable reason for the existence of many consecutive reactions in metabolism is the somewhat limited chemistry that can be performed by a protein molecule (enzyme). To accomplish the net glycolytic reaction given in section 3.3, glucose + $6O_2$ + 38ADP + $38P_i \longrightarrow 6CO_2$ + $6H_2O$ + 38ATP,

in a single step would require a protein molecule and catalytic mechanism of such extreme complexity that it would be unlikely to evolve. Moreover, even if such a protein did evolve, the process of glucose oxidation would be restricted to the formation of ATP under aerobic conditions. In contrast, the breakdown of the overall reaction into several consecutive reactions involving relatively-simple organic transformations not only makes the reaction kinetically feasible, but also, by channelling some of the intermediates into other systems, enables glucose utilisation to proceed anaerobically (with lactate instead of CO_2 as end product) and to supply intermediates for the synthesis of substances such as fats and proteins. Consequently, the consecutive reactions that comprise a metabolic pathway provide both flexibility and economy.

8.5.2 ΔG^o and metabolite concentrations: coupling in parallel

If a metabolic reaction is not at equilibrium, neither the sign nor the magnitude of its standard free energy change, ΔG^o, can be used to predict its state in vivo: the latter state is determined by ΔG, and hence by the ratio K/Γ, whereas ΔG^o involves only K (sections 4.3 and 4.4: see also Banks, 1969). Thus if for the equilibrium, $X \rightleftharpoons Y$, K is chosen in the direction $X \longrightarrow Y$ (section 4.5),

$$\Delta G^o = -RT\ln \left\{ [Y]_{eq}/[X]_{eq} \right\} \qquad \text{(section 4.3)}.$$

If $[Y]_{eq} = [X]_{eq}$, the logarithmic term is zero and ΔG^o is zero: if $[Y]_{eq} > [X]_{eq}$, so that the equilibrium lies to the right, the logarithmic term is positive and ΔG^o is negative: if $[Y]_{eq} < [X]_{eq}$, so that the equilibrium lies to the left, the logarithmic term is negative and ΔG^o is positive. Therefore, the sign of ΔG^o reflects the position of equilibrium relative to the direction chosen for K: it does not indicate whether the reaction is 'energetically' possible or impossible. In particular, a positive value for ΔG^o does not mean that a flux in this direction is impossible; the condition of negativity applies only to ΔG. For example, the reaction catalysed by glycogen phosphorylase,

glycogen$_n$ + $P_i \rightleftharpoons$ glycogen$_{n-1}$ + glucose 1-phosphate,

is far displaced from equilibrium (i.e. kinetically irreversible) in the direction of glycogen breakdown to glucose 1-phosphate in vivo: yet the equilibrium constant is in favour of glycogen synthesis (The value of K is approximately 0.3, and thus ΔG^o is approximately + 2.94 kJ/mole, in the direction of glycogen breakdown). This is because the cellular concentration of P_i (approximately 10mM) is maintained much greater than that of glucose 1-phosphate (approximately 0.03mM). Therefore, since [glycogen$_n$] is approximately equal to [glycogen$_{n-1}$], the mass-action ratio, Γ, in the direction of glycogen breakdown is approximately equal to the ratio, [glucose 1-phosphate]/[P_i], whose value is approximately 0.003. Consequently, the value of K/Γ in this direction is approximately 10^2, so that ΔG is large and negative (approximately - 11.34 kJ/mole) and the reaction is kinetically irreversible in the direction of glycogen breakdown despite a positive ΔG^o (Newsholme & Start, 1973). Some other examples of reactions that proceed in a direction in which ΔG^o is positive are given in Table 2.

Similarly, the magnitude of ΔG^o reflects only the extent of the equilibrium in a given direction, i.e. the size of the ratio $[Y]_{eq}/[X]_{eq}$: it does not indicate displacement from equilibrium and therefore it does not reflect the value of ΔG.

Consequently, it is incorrect to assume that a reaction for which ΔG^o is large and negative in the direction of a metabolic flux must have a value of ΔG that is also large and negative in this direction, and hence must be far displaced from equilibrium in vivo. Nor is it correct to assume that a reaction for which ΔG^o is close to zero must be close to equilibrium in vivo. For example, the cytoplasmic reaction catalysed by lactate dehydrogenase,

pyruvate + NADH $\rightleftharpoons$ lactate + NAD,

is close to equilibrium in vivo; yet the equilibrium constant for this reaction is approximately 10^4 in the direction of lactate formation, i.e. ΔG^o is approximately - 23 kJ/mole in this direction (see Krebs & Veech, 1969). This situation results from large values for the cytoplasmic ratios, [NAD]/[NADH] and [lactate]/[pyruvate], which in rat liver are approximately 10^3 and 10 respectively (Krebs & Veech, 1969); therefore the mass-action ratio in the direction of lactate formation, [lactate][NAD]/[pyruvate][NADH], is approximately 10^4, so that K/Γ is close to unity, ΔG is close to zero and the reaction is close to equilibrium.

Although the value of ΔG^o does not reflect the state of a metabolic reaction in vivo, it is a factor determining the steady-state concentrations of substrates or products. Thus in the system,

$$\longrightarrow X \underset{v_r}{\overset{v_f}{\rightleftharpoons}} Y \longrightarrow \ ,$$

the concentration of Y (for a given steady-state flux) is determined by the reaction removing Y from the system (section 2). This value of [Y] determines the rate of the reverse process (v_r) of reaction $X \rightleftharpoons Y$. [X], and hence v_f, then attains a value such that $(v_f - v_r)$ is equal to the steady-state flux through the system*. Since ΔG for this reaction must be negative, the <u>minimum</u> concentration of X, for a given flux and [Y], is that which gives a value of ΔG just less than zero, i.e. one that just displaces the reaction from equilibrium. Consequently, the effect of ΔG^o on [X] can be examined qualitatively by considering its effect on the minimum value of [X], and hence by considering the reaction as if it were at equilibrium.

In this case,

$[X]_{min}$ = [Y]/K,

where K is the equilibrium constant in the direction $X \longrightarrow Y$. This shows that [X] decreases as K increases, i.e. as ΔG^o in this direction becomes more negative. Therefore, a reaction for which ΔG^o is negative in the direction of the net flux is able to operate at lower substrate concentrations than a reaction for which ΔG^o in this direction is more positive. It is generally advantageous for a cell to avoid excessively-large concentrations of its intermediates, for at least two reasons. First, a high concentration may be precluded because of a limited solubility or because of significant non-enzymic side reactions (see Atkinson, 1969); for example, both pyruvate and oxaloacetate are unstable at high concentrations in aqueous solution. Second, a high concentration may damage the morphology of the cell; for example, high concentrations of long-chain fatty acids uncouple oxidative phosphorylation and damage cell membranes. Therefore it is not surprising that the reactions of a metabolic pathway usually have values of ΔG^o that are either negative or not too far from zero in the direction of the net flux <u>in vivo</u>. This principle is illustrated by the values of ΔG^o for the glycolytic reactions (Table 2): apart from aldolase, all the reactions have either negative ΔG^o values or positive values that are not too far from zero. However, it must be stressed that this principle is only a <u>guide</u> to the likely situation <u>in vivo</u>: some reactions may operate satisfactorily with a large positive value for ΔG^o (e.g. aldolase).

If the value of ΔG^o for the reaction, $X \rightleftharpoons Y$, is too positive to provide a satisfactory concentration of X for the required flux in the direction $X \longrightarrow Y$, this can be overcome by kinetically coupling the reaction in parallel with another reaction for which ΔG is large and negative. Thus in the system,

---

*This shows that the mass-action ratio, [Y]/[X], and hence the value of $\Delta G_{X \longrightarrow Y}$ results from and does not determine the flux in this system: therefore $\Delta G_{X \longrightarrow Y}$ can not be regarded as a 'driving force' for reaction $X \rightleftharpoons Y$ (see also Banks, 1969).

reaction X$\longrightarrow$Y is coupled to reaction A$\longrightarrow$B by an enzyme that catalyses the overall reaction, X+A$\rightleftharpoons$Y+B. If, <u>at the cellular concentrations of A and B</u>, the value of ΔG for the reaction A$\longrightarrow$B (i.e. if it were not coupled to X$\longrightarrow$Y) is ΔG_{AB} and if, <u>at the cellular concentrations of X and Y</u>, the value of ΔG for the reaction X$\longrightarrow$Y (i.e. if it were not coupled to A$\longrightarrow$B) is ΔG_{XY}, the value of ΔG for the coupled reaction is equal to the sum of these component values (since ΔG is a state function): thus,

$$\Delta G_{net} = \Delta G_{XY} + \Delta G_{AB}.$$

(As with reactions coupled together in series, this summation of ΔG values does not reflect, and must therefore not be confused with, the first law of thermodynamics.)

Consequently, if ΔG_{AB} is large and negative, ΔG_{XY} can be <u>positive</u>, provided that the net sum is negative: in such a situation there is a net flux of A and X to B and Y, and hence a net flux of X to Y despite a positive value for ΔG_{XY}. Since ΔG_{XY} is positive, [X] can be smaller than the minimum value in the absence of coupling (i.e. when ΔG_{XY} must be negative); therefore this parallel coupling of the reactions helps to prevent a large steady-state concentration of X that would result from the unfavourable value of $\Delta G°$ for the uncoupled reaction. The role of reactions such as A$\rightarrow$B in 'activating' of 'facilitating' reactions such as X$\rightarrow$Y has been discussed by Atkinson (1969). However it must be stressed that this method for offsetting a positive $\Delta G°$ is not possible unless the reactions are <u>kinetically</u> coupled: for ΔG_{XY} to be positive and to be offset by a negative ΔG_{AB}, there must be an overall reaction, X+A$\rightleftharpoons$Y+B, that is kinetically distinct (i.e. catalysed by a different enzyme) from reactions X$\rightleftharpoons$Y and A$\rightleftharpoons$B. In the absence of kinetic coupling, i.e. if the separate reactions X$\rightleftharpoons$Y and A$\rightleftharpoons$B merely coexist, <u>there is no interaction between ΔG_{XY} and ΔG_{AB}</u>, so that under these conditions, the former must always be negative for a flux in the direction X$\rightarrow$Y: in other words, 'thermodynamic coupling' (i.e. interactions between free energies or 'affinities') is impossible without kinetic coupling (Banks & Vernon, 1970; Keizer, 1975).

A metabolic example of the lowering of a steady-state concentrations by parallel coupling is the conversion of glucose into glucose 6-phosphate in glycolysis. If this reaction proceeded by a direct addition of P_i, i.e.

glucose + $P_i \rightleftharpoons$ glucose 6-phosphate,

for which $\Delta G°$ in the direction of glucose 6-phosphate formation is +12.6 kJ/mole, the minimum (equilibrium) concentration of glucose at cellular concentrations of P_i (approximately 10^{-2}M) and glucose 6-phosphate (approximately 10^{-4}M) would be 1.6M, i.e. 28 g/litre: this is more than three hundredfold greater than the blood glucose concentration and could never be tolerated because of its large osmotic effect. However, in the cell this conversion is accomplished by a parallel coupling to the hydrolysis of ATP to ADP+P_i, in a reaction catalysed by hexokinase:

glucose + P_i + ATP $\rightleftharpoons$ glucose 6-phosphate + ADP + P_i

i.e.

glucose + ATP $\rightleftharpoons$ glucose 6-phosphate + ADP.

The minimum (equilibrium) concentration of glucose is now that corresponding to a zero value for ΔG_{net}. Since the value of ΔG for ATP hydrolysis is ADP + P_i is approximately -50.4 kJ/mole in vivo (see below), the free energy change of the 'component' reaction, glucose + $P_i \rightarrow$ glucose 6-phosphate, is equal to +50.4 kJ/mole when ΔG_{net} is zero. Hence the minimum glucose concentration is that corresponding to a free energy change of +50.4 kJ/mole for this component reaction and it can be calculated from the equation,

$$\Delta G = \Delta G° + RT\ln\Gamma ,$$

where $\Delta G°$ is +12.5 kJ/mole, ΔG is +50.4 kJ/mole and Γ is the ratio, [glucose 6-phosphate]/$[P_i]$[glucose]$_{min}$. Thus,

$$+50.4 = +12.6 + (8.31 \times 300/1000)\ln\left\{ 10^{-4}/10^{-2}[glucose]_{min}\right\},$$

whence,

$$[glucose]_{min} = 3 \times 10^{-9}M.$$

In fact the hexokinase reaction is far displaced from equilibrium and is thus kinetically irreversible in vivo (Table 2), so that the intracellular glucose concentration is much greater than the minimum value: nevertheless, this cellular concentration (approximately $10^{-4}M$) is still very much lower than that which would be required if P_i reacted directly with glucose. This example illustrates one of the roles of ATP hydrolysis in metabolism: the facilitating of a reaction so that it can transmit a flux at physiologically-acceptable substrate concentrations. Other examples of reactions in which ATP hydrolysis acts as a facilitating system are those catalaysed by phosphofructokinase,

fructose 6-phosphate + ATP $\longrightarrow$ fructose bisphosphate + ADP; pyruvate carboxylase,

pyruvate + CO_2 + ATP $\longrightarrow$ oxaloacetate + ADP + P_i; acetyl-CoA synthetase,

acetate + CoA + ATP $\longrightarrow$ acetyl-CoA + AMP + PP_i.

The third example is interesting because the reaction, acetate + CoA $\longrightarrow$ acetyl-CoA, is coupled to the hydrolysis of ATP to AMP and pyrophosphate (PP_i), instead of to ADP and P_i. A possible reason for this is that both [AMP] and [PP_i] are maintained at lower values in vivo than [ADP] and (P_i): for example, in muscle [AMP] and [PP_i] are approximately $2 \times 10^{-5}M$ and $10^{-5}M$, respectively, whereas [ADP] and [P_i] are approximately $2 \times 10^{-4}M$ and $10^{-2}M$, respectively. Furthermore, $\Delta G°$ for the hydrolysis of ATP to AMP + PP_i (-36.2 kJ/mole) is more negative than that for hydrolysis to ADP+ P_i (-29.4 kJ/mole): see Huennekens & Whiteley (1960). Therefore, under these cellular conditions and assuming that [ATP] is $5x10^{-3}$ M,

$$\Delta G_{ATP \rightarrow ADP} = -29.4 + (8.31x300/1000)\ln (10^{-2}x2x10^{-4}/5x10^{-3})$$

$$= -50.4 \text{ kJ/mole};$$

$$\Delta G_{ATP \rightarrow AMP} = -36.2 + (8.31x300/1000)\ln (10^{-5}x2x10^{-5}/5x10^{-3})$$

$$= -79.8 \text{ kJ/mole}.$$

Consequently, the hydrolysis of ATP to AMP provides a more negative ΔG and may thus be a better facilitating reaction in vivo than hydrolysis to ADP. However, the hydrolysis of ATP to AMP is equivalent to the utilisation of two molecules of ATP, since a further molecule of ATP is needed to convert AMP back to ADP and hence to ATP: this occurs in a reaction catalysed by adenylate kinase (myokinase),

AMP + ATP $\rightleftharpoons$ 2ADP.

Therefore, hydrolysis to AMP requires a greater utilisation of fuel, and is thus less economical, than hydrolysis to ADP. Presumably, hydrolysis to ADP will be preferred unless the value of $\Delta G°$ for the reaction to be facilitated is very large and positive. In this connection it is perhaps significant that $\Delta G°$ for the direct reaction of acetate and CoA to form acetyl-CoA is +33.6 kJ/mole (Krebs & Kornberg, 1957), which is much more positive than that for the direct reaction between glucose and P_i (+12.6 kJ/mole).

Finally, the role of ATP hydrolysis in facilitating reactions is shared by several other nucleoside triphosphates. For example GTP (or ITP) hydrolysis is used in the reaction catalysed by phosphoenolpyruvate carboxykinase,

oxaloacetate + G(I)TP $\rightleftharpoons$ phosphoenolpyruvate + G(I)DP + CO_2.

GTP hydrolysis is also used to facilitate several steps in protein synthesis (see Stryer, 1981) as well as the transmission of hormonal effects on cell receptors (see Sibley & Lefkowitz, 1985). UTP hydrolysis facilitates the conversion of glucose 1-phosphate to glycogen and CTP hydrolysis facilitates the formation of some phospholipids (see Stryer, 1981). Other, non nucleoside triphosphate, cofactors are also used to facilitate reactions; for example, lipoic acid in the conversion of pyruvate or 2-oxoglutarate into acetyl-CoA or succinyl-CoA, respectively (see Atkinson, 1969).

8.5.3 Parallel coupling of reactions to drive cyclic fluxes

In the examples discussed in the previous section the large negative free energy change (e.g. ΔG_{AB}) did not drive the coupled reaction (e.g. X$\rightleftharpoons$Y) in a given direction: it merely allowed it to proceed in that direction at a physiologically-acceptable substrate concentration. However, there is one situation where a reaction of large negative ΔG is required to drive (i.e. generate a flux for) the reaction or system to which it is coupled: this is when the latter reaction belongs to a <u>cyclic</u> flux. For example, in the following system,

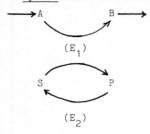

The continuous cyclic flux of S to P cannot occur unless one of the reactions of the cycle (in this case E_1) is kinetically coupled to another flux, $\longrightarrow A \longrightarrow B \longrightarrow$, i.e. to a reaction for which ΔG is negative. This is because, for a continuous cycling of metabolites between S and P, there are net fluxes both from S to P and from P to S; thus the free energy changes of the reactions of the cycle $\Delta G_{S \to P}$ and $\Delta G_{P \to S}$ must both be negative. However, since ΔG is a state function, the net change of free energy around the cycle must be zero, i.e.

$$\Delta G_{S \to P} + \Delta G_{P \to S} = 0.$$

Therefore, in the absence of a coupled flux such as $\longrightarrow A \longrightarrow B \longrightarrow$, $\Delta G_{S \to P}$ and $\Delta G_{P \to S}$ cannot both be negative since their sum must be zero: hence the cyclic flux cannot exist. This is expected because, under these conditions, S and P represent a closed system and thus can only approach equilibrium (section 2). Furthermore, since $\Delta G_{S \to P} = -\Delta G_{P \to S}$, the conversion of S into P under these conditions must be the exact reverse of the conversion of P into S (section 4.5): in other words a closed system between S and P must consist of a <u>single</u>, kinetically-reversible, reaction, $S \rightleftharpoons P$.

However, when the conversion of S into P is kinetically coupled to reaction $A \longrightarrow B$, the value of $\Delta G_{S \to P}$ may be <u>positive</u>: as for reactions that are facilitated by kinetic coupling (section 5.2), a negative ΔG (i.e. $\Delta G_{A \to B}$) may offset a positive ΔG (i.e. $\Delta G_{S \to P}$) to produce a net ΔG (i.e. $\Delta G_{A+S \to B+P}$) that is negative. This enables the net flux of S to P to occur despite a positive value for $\Delta G_{S \to P}$. The condition, $\Delta G_{S \to P} + \Delta G_{P \to S} = 0$, can then be satisfied with a negative value of $\Delta G_{P \to S}$, so that both reactions of the cycle proceed in the directions shown in the diagram and there is a continuous (steady-state) cycling between S and P. Since the cyclic flux could not exist in the absence of the kinetic coupling, flux $\longrightarrow A \longrightarrow B \longrightarrow$ may be considered to drive flux $S \rightleftharpoons P$.

A metabolic example of such a system has been described in section 3.3. This is the continuous cycling between fructose 6-phosphate and fructose bisphosphate which is coupled to, and is thus driven by, the hydrolysis of ATP to ADP and P_i. In turn, ATP is also continuously cycled between ATP and ADP and this cyclic flux of adenine nucleotide is coupled to, and is thus driven by, the metabolism of fuels. Consequently, this particular system consists of two cyclic fluxes coupled in parallel, with the whole system being driven by the flux of a fuel (e.g. glucose, fatty acids) to its end products (e.g. lactate, CO_2):

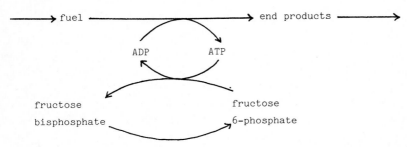

Indeed, in a wide sense this framework must apply to all the metabolic systems of nature. Since the earth has a fixed content of inorganic and organic nutrients, all metabolic systems must be part of wider systems that represent the re-cycling of these nutrients: some of these 'global' cycles, for example the carbon and the nitrogen cycles, are well known. Consequently, all the metabolic fluxes of nature must ultimately be coupled to, and hence driven by, a flux that is continuously supplied to and dissipated from the earth: this fundamental flux is the flux or solar energy and its dissipation as heat. In green plants and some bacteria this flux of radiant energy is kinetically coupled, via the photosynthetic electron transport systems, to the synthesis of ATP and NADPH and thence to the synthesis of glucose from CO_2 (see Mahler & Cordes, 1968; Yudkin & Offord, 1973; Stryer, 1981): glucose then serves as a fuel, and is converted into other fuels such as fatty acids, for the metabolic systems of both photosynthetic and nonphotosynthetic organisms. (Unlike the cellular fluxes and systems discussed in this chapter, the 'global' open systems are not in a steady state. As stated in section 2, each cell or organism effectively insulates a part of the global system and converts it into one in a steady state: this produces a stable metabolic system (homeostasis) within the organism and enables it to develop and regulate its metabolism despite continuous fluctuations in its environment.)

8.5.4 The concept of energy transfer in metabolic systems

The series-coupled system,

$$\longrightarrow S \longrightarrow P \longrightarrow Q \longrightarrow R \longrightarrow$$

can be considered to transfer chemical (internal) energy from S to R, accompanied by the release of $-\Delta U_{S \to R}$ kJ/mole of the internal energy of S. However, the absolute value of the internal energy of one mole of S is unknown, because thermodynamic measurements only give <u>changes</u> in the state functions (see Klotz, 1967); therefore the percentage of the absolute internal energy of S represented by $\Delta U_{S \to R}$, and thus the efficiency of the transfer of internal energy to R, can not be calculated.

The situation is somewhat different for reactions coupled in parallel, for example the system

By the first law of thermodynamics,

$$\Delta U_{X+A \longrightarrow Y+B} = \Delta U_{X \to Y} + \Delta U_{A \to B},$$

where the functions on the right hand side are those for the individual uncoupled reactions (see section 5.2). If $\Delta U_{X \to Y}$ is negative and $\Delta U_{A \to B}$ is positive, ΔU for the coupled reaction is less negative than that for reaction X→Y alone: therefore the latter reaction may be considered to have transferred an amount of chemical energy, equal to the difference between $-\Delta U_{X \to Y}$ amd $\Delta U_{X+A \to Y+B}$ (which, from the above equation, is equal to $\Delta U_{A \to B}$), to reaction A→B. (If $\Delta U_{X \to Y}$ and $\Delta U_{A \to B}$ are both negative, so that their sum is more negative than either, this concept of energy transfer is meaningless.) Since the maximum amount of chemical energy that can be transferred from reaction X→Y is equal to $-\Delta U_{X \to Y}$, it is possible to define a percentage efficiency, η , for the transfer, given by the equation,

$$\eta = \frac{100.\ \text{energy transferred}}{\text{energy available}}$$

$$= \frac{100.\Delta U_{A \to B}}{-\Delta U_{X \to Y}}$$

If the system involves a net change of volume, the energy corresponding to this can be excluded from the efficiency by using ΔH instead of ΔU, in which case,

$$\eta = \frac{100.\Delta H_{A \to B}}{-\Delta H_{X \to Y}}$$

For the synthesis of 38 moles of ATP as a result of the oxidation of one mole of glucose to CO_2 (section 3.3),

$$\longrightarrow \text{glucose} \underset{38(ADP+P_i)}{\overset{6CO_2}{\rightleftharpoons}} 38ATP \longrightarrow \, ,$$

$$\Delta H_{ADP \rightarrow ATP} \quad = \quad -\Delta H_{ATP \rightarrow ADP}$$

$$= \quad +21 \text{ kJ/mole of ADP (Table 1);}$$

$$\Delta H_{glucose \rightarrow CO_2} \quad = \quad -2820 \text{ kJ/mole of glucose (Table 1).}$$

Consequently, for the transfer of chemical energy from glucose oxidation to ATP synthesis,

$$\eta \quad = \quad 100 \times 21 \times 38 / 2820$$

$$= \quad 28\%.$$

It should be noted that this efficiency refers to the transfer of internal energy between reactions, and not between metabolites: efficiencies for the latter process require absolute values for U and H and, as stated above, these are unknown.

The net changes of Gibbs free energy for reactions coupled in parallel, ΔG_{net}, is also the sum of the values of ΔG for the component (uncoupled) reactions (section 5.2): thus, for fixed concentrations of A, B, X and Y,

$$\Delta G_{A+X \rightarrow B+Y} \quad = \quad \Delta G_{X \rightarrow Y} \quad + \quad \Delta G_{A \rightarrow B}.$$

By analogy with the summation of component ΔU or ΔH values, if $\Delta G_{X \rightarrow Y}$ is negative and $\Delta G_{A \rightarrow B}$ is positive (i.e. if reaction X→Y is facilitating reaction A→B) the system behaves as if free energy is transferred from reaction X→Y to reaction A→B with a percentage efficiency given by the equation

$$\eta \quad = \quad \frac{100 . \Delta G_{A \rightarrow B}}{-\Delta G_{X \rightarrow Y}}$$

If the coupled system is at equilibrium, so that $\Delta G_{A+X \rightarrow B+Y}$ is zero,

$$\Delta G_{A \rightarrow B} \quad = \quad -\Delta G_{X \rightarrow Y},$$

and the efficiency of 'free energy transfer' from reaction X→Y to reaction A→B is 100%. As for ΔU and ΔH, the concept of free energy transfer is meaningless if $\Delta G_{X \rightarrow Y}$ and $\Delta G_{A \rightarrow B}$ are both negative.*

This concept of free energy transfer is very common in discussions of metabolism, but is actually very unsatisfactory and can often be dangerously misleading (see Banks, 1969; Banks & Vernon, 1970). This is because, although the summation of component ΔG values to obtain ΔG_{net} is <u>arithmetically</u> analogous to the summation of component ΔU and ΔH values, the underlying reasons for the summations are fundamentally different. The summation of component ΔU or ΔH values results from the first law of thermodynamics (to which the $T\Delta S$ portion of ΔG does not conform: section 3.6), whereas the summation of component ΔG values results from the state function nature of free energy. Consequently, although the summation of component ΔG values appears to indicate that free energy is analogous to enthalpy (or internal energy) and is therefore an entity that can be transferred between reactions or stored in metabolites, this analogy is incorrect because it also implies that free energy is a conventional energy and thus obeys the first law. The common idea that free energy is 'released' when ΔG is negative is also based on this incorrect analogy. Cornish-Bowden (1983) has also pointed out the dangers of regarding free energy as a sort of 'caloric' which can be transferred or stored. One fallacy resulting from the treatment of free energy as a conventional energy was discussed in section 5.1. However, perhaps the best example of a misleading concept arising from the confusion of free energy with conventional energy is the familiar 'high-energy' or 'energy-rich' phosphate bond of ATP and some other phosphate esters: this is considered in the next section.

In a discussion of the physical significance of entropy, Klotz (1967) stated that:- "Ultimately one must realize that entropy is essentially a mathematical function".

It is a concise function of the variables of experience, e.g. temperature, pressure and composition ..... Thus, we ought to view entropy as an index of condition or character (perhaps somewhat analogous to a cost-of-living index or to pH as an index of acidity) rather than as a measure of content of some imaginary fluid. It is an index of capacity for spontaneous change." Because of its $T\Delta S$ component, exactly the same statements apply to Gibs free energy. Most if not all of the problems associated with the misuse of 'free energy' could be avoided if it were referred to as the 'Gibbs function'. This term has been recommended by an international commission (see J. Biol. Chem. 251, 6879-6885) and has been adopted

---

*Other metabolic 'efficiencies' may be defined, but most of these are of dubious significance (see Banks, 1969; Banks & Vernon, 1970). For example, if a system, $\rightarrow X \rightleftharpoons Y \rightarrow$, does w units of physical work per mole of X converted into Y, the system may be considered to produce this work with a percentage efficiency of $100w/(-\Delta G_{X \rightarrow Y})$, because the maximum work possible is equal to $-\Delta G_{X \rightarrow Y}$ (section 3.7). However, the value of $-\Delta G$, and hence this 'efficiency' may be varied by varying the displacement of reaction $X \rightleftharpoons Y$ from equilibrium, i.e. by changing [X] and/or [Y]; but this need not affect w and hence the <u>amount</u> of X required to produce a given amount of work. Consequently, this particular efficiency bears no relation to the economic efficiency (i.e. demand for fuel) and is of little practical value.

in at least one standard physical chemistry text (Atkins, 1982). Its adoption by biochemists will take some time, but when this happens many of the problems raised in this section will become superfluous.

8.5.5 The metabolic importance of ATP

As stated in section 3.3, ATP is synthesised from ADP and P_i in vivo as the result of the metabolism of fuels such as glucose, glycogen and fatty acids. Some of the ATP is produced by relatively simple enzymic reactions, e.g. that catalysed by pyruvate kinase,

phosphoenolpyruvate + ADP $\longrightarrow$ pyruvate + ATP,

but if the fuel is oxidised via the tricarboxylic acid cycle, most of the ATP is produced by the mitochondrial system known as oxidative phosphorylation. The production of ATP serves several important purposes. Hydrolysis of ATP to ADP provides energy for the production of mechanical work in contractile tissues such as muscle. In this situation some of the chemical (internal) energy of the fuels is transferred to ATP and a portion of this energy is converted into work, the remainder being dissipated as heat; thus ATP serves as a 'common currency' through which some of the internal energy of many different types of fuel is directed towards the performance of work. However, ATP hydrolysis to ADP (or AMP) is also kinetically coupled to several metabolic reactions, notably those involved in biosyntheses, and it is this role of ATP that has produced most misunderstanding (see Banks, 1969; Banks & Vernon, 1970).

Biosynthetic (or anabolic) processes generally have positive standard free energy changes ($\Delta G°$), i.e. the position of equilibrium is in the opposite direction to biosynthesis. Since $\Delta G°$ for ATP hydrolysis to ADP or AMP is large and negative (approximately -29 and -36 kJ/mole, respectively), ATP was proposed to transfer much of this 'free energy' to, and thereby drive, these anabolic reactions in the required direction. This led to the concept that the terminal phosphate of ATP was attached to the rest of the molecule by a special 'high-energy' bond which was formed by the input of large amounts of free energy from the fuels that regenerate ATP: thus ATP was proposed to serve as a reservoir of free energy that could be used for biosynthesis (Lipmann, 1941).

However, as pointed out by several workers (see Banks & Vernon, 1970), this concept of a 'high-energy' phosphate bond is meaningless. Indeed it is a very good example of the result of confusing free energy with conventional energy. (This is in addition to the erroneous use of $\Delta G°$ for predicting the state of a reaction in vivo: a positive $\Delta G°$ does not mean that a reaction is 'energetically-impossible' in that direction: section 5.2) As discussed in the previous section, it is certainly possible to consider free energy as something that can be transferred between reactions, and hence from a catabolic reaction to an anabolic reaction via ATP hydrolysis and resynthesis,

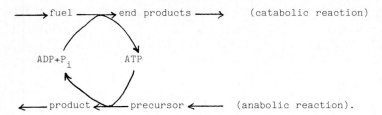

However, this 'transfer' is just an arithmetical consequence of ΔG being a state function and does not actually represent the transfer of an equivalent amount of conventional energy. Therefore, there is no justificatgion for assuming anything energetically special about the phosphate bonds of ATP.

The metabolic significance of the reactions, ATP→ADP + P_i and ATP→AMP + PP_i, is that they are far displaced from equilibrium (i.e. kinetically irreversible) in vivo, so that ΔG is large and negative (section 5.2). This large negative ΔG serves at least three important purposes, all of which have been discussed in previous sections. First, it allows the hydrolysis of ATP to ADP to be associated with the production of mechanical work in muscle: since muscles may perform up to approximately 24 kJ of work per mole of ATP hydrolysed (Wilkie, 1970), the value of $-\Delta G$ for ATP hydrolysis must be at least 24 kJ/mole because $-\Delta G$ is equal to the maximum work that can be done (section 3.7). Second, the large negative ΔG for ATP hydrolysis to ADP or AMP allows these reactions to facilitate other reactions to which they are kinetically coupled (e.g. the reactions catalysed by hexokinase and phosphofructokinase in glycolysis), so that metabolic fluxes can be transmitted at physiologically-acceptable substrate concentrations (section 5.2). This coupling also enables ATP hydrolysis to drive certain cyclic fluxes, e.g. substrate cycles (section 5.3) and the movement of Na^+ and K^+ across cell membranes (section 6). Third, ATP hydrolysis and subsequent resynthesis provides a kinetic coupling between anabolic and catabolic processes (see above) to produce the net coupled system,

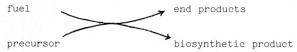

As a result of this coupling by ATP hydrolysis and resynthesis, the large negative ΔG for the metabolism of the fuel to its end products enables the entire anabolic pathway to proceed at physiologically-acceptable concentrations of the precursors. Moreover, if the anabolic system is part of a cyclic flux, i.e.

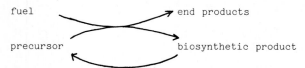

the kinetic coupling provided by ATP hydrolysis and resynthesis enables this cyclic flux to be driven by the flux of fuel to its end products: indeed, without this coupling the cyclic flux could not exist (section 5.3). A metabolic example of such a system is the Cori cycle, in which glucose ($\equiv$ biosynthetic product) synthesised from lactate ($\equiv$ precursor) in mammalian liver is reconverted into lactate in extrahepatic tissues such as skeletal muscle (where glucose serves as a fuel): this cyclic flux of glucose and lactate is coupled, via ATP hydrolysis and resynthesis, to the utilisation of fuels by the liver.

Finally, although ATP hydrolysis plays a central and sometimes unique role in metabolism, this does not result from any special thermodynamic properties of the overall reactions, ATP $\rightarrow$ ADP + P_i or ATP $\rightarrow$ AMP + PP_i. In theory, any reaction could replace ATP hydrolysis, provided that ΔG were sufficiently large and negative. This condition can readily be met by ensuring that the reaction is kinetically irreversible in vivo (e.g. by providing the minimum catalytic activity for transmitting the required net flux). Indeed, other nucleoside triphosphates (GTP, UTP, CTP) and other cofactors are frequently used instead of ATP (section 5.2). Consequently, the reasons why ATP has been selected as the major cellular energy currency and kinetic link between catabolic and anabolic systems will not be found by studying the thermodynamics of the overall reactions, ATP $\rightarrow$ ADP + P_i or ATP $\rightarrow$ AMP + PP_i. Since the transduction of the internal energy of ATP into work, or the coupling of ATP hydrolysis to other reactions, is the result of enzyme activity, the reasons for the selection of ATP will most likely derive from the thermodynamics and kinetics of the interaction of ATP with proteins or protein systems, e.g. the myofibrillar ATPase and the coupling factors of oxidative phosphorylation (see Slater, 1974; Tregear, 1975).

8.6   DISTRIBUTION AND FLUXES OF IONS ACROSS A MEMBRANE

The movement of an ion across a biological membrane (e.g. the cell membrane or inner mitochondrial membrane) may result in an electrical potential difference; for example, the interior of a resting nerve cell is 50-60mV more negative than the exterior, as a result of different permeabilities of the cell membrane to Na^+ and K^+. This potential difference influences the equilibrium or the steady-state distribution of other ions that are transported across the membrane. Thus, in the following system, where E_i and E_o are the electrical potentials on the 'inside' and 'outside' of the membrane, respectively, the positively-charged metabolite, X^+, equilibrates across the membrane by either simple or carrier-mediated

388

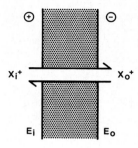

diffusion. If there was no potential difference across the membrane (i.e. if E_i = E_o) the equilibrium concentrations of X^+ on both sides of the membrane would be equal, so that the equilibrium constant for the diffusion, $[X^+]_o/[X^+]_i$, would be unity. However, if E_i is greater than E_o (as in the diagram), so that the inside of the membrane is positive with respect to the outside, the movement of X^+ down the positive potential gradient (i.e. from inside to outside) is facilitated, whereas movement in the direction of the negative gradient (i.e. from outside to inside) is hindered. Consequently, at equilibrium, the concentration of X^+ inside is lower than that outside, and the equilibrium constant for the diffusion, $[X^+]_o/[X^+]_i$, is greater than unity.

The effect of a potential difference on the equilibrium distribution of a charged metabolite across a membrane can be expressed quantitatively as follows:- If a charge of Q coulombs moves from a region of electrical potential, $E_{initial}$, to one of potential, E_{final}, the work done is equal to $Q(E_{initial}-E_{final})$ or $Q\Delta E$, joules (assuming that E is in volts*). If one molecule of an ion carries a charge of z units, one mole of that ion carries a charge of zF coulombs, where F is known as the Faraday constant and has a value of 96,494 coulombs. Therefore, when one mole of an ion moves across a potential difference of ΔE volts the work done is equal to $zF\Delta E$ joules. This provides an additional energy term for the reaction and, since $zF\Delta E$ is real energy, the effect is to produce an enthalpy change of $(\Delta H-zF\Delta E)$, where ΔH is the enthalpy change in the absence of the potential difference. (The negative sign arises because work done represents energy lost to the system:- section 3.1).

---

*Since the terms 'initial' and 'final', and thus the ordering of the potentials in the term ΔE, refer to a flux in the direction of movement of the charge, ΔE defined as $E_{initial}-E_{final}$) is in the same direction as K and Γ for the diffusion process (section 4.5).

Therefore,

$$\Delta G_{net} = \Delta H - T\Delta S - zF\Delta E =$$

$$= -RT\ln(K/\Gamma) - zF\Delta E,$$

where K and Γ are the equilibrium constant and mass-action ratio, respectively, in the direction of the movement of the ion (sections 4.4 and 4.5). There are two ways of proceeding from this equation. One very common way is to express ΔG_{net} as a hypothetical potential difference, ΔP, using the relationship

$$\Delta G_{net} = -zF\Delta P$$

This expresses the chemical $(-RT\ln(K/\Gamma))$ and electrical $(-zF\Delta E)$ components of ΔG_{net} in terms of a single potential difference, ΔP; and examples include 'redox' potentials and 'protomotive force' (Mitchell, 1973; Nicholls, 1982). The use of such potentials simplifies the calculation of energy changes in complex systems. However this is outweighed by a resulting confusion due to the $T\Delta S$ component of ΔG being included. As discussed previously, this component is not a real energy term and therefore any 'potentials' including it are imaginary. Consequently, just as ΔG can easily be confused with real energy (with unfortunate results for metabolism), so these imaginary potentials can easily be confused with real ones (e.g. the membrane potential). This is especially so for redox potentials, and there seems to be no good reason why these should really be used in preference to the equivalent values of ΔG or $\Delta G°$ when discussing metabolic redox systems, for example the mitochondrial and photosynthetic electron transport systems (see also Banks & Vernon, 1970). The same conclusions also apply to 'protonmotive force'.

An alternative approach is to retain the basic form of the equation for ΔG_{net} by combining the electrical component with the equilibrium constant, K. Since $\ln(e^x) = x$, the above equation for ΔG_{net} may be written as;

$$\Delta G_{net} = -RT\ln(K/\Gamma) - RT(zF\Delta E/RT)$$

$$= -RT\ln(K/\Gamma) - RT\ln(e^{zF\Delta E/RT})$$

$$= -RT\ln\left\{ \frac{Ke^{zF\Delta E/RT}}{\Gamma} \right\}$$

$$= -RT\ln(K_E/\Gamma),$$

where,

$$K_E = Ke^{zF\Delta E/RT}.$$

This shows that the effect of ΔE on the distribution of an ion can be expressed as an effect on the equilibrium constant, K_E, whose value is equal to $K.e^{zF\Delta E/RT}$.

Unlike a combined 'potential' K_E actually exists, as it is the concentration ratio (Γ_o) when there is zero net flux across the membrane at potential difference ΔE. Thus, in the hypothetical system where X^+ equilibrates across the membrane, the value of z is +1 and K_E is equal to $K.e.^{F\Delta E/RT}$. If K is defined as $[X]_o/[X]_i$, i.e. in the direction $X_i \rightarrow X_o$, and if the system represents simple or facilitated diffusion so that the value of K is unity,

$$K_E = [X^+]_o[X^+]_i$$
$$= e^{F\Delta E/RT}$$

where ΔE is in the direction $X_i \rightarrow X_o$ and is thus equal to $(E_i - E_o)$. Therefore,

$$[X^+]_o/[X^+]_i = e^{F(E_i - E_o)/RT}$$

or,

$$(E_i - E_o) = (RT/F)\ln\left\{[X^+]_o/[X^+]_i\right\}.$$

This equation is often referred to as a Nernst equation and it shows that, when $(E_i - E_o)$ is positive, the term $\ln\left\{[X^+]_o/[X^+]_i\right\}$ is positive and thus $[X^+]_o \geq [X^+]_i$, in agreement with the qualitative conclusions at the beginning of this section.

If a univalent negatively-charged ion, Y^-, equilibrates across the membrane, so that the value of z is -1,

$$(E_i - E_o] = -(RT/F)\ln\left\{[Y^-]_o/[Y^-]_i\right\}$$
$$= (RT/F)\ln\left\{[Y^-]_i/[Y^-]_o\right\}$$

If both X^+ and Y^- are at equilibrium across the membrane, $(E_i - E_o)$ must be the same for both ions, so that,

$$(RT/F)\ln\left\{[X^+]_o/[X^+]_i\right\} = (RT/F)\ln\left\{[Y^-]_i/[Y^-]_o\right\};$$

consequently,

$$[X^+]_o/[X^+]_i = [Y^-]_i/[Y^-]_o$$

or,

$$[X^+]_o[Y^-]_o = [X^+]_i[Y^-]_i.$$

This is an example of a 'Donnan' relationship between the ions and it applies to any pair of univalent and oppositely-charged ions that are at equilibrium across the membrane, Donnan relationships for ions of valency other than unity can be derived by incorporating the relevant values for z into the above analysis (see also Spanner, 1964; Levin, 1969).

By measuring the value of the membrane potential difference and the concentrations of a diffusible ion, M, on either side, the Nernst equation may be used to determine whether the diffusion system for the ion is at or close to equilibrium. If the measured concentrations and potential difference are not related by the general Nernst equation,

$$(E_i-E_o) \quad = \quad (RT/zF)\ln \left\{ [M]_o/[M]_i \right\} ,$$

the system cannot be at or close to equilibrium. For example, the distributions of Cl^- and K^+ ions across nerve cell membranes are in reasonable agreement with the measured potential difference, indicating that these ions are at or near to equilibrium across the membrane. In contrast, the distribution of Na^+ ions is far from that predicted from the potential difference and the Nernst equation, indicating that the diffusion system for this ion is far from equilibrium. (Since radioactive Na^+ ions have been shown to cross the membrane, a diffusion system for this ion must be present,) Consequently, there must be some system that generates a flux of Na^+ ions across the membrane in the opposite direction to net diffusion, thereby keeping the latter process far-displaced from equilibrium. This system, usually referred to as 'active Na^+/K^+ transport', is a kinetically-irreversible process _in vivo_; it involves the exchange of Na^+ and K^+ across the membrane and it is coupled to the hydrolysis of ATP to ADP $+P_i$. Therefore, the transport of Na^+ and K^+ across the cell membrane can be represented by the following system:

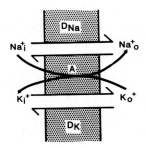

D_{Na} and D_K represent the diffusion of the respective ions and, as stated above, D_K is close to equilibrium whereas D_{Na} is far-displaced from equilibrium: reaction A represents the active transport system which is coupled to ATP hydrolysis (not shown). Thus, in this system, ATP hydrolysis to ADP is used to drive cyclic fluxes of Na^+ and K^+ between the interior and exterior of the cell (see section 5.3). For further details of the transport of ions, including experimental evidence for the above system, see Wood (1968), Davson & Segal (1975) and Harrison & Lunt (1980).

When there is a net flux of ion across the membrane, so that the diffusion system is not at equilibrium, it must be described by a kinetic equation. Thus, in the hypothetical example at the beginning of this section, if there is a net flux of X^+ from inside to outside, the value of this flux, J, is given by the equation,

$$J = k_f[X^+]_i - k_r[X^+]_o,$$

where k_f and k_r are the rate constants for the forward and reverse processes respectively. Since J is zero at equilibrium,

$$k_f[X^+]_{i(eq)} = k_r[X^+]_{o(eq)},$$

so that the equilibrium constant, K_E, in the direction of J, i.e. the ratio $[X^+]_{o(eq)}/[X^+]_{i(eq)}$, is equal to k_f/k_r.

Consequently, since

$$K_E = K e^{zF\Delta E/RT} \qquad \text{(see above)}$$

$$k_f/k_r = (k_f'/k_r')e^{zF\Delta E/RT},$$

where k_f' and k_r' are the rate constants in the absence of a potential difference.

For membrane transport systems resulting from simple or carrier-mediated diffusion the value of K is unity, so that $k_f' = k_r'$. However, this equality does not apply to k_f and k_r: when K is unity these are related by the equation,

$$k_f/k_r = e^{zF\Delta E/RT}$$

so that,

$$k_r = k_f e^{-zF\Delta E/RT}$$

Using this relationship, the kinetic equation for J can be written as,

$$J = k_f \left\{ [X^+]_i - [X^+]_o e^{-zF\Delta E/RT} \right\},$$

in which the value of z would be +1 in this example. Unfortunately, k_f is also a function of ΔE and the relationship between them cannot be derived from thermodynamics. As stated at the beginning of this section, if ΔE is positive the movement of X^+ in the direction of J is facilitated, so that k_f must increase as ΔE increases; however, this is all that can be deduced, from thermodynamics, about the influence of ΔE on k_f. Consequently, to apply the kinetic equation, the value of k_f must either be measured experimentally at the appropriate value of ΔE (e.g. by using radioactive isotopes) or be calculated by another theoretical approach, such as the 'constant-field' theory (Goldman, 1943; Hodgkin & Katz, 1949).

8.7   CONCLUDING REMARKS

This chapter was not intended to provide an exhaustive treatment of the vast subject of thermodynamics as applied to metabolism. Therefore, some topics have been either omitted (e.g. discussion of chemical potentials), or mentioned only in passing (e.g. thermodynamic activities). More detailed treatments of the subject may be found in texts such as Glasstone & Lewis (1963), Spanner (1964) and Klotz (1967).

It is hoped that, by taking a somewhat broad approach and by concentrating on basic principles, the student will have obtained a clearer understanding of the properties and physical significance of the thermodynamic state functions and how they should be applied to metabolic reactions and systems. In particular, he or she should be aware of the danger of confusing Gibbs free energy with conventional energy. Unfortunately this confusion pervades much biochemical literature and thinking, either implicitly (e.g. 'release' or 'transfer' of free energy) or explicitly (e.g. 'dissipation of free energy as heat'), and the student must be on a perpetual alert for this. This situation will almost certainly improve as the term 'free energy' becomes replaced by 'Gibbs function' (see section 5.4), but this change will take some time.

REFERENCES

Atkins, P. W. (1982) Physical Chemistry, 2nd Edn, Oxford University Press.
Atkinson, D. E. (1969) Curr. Top. Cell. Regul. 1, 29-43.
Banks, B. E. C. (1969) Chem. Brit. 5, 514-519.
Banks, B. E. C. & Vernon, C. A. (1970) J. Theor. Biol. 29, 301-326.
Blaxter, K. L. (1962) The Energy Metabolism of Ruminants, Hutchinson, London.
Buecher, Th. & Ruessmann, W. (1964) Angew. Chem. Int. Ed. Engl. 4, 426-439.
Cannon, B. & Nedergaard, J. (1985) Essays Biochem. 20, 110-164.
Cornish-Bowden, A. (1983) Biochem. Soc. Trans. 11, 44-45.
Crabtree, B. & Newsholme, E. A. (1975) in Insect Muscle (Usherwood, P.N.R., ed.), pp.405-500, Academic Press, London and New York.
Crabtree, B. & Newsholme, E. A. (1978) Eur. J. Biochem. 89, 19-22.
Crabtree, B. & Newsholme, E. A. (1985) Curr. Topics Cell Regul. 25, 21-76.
Davson, H. & Segal, M. B. (1975) Introduction to Physiology, vol. 1, pp.116-120, Academic Press, London, New York and San Francisco.
Dawes, E. A. (1972) Quantitative Problems in Biochemistry (5th Edn.) Livingstone Ltd, Edinburgh and London.
Fruton, J. S. & Simmonds, S. (1958) General Biochemistry (2nd Edn.) Wiley and Sons, New York and London.
Glasstone, S. & Lewis, D. (1963) Elements of Physical Chemistry (2nd Edn., revised), MacMillan, London.
Goldman, D. E. (1943) J. Gen. Physiol. 27, 37-60.
Harrison, R. & Lunt, G. G. (1980) Biological Membranes. Their Structure and Function, 2nd Edn, Blackie, Glasgow and London.
Heinrich, B. (1972) Science 175, 185-187; J. Comp. Physiol. 77, 49-64.
Hess, B. & Boiteux, A. (1971) Ann. Rev. Biochem. 40, 237-258.
Hochachka, P. W. (1976) Biochem. Soc. Symp. 41, 3-31.
Hodgkin, A. L. & Katz, B. (1949) J. Physiol. Lond 108, 37-77.
Huennekens, F. M. & Whiteley, H. R. (1960) in Comparative Biochemistry (Florkin, M. & Mason, H. S., eds), vol. 1, pp.107-180, Academic Press, New York and London.
Kacser, H. & Burns, J. A. (1973) Symp. Soc. Exp. Biol. 27, 65-104.
Katchalsky, A. & Curran, P. F. (1965) Nonequilibrium Thermodynamics in Biophysics, Harvard University Press, Cambridge, Mass.
Keizer, J. (1975) J. Theor. Biol. 49, 323-335.
Klotz, I. M. (1967) Energy Changes in Biochemical Reactions, Academic Press, New York and London.
Krebs, H. A. (1963) Adv. Enzyme Regul. 1, 385-400.
Krebs, H. A. & Kornberg, H. L. (1957) Energy Transformations in Living Matter, Springer-Verlag, Berlin, Goettingen and Heidelberg.

394

Krebs, H. A. & Veech, R. L. (1969) in The Energy Level and Metabolic Control in Mitochondria (Papa, S., Tager, J. M., Quagliariello, E. and Slater, E. C., eds.) pp.329-382, Adriatica Editrice, Bari.

Lehninger, A. L. (1970) Biochemistry, Worth Inc., New York.

Levin, R. J. (1969) The Living Barrier, Heinemann, London.

Lipmann, F. (1941) Adv. Enzymol. 1, 99-160.

Mahler, H. R. & Cordes, E. H. (1968) Basic Biological Chemistry, Harper and Row, New York, Evanston and London; Weatherhill, Tokyo.

McDonald, P., Edwards, R. A. & Greenhalgh, J. F. D. (1973) Animal Nutrition (2nd Edn.) Oliver and Boyd, Edinburgh.

Mitchell, P. (1973) J. Bioenergetics 4, 63-91.

Newsholme, E. A. (1970) Essays in Cell Metabolism (Bartley, W., Kornberg, H. L. and Quayle, J. R., eds.), pp.189-223, Wiley-Interscience, London.

Newsholme, E. A. & Crabtree, B. (1973) Symp. Soc. Exp. Biol. 27, 429-460.

Newsholme, E. A. & Crabtree, B. (1976) Biochem. Soc. Symp. 41, 61-110.

Newsholme, E. A. & Gevers, W. (1967) Vitamins and Hormones 25, 1-87.

Newsholme, E. A. & Leech, A. R. (1983) Biochemistry for the Medical Sciences, Wiley, London.

Newsholme, E. A. & Start, C. (1973) Regulation in Metabolism, Wiley and Sons, London, Sydney and Toronto.

Newsholme, E. A., Crabtree, B., Higgins, S. J., Thornton, S. D. & Start, C. (1972) Biochem. J. 128, 89-97.

Nicholls, D. (1982) Bioenergetics, Academic Press, London.

Randle, P. J., England, P. J. & Denton, R. M. (1970) Biochem. J. 117, 677-695.

Rolleston, F. S. (1972) Curr. Top. Cell. Regul. 5, 47-75.

Sibley, D. R. & Lefkowitz, R. J. (1985) Nature 317, 124-129.

Slater, E. C. (1974) Biochem. Soc. Trans. 2, 1149-1163.

Somero, G. N. & Low, P. S. (1976) Biochem. Soc. Symp. 41, 33-42.

Spanner, D. C. (1964) Introduction to Thermodynamics, Academic Press, London and New York.

Stryer, L. (1981) Biochemistry, 2nd Edn, Freeman & Co., San Francisco.

Stucki, J. G. (1983) Biochem. Soc. Trans 11, 45-47.

Tornheim, K. & Lowenstein, J. M. (1976) J. Biol. Chem. 251, 7322-7328.

Tregear, R T. (1975) in Insect Muscle (Usherwood, P. N. R., ed.), pp.357-403, Academic Press, London and New York.

Wilkie, D. R. (1970) Chem. Brit. 6. 475.

Wood, D. W. (1968) Principles of Animal Physiology, pp.137-144, Arnold & Co., London.

Yudkin, M. & Offord, R. E. (1973) Comprehensible Biochemistry, Longmans.

INDEX